AF356601

Zimányi School – Heavy Ion Physics

Zimányi School – Heavy Ion Physics

Editors

Máté Csanád
Péter Kovács
Sándor Lökös
Dániel Kincses

Basel • Beijing • Wuhan • Barcelona • Belgrade • Novi Sad • Cluj • Manchester

Editors

Máté Csanád
Department of
Atomic Physics
ELTE Eötvös
Loránd University
Budapest
Hungary

Péter Kovács
Theoretical Physics
Department
HUN-REN Wigner
Research Centre
Budapest
Hungary

Sándor Lökös
Department of the
Ultrarelativistic Nuclear
Physics and
Hadron Interactions
Institute of Nuclear
Physics PAN
Kraków
Poland

Dániel Kincses
Department of
Atomic Physics
ELTE Eötvös
Loránd University
Budapest
Hungary

Editorial Office
MDPI
St. Alban-Anlage 66
4052 Basel, Switzerland

This is a reprint of articles from the Special Issue published online in the open access journal *Universe* (ISSN 2218-1997) (available at: www.mdpi.com/journal/universe/special_issues/H54ERL6R40).

For citation purposes, cite each article independently as indicated on the article page online and as indicated below:

Lastname, A.A.; Lastname, B.B. Article Title. *Journal Name* **Year**, *Volume Number*, Page Range.

ISBN 978-3-7258-0040-7 (Hbk)
ISBN 978-3-7258-0039-1 (PDF)
doi.org/10.3390/books978-3-7258-0039-1

Contents

About the Editors

Máté Csanád

Máté Csanád is a professor of physics at the Department of Atomic Physics, Eötvös Loránd University, and also the director of the Environmental Science Center, Eötvös Loránd University. His research focus is on femtoscopy, hydrodynamics, and the time evolution of the medium produced in high-energy heavy-ion collisions.

Péter Kovács

Péter Kovács is a senior research fellow at the HUN-REN Wigner Research Centre, Institute for Particle and Nuclear Physics, Theoretical Physics Department, Heavy-ion Physics Research Group. He focuses on effective theories of the strong interaction and their role in high-energy physics and astrophysics.

Sándor Lökös

Sándor Lökös is an Assistant Professor at the Department of the Ultrarelativistic Nuclear Physics and Hadron Interactions of the Institute of Nuclear Physics, Polish Academy of Sciences. His research focuses on the ALICE experiment at the LHC.

Dániel Kincses

Dániel Kincses is a postdoctoral associate at the Department of Atomic Physics, Eötvös Loránd University. His main research focus is on femtoscopy in high-energy heavy-ion physics and investigating the shape of the hadron-emitting source.

Article

Lévy α-Stable Model for the Non-Exponential Low-$|t|$ Proton–Proton Differential Cross-Section

Tamás Csörgő [1,2], **Sándor Hegyi** [2] and **István Szanyi** [1,2,3,*]

1. MATE Institute of Technology, Károly Róbert Campus, Mátrai út 36, H-3200 Gyöngyös, Hungary; tcsorgo@cern.ch
2. Wigner Research Center for Physics, P.O. Box 49, H-1525 Budapest, Hungary; hegyi.physics@gmail.com
3. Department of Atomic Physics, Eötvös University, Pázmány P. s. 1/A, H-1117 Budapest, Hungary
* Correspondence: iszanyi@cern.ch

Abstract: It is known that the Real Extended Bialas–Bzdak (ReBB) model describes the proton–proton (pp) and proton–antiproton ($p\bar{p}$) differential cross-section data in a statistically non-excludible way, i.e., with a confidence level greater than or equal to 0.1% in the center of mass energy range $546\,\mathrm{GeV} \leq \sqrt{s} \leq 8\,\mathrm{TeV}$ and in the squared four-momentum transfer range $0.37\,\mathrm{GeV}^2 \leq -t \leq 1.2\,\mathrm{GeV}^2$. Considering, instead of Gaussian, a more general Lévy α-stable shape for the parton distributions of the constituent quark and diquark inside the proton and for the relative separation between them, a generalized description of data is obtained, where the ReBB model corresponds to the $\alpha = 2$ special case. Extending the model to $\alpha < 2$, we conjecture that the validity of the model can be extended to a wider kinematic range, in particular, to lower values of the four-momentum transfer $-t$. We present the formal Lévy α-stable generalization of the Bialas–Bzdak model and show that a simplified version of this model can be successfully fitted, with $\alpha < 2$, to the non-exponential, low $-t$ differential cross-section data of elastic proton–proton scattering at $\sqrt{s} = 8\,\mathrm{TeV}$.

Keywords: elastic scattering; proton-proton; scattering amplitude

1. Introduction

The Bialas–Bzdak (BB) model considers the proton as a bound state of a quark and a diquark, $p = (q, d)$ for short [1]. The diquark in the proton may also be considered to be a weakly bound state of two constituent quarks, leading to the $p = (q, (q, q))$ variant of the BB model; however, in Ref. [2], it was shown that the $p = (q, (q, q))$ variant of the BB model gives two many diffractive minima, whereas, experimentally, only a single minimum is observed in the differential cross-section of proton–proton (pp) collisions. Thus, in recent studies, in Refs. [3,4], the $p = (q, d)$ version of the model was utilized.

Originally, the BB model considers Gaussian shapes for the parton distributions of constituent quarks and diquarks inside the proton and for the relative separation between them. By these considerations based on R. J. Glauber's multiple scattering theory [5,6], the inelastic scattering cross-section of protons at a fixed $\sqrt{s}$ energy and a fixed b impact parameter value is constructed and denoted as $\tilde{\sigma}_{in}(s, \vec{b})$.

The elastic scattering amplitude in the impact parameter representation is written in terms of $\tilde{\sigma}_{in}(s, \vec{b})$ as a solution of the unitarity equation. The imaginary part of the elastic scattering amplitude is the dominant part, whereas the real part can be considered as a smaller correction. Bialas and Bzdak in Ref. [1] neglected the real part of the amplitude and used a fully imaginary amplitude,

$$\tilde{t}_{el}(s, \vec{b}) = i\left(1 - \sqrt{1 - \tilde{\sigma}_{in}(s, \vec{b})}\right), \tag{1}$$

for the calculations of the scattering cross sections. However, in a model where the amplitude does not have a real part, the characteristic minimum–maximum region of the pp

Citation: Csörgő, T.; Hegyi, S; Szanyi, I. Lévy α-Stable Model for the Non-Exponential Low-$|t|$ Proton–Proton Differential Cross-Section. *Universe* **2023**, *9*, 361. https://doi.org/10.3390/universe9080361

Academic Editor: Jun Xu

Received: 31 May 2023
Revised: 13 July 2023
Accepted: 25 July 2023
Published: 3 August 2023

differential cross-section can not be described properly. In Ref. [7], the elastic scattering amplitude was extended with a real part in a way that the unitarity constraint is fulfilled. This amplitude reads as:

$$t_{el}(s,\vec{b}) = i\left(1 - e^{i\,\alpha\,\tilde{\sigma}_{in}(s,\vec{b})}\sqrt{1 - \tilde{\sigma}_{in}(s,\vec{b})}\right),$$ (2)

where α is a free parameter to be fitted to the data. In the case of $\alpha = 0$, Equation (2) reduces to Equation (1), i.e., to a scattering amplitude that has a vanishing real part.

The model for the elastic proton–proton scattering amplitude, as defined by Equation (2), with $\tilde{\sigma}_{in}(s,\vec{b})$, as defined in Ref. [1], is called the Real Extended Bialas–Bzdak (ReBB) model. In recent studies [3,4], it was shown that the ReBB model describes pp and proton–antiproton ($p\bar{p}$) differential cross-section data in the center of mass energy range of 0.546 TeV $\leq \sqrt{s} \leq$ 8 TeV and in the squared four-momentum transfer range of 0.37 GeV$^2 \leq -t \leq$ 1.2 GeV^2 in a statistically non-excludible manner, i.e., with a confidence level greater than or equal to 0.1%.

The free parameters of the ReBB model are the Gaussian radii of the quark, the diquark, and the separation between them (correspondingly, R_q, R_d, and R_{qd}) and also, the α parameter regulating the real part of the scattering amplitude. Two additional fit parameters could be present: λ, the ratio of the quark and diquark masses, and A_{qq}, the normalization parameter appearing in the inelastic quark–quark cross-section. However, it was shown in Ref. [2] and later confirmed in Ref. [3] that A_{qq} can be fixed at a value of 1.0, whereas λ can be fixed at a value of $1/2$.

The energy dependence of the ReBB model parameters for pp and $p\bar{p}$ scattering were determined in Ref. [3]. It was found that the energy dependencies of the radius parameters are the same for pp and $p\bar{p}$ scattering, whereas the energy dependencies of the α parameter for pp and $p\bar{p}$ scattering are different, i.e., there are different α^{pp} and $\alpha^{p\bar{p}}$ parameters. The energy dependencies of all the five parameters in the energy range of $0.546 \leq \sqrt{s} \leq 8$ TeV are determined by linear logarithmic functions [3,4].

Considering, instead of Gaussian, a more general Lévy α-stable shape for the parton distributions of the constituent quark and diquark inside the proton and for the relative separation between them, an improved description to the data in a wider kinematic range ($\sqrt{s} < 0.546$ TeV, $\sqrt{s} > 8$ TeV, $-t < 0.37$ GeV^2, $-t > 1.2$ GeV^2) is anticipated.

The 0.37 GeV$^2 \leq -t \leq$ 1.2 GeV^2 interval at LHC energies includes the region of the characteristic minimum–maximum structure of the pp elastic differential cross-section. In the 0.01 GeV$^2 \lesssim -t \lesssim$ 0.15 GeV^2 interval, another characteristic structure, a non-exponential behavior is observed. A significant non-exponential behavior was measured by TOTEM at CERN LHC at 8 and 13 TeV center of mass energies [8,9]. Similar behavior was observed also at the CERN ISR accelerator in the 1970s [10], where measurements were made in the 20 GeV $\lesssim \sqrt{s} \lesssim$ 60 GeV energy region.

In Ref. [11], the model-independent Lévy imaging method is successfully employed to describe the pp and $p\bar{p}$ differential cross-section data both at the low and the high $-t$ region simultaneously. In Ref. [12], the model-independent Lévy imaging method was employed to reconstruct the proton inelasticity profile function. This method established a statistically significant proton hollowness effect [13–17], well beyond the 5σ discovery limit at $\sqrt{s} = 13$ TeV. These results suggest that Lévy α-stable models are efficient tools in describing pp and $p\bar{p}$ differential cross-section data, and the ReBB model needs to be Lévy α-stable generalized to have a stronger non-exponential feature at low $-t$ and to accommodate the new features of the differential cross-section data such the hollowness effect at $\sqrt{s} = 13$ TeV or larger energies. In the present work, we complete the formal Lévy α-stable generalization of the Bialas–Bzdak model.

This paper is organized as follows. In Section 2, we deduce the formal Lévy α-stable generalization of the Bialas–Bzdak model and discuss the technical difficulties preventing us to perform an efficient fitting procedure of the model parameters to the experimental data with the full Lévy α-stable generalized Bialas-Bzdak model. In Section 3, we show successful fits to the low $-t$ differential cross-section data at LHC energies with a simple Lévy α-stable model deduced by approximations from the Lévy α-stable generalized Bialas–Bzdak (LBB) model. In Section 4, the parameters of the LBB model is related to the $t = 0$ measurable quantities and to the parameters of the simple Lévy α-stable model. Finally, we summarize and conclude in Section 5.

2. From Gaussian to Lévy α-Stable $p = (q, d)$ BB Model

First, we recapitulate the BB model using normalized Gaussian distributions and introduce some reinterpretations of some of its parts. Then, we change the normalized Gaussian distributions to normalized Lévy α-stable distributions, resulting in the Lévy α-stable generalized BB model.

The inelastic scattering cross-section at a fixed $\vec{b}$ impact parameter value is given as [1]:

$$\tilde{\sigma}_{in}(\vec{b}) = \int_{-\infty}^{+\infty} \cdots \int_{-\infty}^{+\infty} d^2\vec{s}_q d^2\vec{s}'_q d^2\vec{s}_d d^2\vec{s}'_d D(\vec{s}_q, \vec{s}_d) D(\vec{s}'_q, \vec{s}'_d) \sigma(\vec{s}_q, \vec{s}_d; \vec{s}'_q, \vec{s}'_d; \vec{b}), \tag{3}$$

where $D(\vec{s}'_q, \vec{s}'_d)$ is the quark–diquark distribution inside one of the colliding protons, $\sigma(\vec{s}_q, \vec{s}_d; \vec{s}'_q, \vec{s}'_d; \vec{b})$ is the probability of inelastic collision, and the variables we integrate over are the transverse positions of the quarks and diquarks inside the two colliding protons. Note that the energy dependence of $\tilde{\sigma}_{in}(\vec{b})$ is not written out here for clarity reasons; however, through the $\sqrt{s}$ dependence of the model parameters, $R_q(s)$, $R_d(s)$, and $R_{qd}(s)$, $\tilde{\sigma}_{in}(\vec{b})$ has an $\sqrt{s}$ dependence too.

The quark–diquark distribution is considered to be Gaussian:

$$D(\vec{s}_q, \vec{s}_d) = \frac{1 + \lambda^2}{R_{qd}^2 \pi} e^{-(s_q^2 + s_d^2)/R_{qd}^2} \delta^2(\vec{s}_d + \lambda\vec{s}_q), \tag{4}$$

where $\lambda = m_q/m_d$, the ratio of the quark and diquark masses, and R_{qd} are free parameters of the model. The two-dimensional Dirac δ function fixes the center-of-mass of the proton and reduces the dimension of the integral in Equation (3) from 8 to 4. Accordingly, the diquark positions can be expressed by that of the quarks:

$$\vec{s}_d = -\lambda\,\vec{s}_q, \;\; \vec{s}'_d = -\lambda\,\vec{s}'_q. \tag{5}$$

After integration over $\vec{s}_d$, $D(\vec{s}_q, \vec{s}_d)$ becomes a Gaussian in $\vec{s}_q$; then, after the integration, also over $\vec{s}_q$, we obtain unity:

$$\int d^2\vec{s}_d D(\vec{s}_q, \vec{s}_d) \;=\; G\left(\vec{s}_q | R_{qd}/\sqrt{2(1 + \lambda^2)}\right), \tag{6}$$

$$\int d^2\vec{s}_q d^2\vec{s}_d D(\vec{s}_q, \vec{s}_d) \;=\; 1, \tag{7}$$

where:

$$G(\vec{x}|R_G) = \frac{1}{(2\pi)^2} \int d^2q\, e^{i\vec{q}^T\vec{x}} e^{-\frac{1}{2}q^2 R_G^2} = \frac{1}{2\pi R_G^2} e^{-\frac{x^2}{2R_G^2}} \tag{8}$$

is the normalized bivariate Gaussian distribution.

We may reinterpret $D(\vec{s}_q, \vec{s}_d)$ as the distribution of the relative separation between the quark and the diquark in a single proton, namely:

$$D(\vec{s}_q, \vec{s}_d) = (1 + \lambda)^2 G\left(\vec{s}_q - \vec{s}_d | R_{qd}/\sqrt{2}\right) \delta^2(\vec{s}_d + \lambda\vec{s}_q) \tag{9}$$

which is correctly normalized as follows:

$$\int d^2\vec{s}_d D(\vec{s}_q, \vec{s}_d) \;=\; G\!\left(\vec{s}_q | R_{qd*}/\sqrt{2}\right), \tag{10}$$

$$\int d^2\vec{s}_q d^2\vec{s}_d D(\vec{s}_q, \vec{s}_d) \;=\; 1, \tag{11}$$

where:

$$R_{qd*} = \frac{R_{qd}}{1+\lambda}. \tag{12}$$

Here, we have rescaled the parameter R_{qd} of the original Bialas–Bzdak model to the parameter that characterizes the uncertainty of the location of a dressed quark inside the proton. The advantage of this interpretation is that we prepare the ground for the generalization to the case of Levy α-stable distributions and instead of taking the product of two Gaussians, as in Equation (4), we had an equivalent rewrite where the relative coordinate distribution of a quark and a diquark is Gaussian, with rescaled parameters. This rewrite is very advantageous, as the product of two Levy distributions is not a Levy distribution, with the exception of the $\alpha_L = 2$ Gaussian case. As such, to have only one Gaussian in the relative coordinate avoids the problem of having products of Levy α-stable distributions in the formulas.

The term $\sigma(\vec{s}_q, \vec{s}_d; \vec{s}_q', \vec{s}_d'; \vec{b})$ is the probability of inelastic interactions at a fixed impact parameter and transverse positions of all constituents and given by a Glauber expansion as follows:

$$\sigma(\vec{s}_q, \vec{s}_d; \vec{s}_q', \vec{s}_d'; \vec{b}) = 1 - \left[1 - \sigma_{qq}(\vec{s}_q, \vec{s}_q'; \vec{b})\right]\left[1 - \sigma_{qd}(\vec{s}_q, \vec{s}_d'; \vec{b})\right] \times \tag{13}$$

$$\times \left[1 - \sigma_{dq}(\vec{s}_q', \vec{s}_d; \vec{b})\right]\left[1 - \sigma_{dd}(\vec{s}_d, \vec{s}_d'; \vec{b})\right],$$

where:

$$\sigma_{qq}(\vec{s}_q, \vec{s}_q'; \vec{b}) \equiv \sigma_{qq}(\vec{b} + \vec{s}_q' - \vec{s}_q),$$

$$\sigma_{qd}(\vec{s}_q, \vec{s}_d'; \vec{b}) \equiv \sigma_{qd}(\vec{b} + \vec{s}_d' - \vec{s}_q),$$

$$\sigma_{dq}(\vec{s}_d, \vec{s}_q'; \vec{b}) \equiv \sigma_{dq}(\vec{b} + \vec{s}_q' - \vec{s}_d),$$

and:

$$\sigma_{dd}(\vec{s}_d, \vec{s}_d'; \vec{b}) \equiv \sigma_{dd}(\vec{b} + \vec{s}_d' - \vec{s}_d)$$

are the inelastic differential cross-sections of the binary collisions of the constituents. They have Gaussian shapes:

$$\sigma_{ab}(\vec{x}) = A_{ab} e^{-\vec{s}^2/S_{ab}^2} \tag{14}$$

with $S_{ab}^2 = R_a^2 + R_b^2$ and $a,b \in \{q,d\}$. Equation (14) can be rewritten in terms of normalized bivariate Gaussian distribution:

$$\sigma_{ab}(\vec{s}) = A_{ab}\pi S_{ab}^2 G\!\left(\vec{s}|S_{ab}/\sqrt{2}\right). \tag{15}$$

We can reinterpret the inelastic constituent–constituent collisions by assuming that the constituent quark and the constituent diquark have Gaussian parton distributions, characterized by $G(\vec{s}_q|R_q/\sqrt{2})$ and $G(\vec{s}_d|R_d/\sqrt{2})$. Then, the probability of inelastic collisions at a given impact parameter b is proportional to their convolution:

$$\sigma_{ab}(\vec{s}) = A_{ab}\pi S_{ab}^2 \int d^2 s_a G(\vec{s}_a|R_a/\sqrt{2})G(\vec{s} - \vec{s}_a|R_b/\sqrt{2}) \tag{16}$$

$$\equiv A_{ab}\pi S_{ab}^2 G\!\left(\vec{s}|S_{ab}/\sqrt{2}\right).$$

The inelastic quark–quark, quark–diquark, and diquark–diquark cross-sections are obtained by integration:

$$\sigma_{ab,\text{inel}} = \int\limits_{-\infty}^{+\infty}\int\limits_{-\infty}^{+\infty} \sigma_{ab}(\vec{s})\,\mathrm{d}^2 s = A_{ab}\pi S_{ab}^2\,. \tag{17}$$

The number of the free parameters of the model can be reduced by demanding that the ratios of the cross-sections are:

$$\sigma_{qq,\text{inel}} : \sigma_{qd,\text{inel}} : \sigma_{dd,\text{inel}} = 1 : 2 : 4\,, \tag{18}$$

expressing the idea that the constituent diquark contains twice as many partons than the constituent quark and also that the colliding constituents do not "shadow" each other.

Then, the probabilities of inelastic constituent–constituent collisions can be written in the following form:

$$\sigma_{qq}(\vec{s}_q, \vec{s}_q'; \vec{b}) = 2\pi A_{qq} R_q^2 G(\vec{b} + \vec{s}_q' - \vec{s}_q | R_q), \tag{19}$$

$$\sigma_{qd}(\vec{s}_q, \vec{s}_d'; \vec{b}) = 4\pi A_{qq} R_q^2 G\left(\vec{b} + \vec{s}_d' - \vec{s}_q \left| \sqrt{\frac{R_q^2 + R_d^2}{2}}\right.\right), \tag{20}$$

$$\sigma_{dq}(\vec{s}_q', \vec{s}_d; \vec{b}) = 4\pi A_{qq} R_q^2 G\left(\vec{b} + \vec{s}_q' - \vec{s}_d \left| \sqrt{\frac{R_q^2 + R_d^2}{2}}\right.\right), \tag{21}$$

$$\sigma_{dd}(\vec{s}_d, \vec{s}_d'; \vec{b}) = 8\pi A_{qq} R_q^2 G(\vec{b} + \vec{s}_d' - \vec{s}_d | R_d). \tag{22}$$

Substituting these into Equation (13), then substituting $\sigma(\vec{s}_q, \vec{s}_d; \vec{s}_q', \vec{s}_d'; \vec{b})$ into Equation (3), we get a sum of eleven integral terms (with proper sign) for $\tilde{\sigma}_{in}(\vec{b})$:

$$\tilde{\sigma}_{in}(\vec{b}) = \tilde{\sigma}_{in}^{qq}(\vec{b}) + 2\tilde{\sigma}_{in}^{qd}(\vec{b}) + \tilde{\sigma}_{in}^{dd}(\vec{b}) - [2\tilde{\sigma}_{in}^{qq,qd}(\vec{b}) + \tilde{\sigma}_{in}^{qd,dq}(\vec{b}) + \tilde{\sigma}_{in}^{qq,dd}(\vec{b}) + \tag{23}$$
$$+ 2\tilde{\sigma}_{in}^{qd,dd}(\vec{b})] + [\tilde{\sigma}_{in}^{qq,qd,dq}(\vec{b}) + 2\tilde{\sigma}_{in}^{qq,qd,dd}(\vec{b}) + \tilde{\sigma}_{in}^{dd,qd,dq}(\vec{b})] - \tilde{\sigma}_{in}^{qq,qd,dq,dd}(\vec{b}).$$

Let us have a look for the most general fourth-order term, $\tilde{\sigma}_{in}^{qq,qd,dq,dd}(\vec{b})$. After making use of the presence of the Dirac δ function in Equation (9), we have to calculate a four-dimensional integral of products of normalized bivariate Gaussian distributions:

$$\tilde{\sigma}_{in}^{qq,qd,dq,dd}(\vec{b}) = \int d^2 s_q d^2 s_q' G(\vec{s}_q | R_{qd*}/\sqrt{2}) G(\vec{s}_q' | R_{qd*}/\sqrt{2}) \times \tag{24}$$
$$\times \sigma_{qq}(\vec{s}_q, \vec{s}_q'; \vec{b})\sigma_{qd}(\vec{s}_q, -\lambda\vec{s}_q'; \vec{b})\sigma_{dq}(\vec{s}_q', -\lambda\vec{s}_q; \vec{b})\sigma_{dd}(-\lambda\vec{s}_q, -\lambda\vec{s}_q'; \vec{b}).$$

Such an integral results in an expression having a Gaussian shape. The lower-order terms can be obtained from Equation (24) by excluding the proper σ_{ab} term/terms from the integrand. Thus, after computing the integrals in all order, we get the sum of eleven different Gaussian-shaped terms, i.e., the BB model as introduced in Ref. [1].

Now, we perform the Lévy α-stable generalization of the BB model.

Let us introduce the normalized bivariate symmetric Lévy α-stable distribution,

$$L(\vec{x}|\alpha_L, R_L) = \frac{1}{(2\pi)^2}\int d^2 q\, e^{i\vec{q}^T \vec{x}} e^{-|q^2 R_L^2|^{\alpha_L/2}}, \tag{25}$$

which, for $\alpha_L = 2$, gives exactly the bivariate Gaussian distribution:

$$L(\vec{x}|\alpha_L = 2, R_L = R_G/\sqrt{2}) \equiv G(\vec{x}|R_G). \tag{26}$$

Note that the Lévy index of stability α_L, that controls the power-law tails of the inelastic cross-sections, is a different parameter from the α parameter of the ReBB model, that controls the opacity or the real part of the scattering amplitude. Due to historic reasons, both were denoted by α originally, but in this work, we add a subscripted L to distinguish the Lévy parameter α_L from the opacity parameter α.

Since we work with here with symmetric Lévy α-stable distribution, the skewness parameter $\beta_L = 0$ of the Lévy stable source distributions is implicit and are assumed to have zero values. The shift parameter δ_L of the Lévy stable source distribution is explicitely written out when considering the impact parameter picture, while the overall shift of the impact parameter cancels from the final results hence it is assumed to have a vanishing value.

We then consider that the relative separation between the quark and the diquark in a single proton follows Lévy α-stable distribution:

$$D(\vec{s}_q, \vec{s}_d) = (1+\lambda)^2 L\left(\vec{s}_q - \vec{s}_d|\alpha_L, R_L = R_{qd}/2\right)\delta^2(\vec{s}_d + \lambda\vec{s}_q) \tag{27}$$

with:

$$\int d^2\vec{s}_d D(\vec{s}_q, \vec{s}_d) = L\left(\vec{s}_q|\alpha_L, R_{qd*}/2\right), \tag{28}$$

$$\int d^2\vec{s}_q d^2\vec{s}_d D(\vec{s}_q, \vec{s}_d) = 1, \tag{29}$$

similarly to the original case with Gaussian distributions.

As the next step in the generalization, we consider, instead of Gaussian, Lévy α-stable parton distributions for the constituent quark and the constituent diquark: $L(\vec{s}_q|\alpha_L, R_q/2)$ and $L(\vec{s}_d|\alpha_L, R_d/2)$. Then, as in the Gaussian case above, the probability of inelastic collisions at a given impact parameter b is proportional to their convolution:

$$\sigma_{ab}(\vec{s}) = A_{ab}\pi S_{ab}^2 \int d^2s_a L(\vec{s}_a|\alpha_L, R_a/2)L(\vec{s} - \vec{s}_a|\alpha_L, R_b/2) \tag{30}$$

$$= A_{ab}\pi S_{ab}^2 L(\vec{s}|\alpha_L, S_{ab}/2),$$

where now:

$$S_{ab} = \left(R_a^{\alpha_L} + R_b^{\alpha_L}\right)^{1/\alpha_L}, \tag{31}$$

i.e., in this case, after making use of the convolution theorem, the radii add up not quadratically, but at the power of α_L.

Then:

$$\sigma_{qq}(\vec{s}_q, \vec{s}_q'; \vec{b}) = \pi A_{qq}\left(2R_q^{\alpha_L}\right)^{2/\alpha_L} L\left(\vec{b} + \vec{s}_q' - \vec{s}_q|\alpha_L, \left(2R_q^{\alpha_L}\right)^{1/\alpha_L}/2\right), \tag{32}$$

$$\sigma_{qd}(\vec{s}_q, \vec{s}_d'; \vec{b}) = 2\pi A_{qq}\left(2R_q^{\alpha_L}\right)^{2/\alpha_L} L\left(\vec{b} + \vec{s}_d' - \vec{s}_q\bigg|\alpha_L, \left(R_q^{\alpha_L} + R_d^{\alpha_L}\right)^{1/\alpha_L}/2\right), \tag{33}$$

$$\sigma_{dq}(\vec{s}_q', \vec{s}_d; \vec{b}) = 2\pi A_{qq}\left(2R_q^{\alpha_L}\right)^{2/\alpha_L} L\left(\vec{b} + \vec{s}_q' - \vec{s}_d\bigg|\alpha_L, \left(R_q^{\alpha_L} + R_d^{\alpha_L}\right)^{1/\alpha_L}/2\right), \tag{34}$$

and

$$\sigma_{dd}(\vec{s}_d, \vec{s}_d'; \vec{b}) = 4\pi A_{qq}\left(2R_q^{\alpha_L}\right)^{2/\alpha_L} L\left(\vec{b} + \vec{s}_d' - \vec{s}_d)|\alpha_L, \left(2R_d^{\alpha_L}\right)^{1/\alpha_L}/2\right). \tag{35}$$

Equation (3) with Equation (13), Equation (27), and Equations (32)–(35) define the Lévy α_L-stable generalized Bialas–Bzdak (LBB) model for $\tilde{\sigma}_{in}(b)$. Now, in Equation (23), instead of a sum of integrals of products of normalized Gaussian distributions, there are a sum of integrals of products of normalized Lévy α_L-stable distributions. Though integrals of products of Gaussian distributions can be calculated, the calculation of integrals of products of Lévy α_L-stable distributions is an issue. Integrals of products of Lévy α_L-stable distributions can be easily calculated if the integral can be written in a convolution form. This is the case for the first three terms in Equation (23). The results can be written in terms of Lévy α_L-stable distributions:

$$\tilde{\sigma}_{in}^{qq}(\vec{b}) = \pi A_{qq}\left(2R_q^{\alpha_L}\right)^{2/\alpha_L} \times \tag{36}$$

$$\times \int d^2 s_q d^2 s'_q L(\vec{s}_q|\alpha_L, R_{qd*}/2) L(\vec{s}'_q|R_{qd*}/2) L\left(\vec{b} + \vec{s}'_q - \vec{s}_q \Big| \left(2R_q^{\alpha_L}\right)^{1/\alpha_L}/2\right)$$

$$= \pi A_{qq}\left(2R_q^{\alpha_L}\right)^{2/\alpha_L} L\left(\vec{b}\Big|\alpha_L, \left(2R_{qd*}^{\alpha_L} + 2R_q^{\alpha_L}\right)^{1/\alpha_L}/2\right),$$

$$\tilde{\sigma}_{in}^{qd}(\vec{b}) = 2\pi A_{qq}\left(2R_q^{\alpha_L}\right)^{2/\alpha_L} \times \tag{37}$$

$$\times \int d^2 s_q d^2 s'_q L(\vec{s}_q|R_{qd*}/2) L(\vec{s}'_q|R_{qd*}/2) L\left(\vec{b} - \lambda\vec{s}'_q - \vec{s}_q \Big|\alpha_L, \left(R_q^{\alpha_L} + R_d^{\alpha_L}\right)^{1/\alpha_L}/2\right)$$

$$= 2\pi A_{qq}\left(2R_q^{\alpha_L}\right)^{2/\alpha_L} L\left(\vec{b}\Big|\alpha_L, \left((1 + \lambda^{\alpha_L})R_{qd*}^{\alpha_L} + R_q^{\alpha_L} + R_d^{\alpha_L}\right)^{1/\alpha_L}/2\right),$$

$$\tilde{\sigma}_{in}^{dd}(\vec{b}) = 4\pi A_{qq}\left(2R_q^{\alpha_L}\right)^{2/\alpha_L} \times \tag{38}$$

$$\times \int d^2 s_q d^2 s'_q L(\vec{s}_q|R_{qd*}/2) L(\vec{s}'_q|R_{qd*}/2) L\left(\vec{b} + \lambda(\vec{s}_q - \vec{s}'_q)|\alpha_L, (2R_d^{\alpha_L})^{1/\alpha_L}/2\right)$$

$$= 4\pi A_{qq}\left(2R_q^{\alpha_L}\right)^{2/\alpha_L} L\left(\vec{b}\Big|\alpha_L, \left(2\lambda^{\alpha_L}R_{qd*}^{\alpha_L} + 2R_d^{\alpha_L}\right)^{1/\alpha_L}/2\right).$$

The results of the remaining eight integrals, corresponding to higher-order terms in the BB model, are yet to be determined in terms of analytic formulas.

Whereas univariate and multivariate Gaussian distributions have closed forms in terms of elementary functions, univariate and multivariate Lévy α_L-stable distributions have forms in terms of special functions. This makes it hard to perform a numerical fitting procedure of the model parameters to the experimental data. To complete this work in the future, a relatively high computing capacity or improved analytic insight will be needed. In this work, we have chosen another approach, limiting the domain of the applicability of the calculations in the squared four-momentum transfer $-t$. This allows for certain simplifications and results in an increased analytic insight to certain properties of the LBB model.

A possible alternative to the Lévy α-stable generalization of the BB model could be its Tsallis or q-exponential generalization, since data from high-energy collisions have shown such distribution. The presence of the Tsallis distribution was explained in Ref. [18] using the fractal approach to the non-perturbative QCD, and also, the q index was expressed in terms of the number of colors and the number of flavors. The validity of the derived relation was reinforced later in Ref. [19]. These results suggest that the investigation of the Tsallis generalization of the BB model is worthwhile. This will be done in a future study. In this manuscript, we investigate the Lévy α-stable generalization of the BB model.

3. A Simple Lévy α-Stable Model

Now, we check if the Lévy α-stable generalization of the BB model has an enhanced potential, as compared to the ReBB model, or not. The mathematical and computing difficulties discussed in the previous section can be bypassed by introducing new approximations that are valid at low $-t$, in the domain where the original ReBB model had difficulties to describe the strongly non-exponential features of the experimental data of elastic proton-proton scattering at the TeV energy scale. Our aim is, thus, to deduce a model for the differential cross-section which is valid at the low-$|t|$ region.

Since the elastic scattering amplitude is predominantly imaginary in this kinematic region, we approximate it by an imaginary part, as given by Equation (1). Low-$|t|$ scattering corresponds to high-b scattering and, at high b values $\tilde{\sigma}_{in}(s, b)$, is small. Thus, the leading order term in the Taylor expansion of Equation (1), i.e.,

$$\tilde{t}_{el}(s, \vec{b}) = \frac{i}{2}\tilde{\sigma}_{in}(s, \vec{b}), \tag{39}$$

should be a reasonable approximation at low $-t$ values in the $\alpha = 0$ (vanishing real part) case.

As discussed in Section 3, low-$|t|$ scattering corresponds to high-b scattering and, at high b values $\tilde{\sigma}_{in}(s, b)$, is small. Thus, the leading order term in the Taylor expansion of Equation (2), i.e.,

$$\tilde{t}_{el}(s, \vec{b}) = \left(\alpha + \frac{i}{2}\right)\tilde{\sigma}_{in}(s, \vec{b}), \tag{40}$$

should be a reasonable approximation at low $-t$ values if the opacity parameter α is small.

In Section 2, we discussed that in the Lévy α-stable generalized case of the BB model, the leading order terms in $\tilde{\sigma}_{in}(s, b)$ are Lévy-α-stable-shaped terms. Motivated by this fact in our simplified model, we approximate $\tilde{\sigma}_{in}(s, b)$ with a single Lévy-αstable-shaped term, i.e.,

$$\tilde{\sigma}_{in}(s, \vec{b}) = \tilde{c}(s)L(\vec{b}|\alpha_L(s), r(s)) \tag{41}$$

where $\tilde{c}(s)$ is in general a complex-valued and s dependent function, while $\alpha_L(s)$, and $r(s)$ are adjustable parameters determined at a given $\sqrt{s}$ energy,

Then, by Equation (39), we have:

$$\tilde{t}_{el}(s, \vec{b}) = ic(s)L(\vec{b}|\alpha_L(s), r(s)), \tag{42}$$

where $c(s) = \tilde{c}(s)/2$ is a rescaled and complex valued parameter. Now, we transform the impact parameter amplitude into momentum space:

$$t(s, t) = \int d^2 b\, e^{i\vec{\Delta}^T \vec{b}}\tilde{t}(s, \vec{b}) = ic(s)e^{-|tr^2(s)|^{\alpha_L(s)/2}}, \tag{43}$$

where $|\vec{\Delta}| \simeq \sqrt{-t}$. The resulting differential cross-section is:

$$\frac{d\sigma}{dt}(s, t) = \frac{1}{4\pi}|t(s, t)|^2 = a(s)e^{-|tb(s)|^{\alpha_L(s)/2}}, \tag{44}$$

where $a(s) = \frac{|c(s)|^2}{4\pi}$ and $b(s) = 2^{2/\alpha_L(s)}r^2(s)$. Thus, finally, this simple model for the differential cross-sections has three adjustable parameters, α_L, a, and b, to be determined at a given energy.

The result of a fit to the TOTEM pp elastic differential cross-section data at $\sqrt{s} = 8$ TeV by the model defined by Equation (44) is shown in Figure 1. One can see that the non-exponential model with $\alpha_L = 1.953 \pm 0.004$ represents the low-$|t|$ differential cross-section data with a confidence level of 55%.

Figure 2 shows the ratio, $(d\sigma/dt - ref)/ref$, with $ref = Ae^{-Bt}$, used by the TOTEM collaboration [8] to make the relatively small, but significant, low-$|t|$ non-exponential behavior visible. One can clearly see that our model successfully describes the low-$|t|$ data.

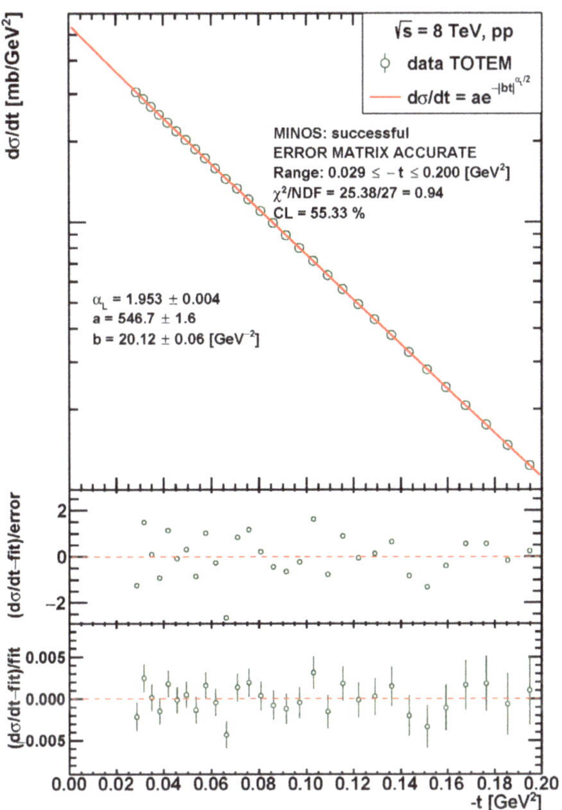

Figure 1. Fit-to-the-TOTEM pp elastic differential cross-section data at $\sqrt{s} = 8$ TeV [8] by the model defined by Equation (44).

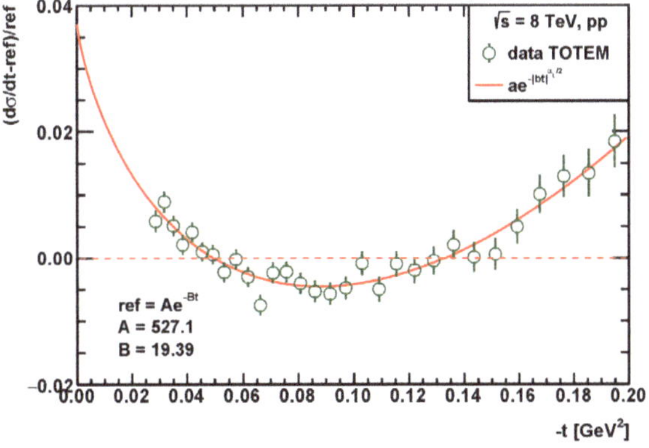

Figure 2. The ratio, $(d\sigma/dt - ref)/ref$, evaluated from the TOTEM pp elastic differential cross-section data at $\sqrt{s} = 8$ TeV [8]. The curve corresponds to the fitted model defined by Equation (44).

4. The $t = 0$ Measurable Quantities and the BB Model Parameters

In this section, we relate the $t = 0$ measurable quantities and the LBB parameters. First, we work with the original BB model with Gaussian distributions and then derive the formulas for the Levy α-stable generalized case. We note again that to avoid confusion with the α parameter of the ReBB model regulating the real part of the amplitude and that of the Lévy α-stable distribution, the latter we have denoted in this manuscript as α_L. For the $\alpha_L = 2$ limiting case, the relations obtained in the original BB model are recovered.

Now, we consider only the leading order terms in $\tilde{\sigma}_{in}(s, b)$, i.e., $\tilde{\sigma}_{in}^{qq}(\vec{b})$, $\tilde{\sigma}_{in}^{qd}(\vec{b})$, and $\tilde{\sigma}_{in}^{dd}(\vec{b})$, which give the dominant contribution at $t = 0$. We get the amplitude in momentum space by Fourier transformation as in Equation (43). As discussed in the Introduction, the parameter A_{qq} can be fixed at a value of 1.0, whereas λ can be fixed at a value of $1/2$. We use these specific values below.

With Gaussian distributions in the BB model, in the low-$|t|$ approximation, σ_{tot} is related to the square of the quark radius R_q,

$$\sigma_{tot} = 2Imt(s, t = 0) = 18\pi R_q^2, \tag{45}$$

whereas the ratio of the real to the imaginary part of the forward scattering amplitude is related to the α parameter of the ReBB model,

$$\rho_0 = \frac{Ret(s, t = 0)}{Imt(s, t = 0)} = 2\alpha. \tag{46}$$

Note that this result for ρ_0 holds also in the Levy α-stable generalized case.

The low-$|t|$ pp differential cross-section is in the form [8]:

$$\frac{d\sigma}{dt} = \frac{1}{4\pi}|t(s, t)|^2 = ae^{-b_1 t + b_2 t^2} \tag{47}$$

where:

$$a = \frac{d\sigma}{dt}\bigg|_{t=0} \tag{48}$$

is the optical point,

$$b_1 = \left(\frac{d}{dt}\ln\frac{d\sigma}{dt}\right)\bigg|_{t=0} \tag{49}$$

is the slope parameter, and

$$b_2 = \frac{1}{2}\left(\frac{d}{dt^2}\ln\frac{d\sigma}{dt}\right)\bigg|_{t=0} \tag{50}$$

is the curvature parameter. These measurable quantities can be expressed in terms of the ReBB model parameters:

$$a = \frac{81}{4}\pi Rq^4\left(1 + 4\alpha^2\right), \tag{51}$$

$$b_1 = \frac{2}{9}R_{qd}^2 + \frac{2}{3}R_d^2 + \frac{1}{3}R_q^2, \tag{52}$$

and

$$b_2 = \frac{1}{324}\left(R_{qd}^2 - 3R_d^2 + 3R_q^2\right)^2. \tag{53}$$

Now, we turn to the LBB model. Using the Levy α-stable generalized forms of the leading order terms in $\tilde{\sigma}_{in}(s, b)$, i.e., Equations (36)–(38), the total cross-section is:

$$\sigma_{tot} = 9\pi\left(2R_q^{\alpha_L}\right)^{2/\alpha_L}. \tag{54}$$

Furthermore, we consider now that the differential cross-section has the form as written in Equation (44). Now, the optical point is:

$$a = \frac{81}{16}\pi(2Rq^{\alpha_L})^{4/\alpha_L}\left(1+4\alpha^2\right),\tag{55}$$

whereas the slope parameter is:

$$b = \frac{1}{36}\left(\frac{4}{3}\right)^{2/\alpha_L}\left((2+2^{\alpha_L})R_{qd}^{\alpha_L}+3^{\alpha_L}\left(2R_d^{\alpha_L}+R_q^{\alpha_L}\right)\right)^{2/\alpha_L}.\tag{56}$$

One can easily check that for $\alpha_L = 2$, Equation (54) reduces to Equation (45), Equation (55) to Equation (51), and Equation (56) to Equation (52). Since the function in Equation (44) is not an analytic function of t at $t = 0$, Equation (56) was obtained by a Taylor expansion in $t^{\alpha_L/2}$ around zero and by keeping only the leading order term.

As discussed in Section 3, the Levy scale parameter r in our simple Lévy α-stable model is related to the slope parameter. The relation can be rewritten as $r = \sqrt{b}/2^{1/\alpha_L}$. Then, this r parameter can be expressed in terms of the LBB model parameters:

$$r = \frac{1}{6}\left(\frac{2}{3}\right)^{1/\alpha_L}\left((2+2^{\alpha_L})R_{qd}^{\alpha_L}+3^{\alpha_L}\left(2R_d^{\alpha_L}+R_q^{\alpha_L}\right)\right)^{1/\alpha_L}.\tag{57}$$

Thus, we have shown that the parameters of our simple Lévy α-stable model, namely, a and b (or equivalently, r), can be approximately expressed in terms of those of the LBB model.

In Ref. [20], the three-dimensional radius of the proton is defined and its relation to the slope parameter is derived. In our work, we related the Levy scale parameter r in our simple Lévy α-stable model to the elastic slope parameter and expressed it in terms of the radii of the constituents of the proton (R_q and R_d) and their typical separation (R_{qd}).

Finally, we note that there are five measurable parameters at the forward region: the total cross-section, the ratio of the real to the imaginary part of the forward scattering amplitude, the optical point, the slope parameter, and the curvature parameter. The ReBB model has four free parameters, whereas the LBB model has five. This naturally suggests that the LBB can give a better description to the data than the ReBB model.

5. Summary

The ReBB model turned out to be an efficient tool in describing pp and $p\bar{p}$ differential cross-section data, but in a limited $\sqrt{s}$ and $-t$ range. The validity range of the ReBB model in $\sqrt{s}$ does not include 13 TeV, possibly due to the significant hollowness effect observed at that energy. The validity range of the ReBB model in $-t$ includes the minimum–maximum structure of the differential cross-section, but does not include the significant non-exponential behavior at low $-t$ values. To overcome these shortcomings of the ReBB model, in this paper, we introduce the Lévy α-stable generalized Real Extended Bialas–Bzdak (LBB) model. The fitting of the parameters of the LBB model to the experimental data, however, requires the solution of difficult and complex technical (mathematical and computational) problems. However, in the low four-momentum transfer region, based on our novel approximations and the idea of the Levy-α-stable-shaped inelastic scattering probability suggested by the LBB model, we deduced and fitted a highly simplified Levy α-stable model of the pp differential cross section to the measured data at $\sqrt{s} = 8$ TeV. The results show that our simple model represents the low-$|t|$ experimental data in a statistically acceptable manner. This is a promising prospect for the future utility of the Lévy α-stable generalized Real Extended Bialas–Bzdak (LBB) model.

We have shown also that the parameters of our simple Lévy α-stable model, namely, a and b (or equivalently, r), can be approximately expressed in terms of those of the LBB model, which is based on R. J. Glauber's multiple diffractive scattering theory. We emphasize that there are five measurable parameters at the forward region, whereas the

ReBB model has only four free parameters. Since the LBB model has five free parameters, it is natural to expect that it can give a better description to the data than the ReBB model.

In the next steps of our research, we are planning to extend the fits with our simple model for all the energies where low-$|t|$ experimental data exist, and after solving the technical issues, to fit the full LBB model to all the existing experimental pp and $p\bar{p}$ differential cross-section data.

Author Contributions: Conceptualization, T.C.; methodology, T.C., S.H. and I.S.; investigation, I.S.; writing—original draft preparation, I.S.; writing—review and editing, T.C. and S.H.; supervision, T.C. All authors have read and agreed to the published version of the manuscript.

Funding: NKFIH Grants no. K133046 and 2020-2.2.1-ED-2021-00181; ÚNKP-22-3 New National Excellence Program of the Ministry for Innovation and Technology from the source of the National Research, Development and Innovation Fund.

Institutional Review Board Statement: Not applicable.

Informed Consent Statement: Not applicable.

Conflicts of Interest: The authors declare no conflict of interest.

References

1. Bialas, A.; Bzdak, A. Constituent quark and diquark properties from small angle proton-proton elastic scattering at high energies. *Acta Phys. Polon.* **2007**, *38* , 159–168.
2. Nemes, F.; Csörgő, T. Detailed Analysis of $p^{+}p$ Elastic Scattering Data in the Quark-Diquark Model of Bialas and Bzdak from $\sqrt{s} = 23.5$ GeV to 7 TeV. *Int. J. Mod. Phys.* **2012**, *27*, 1250175. [CrossRef]
3. Csörgő, T.; Szanyi, I. Observation of Odderon effects at LHC energies: A real extended Bialas–Bzdak model study. *Eur. Phys. J. C* **2021**, *81*, 611. [CrossRef]
4. Szanyi, I.; Csörgő, T. The ReBB model and its H(x) scaling version at 8 TeV: Odderon exchange is a certainty. *Eur. Phys. J. C* **2022**, *82*, 827. [CrossRef]
5. Glauber, R.J.; Matthiae, G. High-energy scattering of protons by nuclei. *Nucl. Phys. B* **1970**, *21*, 135–157. [CrossRef]
6. Glauber, R.J.; Velasco, J. Multiple Diffraction Theory of $p\bar{p}$ Scattering at 546-GeV. *Phys. Lett. B* **1984**, *147*, 380–384. [CrossRef]
7. Nemes, F.; Csörgő, T.; Csanád, M. Excitation function of elastic pp scattering from a unitarily extended Bialas–Bzdak model. *Int. J. Mod. Phys.* **2015**, *30*, 1550076. [CrossRef]
8. Antchev, G.; Aspell, F.; Atanassov, I.; Avati, V.; Baechler, J.; Baldenegro Barrera, C.; Berardi, V.; Berretti, M.; Bossini, E.; Bottigli, U.; et al. Evidence for non-exponential elastic proton–proton differential cross-section at low $|t|$ and $\sqrt{s} = 8$ TeV by TOTEM. *Nucl. Phys. B* **2015**, *899*, 527–546. [CrossRef]
9. Antchev, G.; Aspell, F.; Atanassov, I.; Avati, V.; Baechler, J.; Baldenegro Barrera, C.; Berardi, V.; Berretti, M.; Bossini, E.; Bottigli, U.; et al. Elastic differential cross-section measurement at $\sqrt{s} = 13$ TeV by TOTEM. *Eur. Phys. J. C* **2019**, *79*, 861. [CrossRef]
10. Barbiellini, G.; Bozzo, M.; Darriulat, P.; Palazzi, G.; De Zorzi, G.; Fainberg, A.; Ferrero, M.; Holder, M.; McFarland, A.; Maderni, G.; et al. Small-angle proton-proton elastic scattering at very high energies (460 GeV^{2} < s < 2900 GeV^{2}). *Phys. Lett. B* **1972**, *39*, 663–667. [CrossRef]
11. Csörgő, T.; Pasechnik, R.; Ster, A. Odderon and proton substructure from a model-independent Lévy imaging of elastic pp and $p\bar{p}$ collisions. *Eur. Phys. J.* **2019**, *79*, 62. [CrossRef] [PubMed]
12. Csörgő, T.; Pasechnik, R.; Ster, A. Proton structure and hollowness from Lévy imaging of pp elastic scattering. *Eur. Phys. J.* **2020**, *80*, 126. [CrossRef]
13. Dremin, I.M. Critical regime of proton elastic scattering at the LHC. *JETP Lett.* **2014**, *99*, 243–245. [CrossRef]
14. Dremin, I.M. Interaction region of high energy protons. *Phys. Usp.* **2015**, *58*, 61–70. [CrossRef]
15. Albacete, J.L.; Soto-Ontoso, A. Hot spots and the hollowness of proton–proton interactions at high energies. *Phys. Lett.* **2017**, *770*, 149–153. [CrossRef]
16. Troshin, S.M.; Tyurin, N.E. Experimental signatures of hadron asymptotics at the LHC. *Int. J. Mod. Phys.* **2017**, *32*, 1750103. [CrossRef]
17. Broniowski, W.; Jenkovszky, L.; Ruiz Arriola, E.; Szanyi, I. Hollowness in pp and $p\bar{p}$ scattering in a Regge model. *Phys. Rev.* **2018**, *98*, 074012. [CrossRef]
18. Deppman, A.; Megias, E.; Menezes, D.P. Fractals, nonextensive statistics, and QCD. *Phys. Rev. D* **2020**, *101*, 034019. [CrossRef]

19. Megias, E.; Deppman, A.; Pasechnik, R.; Tsallis, C. Comparative study of the heavy-quark dynamics with the Fokker-Planck Equation and the Plastino-Plastino Equation. *arXiv* **2023**, arXiv:2303.03819.
20. Petrov, V.A.; Okorokov, V.A. The size seems to matter or where lies the "asymptopia"? *Int. J. Mod. Phys. A* **2018**, *33*, 1850077. [CrossRef]

 *universe*

Article

Coulomb Corrections for Bose–Einstein Correlations from One- and Three-Dimensional Lévy-Type Source Functions

Bálint Kurgyis, Dániel Kincses, Márton Nagy and Máté Csanád *

Department of Atomic Physics, Eötvös Loránd University (ELTE), Pázmány Péter Stny. 1/A, H-1117 Budapest, Hungary; kincses@ttk.elte.hu (D.K.); nmarci@elte.hu (M.N.)
* Correspondence: csanad@elte.hu

Abstract: In the study of femtoscopic correlations in high-energy physics, besides Bose–Einstein correlations, one has to take final-state interactions into account. Amongst them, Coulomb interactions play a prominent role in the case of charged particles. Recent measurements have shown that in heavy-ion collisions, Bose–Einstein correlations can be best described by Lévy-type sources instead of the more common Gaussian assumption. Furthermore, three-dimensional measurements have indicated that, depending on the choice of frame, a deviation from spherical symmetry observed under the assumption of Gaussian source functions persists in the case of Lévy-type sources. To clarify such three-dimensional Lévy-type correlation measurements, it is thus important to study the effect of Coulomb interactions in the case of non-spherical Lévy sources. We calculated the Coulomb correction factor numerically in the case of such a source function for assorted kinematic domains and parameter values using the Metropolis–Hastings algorithm and compared our results with previous methods to treat Coulomb interactions in the presence of Lévy sources.

Keywords: femtoscopy; Bose–Einstein correlations; Lévy distribution; anomalous diffusion; heavy-ion collisions; high-energy physics

Citation: Kurgyis, B.; Kincses, D.; Nagy, M.; Csanád, M. Coulomb Corrections for Bose–Einstein Correlations from One- and Three-Dimensional Lévy-Type Source Functions. *Universe* **2023**, *9*, 328. https://doi.org/10.3390/universe9070328

Academic Editor: Maxim Chernodub

Received: 12 June 2023
Revised: 3 July 2023
Accepted: 8 July 2023
Published: 10 July 2023

1. Introduction

The investigation of Bose–Einstein or HBT correlations offers a way to gain information about the space-time dynamics of heavy-ion collisions on the femtometer scale. Such information can lead to a better understanding of the space-time geometry of the collision and particle production mechanisms and could even indicate critical phenomena [1–4].

For the study of Bose–Einstein correlation functions, one usually makes an assumption for the source function. There is now a large amount of evidence showing that in heavy-ion collisions, there is indeed a significant deviation from Gaussian shape, such as source imaging results showing a long-range, power-law-type component [5–7]. It turns out that a suitable choice is a Lévy-type source function [4,8,9].

In one-dimensional correlation measurements, the correlation function is measured as a function of only one relative momentum variable, the magnitude of the momentum difference. This type of measurement assumes the spherical symmetry of the spatial source (and thus the momentum space correlation) and is more suitable for situations wherein a lack of experimental statistics prevents the detailed mapping of the momentum space. Three-dimensional measurements, on the other hand, can yield further information about the space-time geometry of the source, so whenever experimental statistics make it possible, it is desirable to perform such measurements. Building on the substantial progress that three-dimensional Gaussian measurements have made in the understanding of the space-time structure of particle production in heavy-ion collisions, it is also of interest to perform Lévy-type measurements in a three-dimensional setting. The first such measurement has already been reported [10].

In the present paper, we aimed at developing a methodology for such a measurement. Coulomb corrections are an essential ingredient of all HBT correlation measurements that use identical charged particles, as most do: the final-state Coulomb repulsion of the outgoing particles modifies the shape of the observed correlation function in a complicated manner, and in experimental analyses, one usually applies a correction factor, the Coulomb correction, to account for this effect [11,12]. At present, the Coulomb correction for Lévy distributions is available only in the spherically symmetric case [13]. Our goal was two-fold: first, we investigated the Coulomb correction for three-dimensional Lévy sources and determined a sound method for its use in experimental work. Second, in doing so we encountered the question of the proper choice of coordinate frame, namely the longitudinally co-moving system (LCMS) and pair center of mass system (PCMS) of the particle pair. We thus investigated the implications of using these coordinate frames for the measurements and calculations.

1.1. Two-Particle Correlation Functions

The n-particle correlation functions are defined as

$$C_n = \frac{N_n(k_1, \cdots k_n)}{\prod_{i=1}^{n} N_1(k_i)},\tag{1}$$

where N_n is the n-particle invariant momentum distribution. In a statistical picture, one introduces the source function $S(x, k)$ that characterizes the particle production at a given space-time point x and momentum k, and writes up the $N_n(k_1, \ldots, k_n)$ distribution using this function and the n-particle wave function $\psi_n(x_1, \cdots x_n, k_1, \cdots k_n)$ as

$$N_n(k_1, \cdots k_n) = \int |\psi_n(x_1, \cdots x_n, k_1, \cdots k_n)|^2 \prod_{i=1}^{n} S(x_i, k_i)\mathrm{d}x_i.\tag{2}$$

In particular, for single-particle distributions, we have $|\psi_1|^2 = 1$. Thus,

$$N_1(k) = \int \mathrm{d}x\, S(x, k),\tag{3}$$

and so one obtains the two-particle correlation function as

$$C_2(k_1, k_2) = \frac{\int \mathrm{d}x_1 \mathrm{d}x_2\, S(x_1, k_1) S(x_2, k_2) |\psi_2(x_1, k_1, x_2, k_2)|^2}{\int \mathrm{d}x_1\, S(x_1, k_1) \int \mathrm{d}x_2\, S(x_2, k_2)}.\tag{4}$$

We introduced the average and relative space-time and momentum variables as

$$q \equiv k_1 - k_2, \qquad K \equiv \frac{k_1 + k_2}{2}, \qquad \rho \equiv x_1 - x_2, \qquad R = \frac{x_1 + x_2}{2}.\tag{5}$$

Although the two-particle wave function appearing in the above formula itself is a function of all relevant variables, its modulus square depends only on ρ and q, as shown below. We assumed this in advance and used a notation that reflected this by suppressing the K and R variables in the notation of $|\psi_2|^2$.

The overall normalization of the $S(x, k)$ source function cancels from the C_2 correlation function, so from now on we treat it as unity, i.e., we write

$$C_2(q, K) = \int |\psi_2(q, \rho)|^2 S(R - \tfrac{1}{2}\rho, K - \tfrac{1}{2}q) S(R + \tfrac{1}{2}\rho, K + \tfrac{1}{2}q)\, \mathrm{d}\rho\, \mathrm{d}R.\tag{6}$$

A usual approximation is that one neglects the $\pm\tfrac{1}{2}k$ in the arguments of the source functions; in other words, one approximates the k_1 and k_2 momenta in the source functions as $k_1 \approx k_2 \approx K$. In doing so, we noted that it was useful to introduce the relative coordinate distribution, also called pair distribution, $D(\rho, K)$, which is the auto-convolution of the source function in the first variable:

$$D(\rho, K) = \int \mathrm{d}R\, S(R + \tfrac{1}{2}\rho, K) S(R - \tfrac{1}{2}\rho, K),\tag{7}$$

with which the two-particle correlation function could then be expressed as

$$C_2(q,K) = \int |\psi_2(q,\rho)|^2 D(\rho,K)\mathrm{d}\rho. \tag{8}$$

1.2. Lévy Sources

For the source function, we assumed a symmetric Lévy distribution [14]:

$$S(r,K) = \mathcal{L}^{(4D)}\left(r^\mu, \alpha(K), R^2_{\sigma v}(K)\right) = \int \frac{\mathrm{d}^4 q}{(2\pi)^4} e^{iq_\mu r^\mu} e^{-\frac{1}{2}|q^\sigma R^2_{\sigma v} q^v|^{\alpha/2}}, \tag{9}$$

where α is the Lévy exponent, and $R^2_{\sigma v}$ is a two-index symmetric tensor containing the squares of the Lévy scale parameters. The momentum dependence of the source was assumed to manifest itself through the momentum dependence of these parameters. For such a Lévy-type source, the relative coordinate distribution $D(r,K)$ is itself a Lévy distribution with the same α but with scale parameters modified as $R^2 \to 2^{2/\alpha} R^2$.

By choosing a reference frame and making some assumptions, we constrained the form of the $R^2_{\sigma v}$ matrix. In the case of Bose–Einstein correlation measurements in high-energy heavy-ion collisions, one sets the laboratory frame as the center-of-mass frame of the colliding nuclei. Most two-particle Bose–Einstein measurements are carried out with respect to the so-called longitudinally co-moving system (LCMS) (see, e.g., Refs. [3,4]), which is defined as the frame that is connected to the laboratory frame by a Lorentz boost along the collision axis (z axis), with the criterion that the longitudinal component of the average momentum of the particle pair K_z vanishes in this frame.

We made the assumption that our source could be described by a spatially three-dimensional symmetric Lévy shape with only diagonal terms in the scale parameter matrix $R^2_{\sigma v}$, and that the freeze-out was simultaneous in the LCMS frame. We saw that the momentum variable of the source function translated essentially as the average momentum of the particle pair (after the $k_1 \approx k_2 \approx K$ approximation), so for $S(x,K)$, we determined the mean LCMS frame as the frame wherein its K variable has no K_z component. Therefore, the $R^2_{\sigma v}$ tensor has the following form:

$$R^2_{\sigma v} = \begin{pmatrix} 0 & 0 & 0 & 0 \\ 0 & R^2_{\text{out}} & 0 & 0 \\ 0 & 0 & R^2_{\text{side}} & 0 \\ 0 & 0 & 0 & R^2_{\text{long}} \end{pmatrix}, \tag{10}$$

where out, side, and long indicate that we used Bertsch–Pratt coordinates [15,16]. We could then simplify the four-dimensional Lévy distribution as a product of a Dirac delta function and a three-dimensional symmetric Lévy distribution:

$$\mathcal{L}^{(4D)} = \delta(t^L)\mathcal{L}^{(3D)}\left(\vec{r}^L, \alpha, R_{\text{out}}, R_{\text{side}}, R_{\text{long}}\right), \tag{11}$$

$$\mathcal{L}^{(3D)}\left(\vec{r}^L, \alpha, R_{\text{out}}, R_{\text{side}}, R_{\text{long}}\right) = \int \frac{\mathrm{d}^3 q}{(2\pi)^3} e^{-i\vec{q}\vec{r}^L} e^{-\frac{1}{2}|q^2_{\text{out}} R^2_{\text{out}} + q^2_{\text{side}} R^2_{\text{side}} + q^2_{\text{long}} R^2_{\text{long}}|^{\alpha/2}}, \tag{12}$$

where the L superscript indicates that these coordinates are in the LCMS. As noted above, for such a source function, the pair distribution is a Lévy distribution with modified scale parameters:

$$D(\vec{r}^L, K) = \delta(t^L)\mathcal{L}^{(3D)}\left(\vec{r}^L, \alpha, 2^{\frac{1}{\alpha}} R_{\text{out}}, 2^{\frac{1}{\alpha}} R_{\text{side}}, 2^{\frac{1}{\alpha}} R_{\text{long}}\right), \tag{13}$$

From this, in the case when final-state interactions were neglected and thus the final-state wave function was a symmetrized plane wave, we could easily obtain the form of the two-particle correlation function in the LCMS with the above-mentioned source by means of an (inverse) Fourier transform [14]:

$$C_2^{(0)}(\vec{q}, \alpha, R_{\text{out}}, R_{\text{side}}, R_{\text{long}}) = 1 + e^{-|q_{\text{out}}^2 R_{\text{out}}^2 + q_{\text{side}}^2 R_{\text{side}}^2 + q_{\text{long}}^2 R_{\text{long}}^2|^{\alpha/2}}. \tag{14}$$

2. Methodology

2.1. Coulomb Interaction

To take into account the Coulomb interaction, one has to use the Coulomb interacting two-particle wave function. This is the solution of the two-particle Schrödinger equation with a repulsive Coulomb force and the appropriate boundary conditions at infinity.

Utilizing the Schrödinger equation implies a non-relativistic treatment, which is a justifiable approximation in the PCMS (pair co-moving system) frame, i.e., the center-of-mass frame of the two particles. The solution of the Schrödinger equation of interest to us is written as [11,17]

$$\psi(\vec{R}^P, \vec{r}^P, \vec{K}^P, \vec{k}^P) = \frac{N}{\sqrt{2}} e^{-i2\vec{K}\vec{R}} \left[e^{i\vec{k}\vec{r}} F(-i\eta, 1, i(kr - \vec{k}\vec{r})) + \right.$$
$$\left. + e^{-i\vec{k}\vec{r}} F(-i\eta, 1, i(kr + \vec{k}\vec{r})) \right], \tag{15}$$

where a symmetrization has been performed, as required for pairs of identical bosons. In this expression, $F(a, b, z)$ is the confluent hypergeometric function, $\vec{k} = \vec{q}/2$, $k = |\vec{k}|$, and

$$\eta = \frac{mc^2 \alpha}{2\hbar ck}, \qquad\qquad N = e^{-\frac{\pi\eta}{2}} \Gamma(1 + i\eta), \tag{16}$$

where α is the fine-structure constant; m is the particle mass (e.g., pion mass); and $\Gamma(z)$ is the gamma function.

To evaluate the two-particle correlation function, we needed the modulus square of the wave function, with which the $\vec{R}$ and $\vec{K}$ dependence was lost (as mentioned earlier):

$$|\psi(\vec{r}^P, \vec{k}^P)|^2 = \frac{2\pi\eta}{e^{2\pi\eta} - 1} \cdot \frac{1}{2} \cdot \left[|F(-i\eta, 1, i(kr + \vec{k}\vec{r}))|^2 + \right.$$
$$\left. + e^{2i\vec{k}\vec{r}} F(-i\eta, 1, i(kr - \vec{k}\vec{r})) F(i\eta, 1, -i(kr + \vec{k}\vec{r})) \right] + (\vec{r} \leftrightarrow -\vec{r}). \tag{17}$$

To arrive at the two-particle correlation function, one has to evaluate a $d^4 r$ integral over the whole space-time. This can be performed in any coordinate frame, but our approximations detailed above made strong arguments in favor of some preferred coordinate systems. However, even with these in mind, we had several options to explore:

1. We could assume that the $R_{\sigma\nu}^2$ matrix, and thus the whole source function, is the same in the PCMS and the LCMS frames. This is essentially an approximation where $\vec{K} \approx 0$. However, this is a rather strong approximation, and one of the goals of HBT measurements is indeed to explore the average momentum (or transverse mass) dependence of the parameters that describe the source.

2. There are two objects, one in the PCMS (the wave function) and the other in the LCMS (the source function). We could try to transform the wave function from the PCMS to the LCMS and then use the simple form of the source function and obtain the result in LCMS coordinates. However, the two-particle wave function of Equation (17) is not a relativistic expression; thus, we refrained from trying to come up with the right transformation of this object.

3. The third option was to evaluate the integral in the PCMS, as the two-particle Coulomb wave function is only known in the PCMS. This meant that the Lévy source had to be transformed from the LCMS to the PCMS.

Below, we proceed with the third option listed above. We introduce some further notations: the average transverse momentum in the LCMS K_T, the transverse mass $m_T = \sqrt{m^2 + K_T^2}$, and the $\beta_T = K_T / m_T$ factor. The Lorentz boost from the LCMS to the PCMS is then

$$\Lambda^\nu_\mu = \frac{1}{m} \begin{pmatrix} m_T & -K_T & 0 & 0 \\ -K_T & m_T & 0 & 0 \\ 0 & 0 & m & 0 \\ 0 & 0 & 0 & m \end{pmatrix}. \tag{18}$$

The Lévy distribution then transforms as a scalar from the LCMS to the PCMS, meaning we had to evaluate Equation (11) at the coordinates $r' = \Lambda^{-1} r$, where the transformation is the following:

$$\begin{pmatrix} t^L \\ \vec{r}^L \end{pmatrix} = \frac{1}{m} \begin{pmatrix} m_T t^P + K_T r^P_{\text{out}} \\ K_T t^P + m_T r^P_{\text{out}} \\ m r^P_{\text{side}} \\ m r^P_{\text{long}} \end{pmatrix}. \tag{19}$$

The temporal integral could then be easily evaluated and, subsequently, knowing the form of $D(\vec{r}, K)$ from Equation (13) above, we were left with the following expression (where $2\vec{k} = \vec{q}$):

$$C_2^{(C)}(\vec{q}) = \int d^3 r |\psi(\vec{k}, \vec{r})|^2 \mathcal{L}^{(3D)}\left(\sqrt{1 - \beta_T^2} r_{\text{out}}, r_{\text{side}}, r_{\text{long}}, \alpha, 2^{\frac{1}{\alpha}} R_{\text{out}}, 2^{\frac{1}{\alpha}} R_{\text{side}}, 2^{\frac{1}{\alpha}} R_{\text{long}}\right), \tag{20}$$

where we dropped the P superscripts for simplicity, but every momentum and spatial coordinate is in the PCMS. Furthermore, we could utilize a simple scaling relation of the three-dimensional Lévy distribution:

$$\mathcal{L}^{(3D)}\left(\sqrt{1 - \beta_T^2} r_{\text{out}}, r_{\text{side}}, r_{\text{long}}, \alpha, 2^{\frac{1}{\alpha}} R_{\text{out}}, 2^{\frac{1}{\alpha}} R_{\text{side}}, 2^{\frac{1}{\alpha}} R_{\text{long}}\right) \sim$$
$$\sim \mathcal{L}^{(3D)}\left(\vec{r}, \alpha, 2^{\frac{1}{\alpha}} R_{\text{out}} / \sqrt{1 - \beta_T^2}, 2^{\frac{1}{\alpha}} R_{\text{side}}, 2^{\frac{1}{\alpha}} R_{\text{long}}\right), \tag{21}$$

a relation that could be derived by scaling the $\vec{q}$ integration variable in the definition of $\mathcal{L}^{(3D)}$, Equation (11). In this equation, $\sim$ stands for proportionality, and constant factors in $S(x, k)$ cancel from the two-particle correlation function. Thus, the integral we intended to calculate was

$$C_2^{(C)}(\vec{q}, \alpha, R_1, R_2, R_3) = \int d^3 r |\psi(\vec{k}, \vec{r})|^2 \mathcal{L}^{(3D)}(\vec{r}, \alpha, R_1, R_2, R_3), \tag{22}$$

where $R_1 = 2^{\frac{1}{\alpha}} R_{\text{out}} / \sqrt{1 - \beta_T^2}$, $R_2 = 2^{\frac{1}{\alpha}} R_{\text{side}}$, $R_3 = 2^{\frac{1}{\alpha}} R_{\text{long}}$. This expression could be evaluated numerically.

2.2. Numerical Simulations

For the evaluation of the integral, we utilized the Metropolis–Hastings algorithm. This algorithm can be used to evaluate integrals of the form

$$I = \int_\Omega dx f(x) \cdot g(x), \tag{23}$$

where $f(x)$ can be thought of as a probability distribution, and $g(x)$ is the function of interest [18,19]. In our case, the three-dimensional symmetric Lévy distribution was the probability distribution, and the function of interest was from Equation (17):

$$f(x) dx := \mathcal{L}^{(3D)}(\vec{r}, \alpha, R_1, R_2, R_3) d^3 r, \tag{24}$$
$$g(x) := |\psi(\vec{k}, \vec{r})|^2. \tag{25}$$

We could utilize two transformations. First, with the reflection relations of the confluent hypergeometric functions, we used the second term in Equation (17):

$$e^{2i\vec{k}\vec{r}}F(-i\eta,1,i(kr-\vec{k}\vec{r}))F(i\eta,1,-i(kr+\vec{k}\vec{r})) =$$
$$= F(1+i\eta,1,-i(kr-\vec{k}\vec{r}))F(1-i\eta,1,-i(kr+\vec{k}\vec{r})). \tag{26}$$

Additionally, we could transform the 3D symmetric Lévy distribution:

$$\mathcal{L}^{(3D)}(\vec{r},\alpha,R_1,R_2,R_3) = \frac{1}{R_1 R_2 R_3}\mathcal{L}^{(1D)}(s(\vec{r}),\alpha,1), \tag{27}$$

where

$$s(\vec{r}) = \sqrt{\frac{r_{\text{out}}^2}{R_1^2} + \frac{r_{\text{side}}^2}{R_2^2} + \frac{r_{\text{long}}^2}{R_3^2}}, \tag{28}$$

and $\mathcal{L}^{(1D)}$ is the spherically symmetric version of the three-dimensional Lévy distribution, as a function of the radial variable. Its its expression is then

$$\mathcal{L}^{(1D)}(x,\alpha,1) = \frac{1}{2\pi^2 x}\int_0^\infty dq\, q^2 \sin(qx)e^{-\frac{1}{2}q^\alpha}. \tag{29}$$

We could thus perform the integral in Equation (22); this was carried out using spherical coordinates on the domain $\Omega = [0,r_{\text{max}}] \times [0,2\pi] \times [0,\pi]$, with an r_{max} chosen so that the integral of the Lévy distribution ($I = \int \mathcal{L}$) was a maximum of 1% less than 1 ($I \geq 0.99$).

3. Results

First, we compared our three-dimensional calculations and other available, spherically symmetric calculations for Lévy sources. Then, we investigated the implications of the fact that most measurements are in the LCMS, and the source is assumed to be spherical there for one-dimensional analyses, but the integral of Equation (22) is in the PCMS.

3.1. Three-Dimensional Calculations

Three-dimensional calculations are rather time-consuming, and their numerical precision could also be problematic for implementation when investigating experimental data. Instead, we aimed to find an approximation that was precise and fast enough to be utilized in actual experimental analyses. Our approach here was that we fixed a set of parameters (α, R_1, R_2, R_3) and evaluated the integral at 100^3 points in momentum space. This gave us a fine enough resolution in momentum space for comparison purposes. First, let us compare the two-particle correlation functions in the PCMS. In Figure 1, we can see the Bose–Einstein correlation functions with Coulomb interactions (full BEC) and without any final-state interactions (free BEC) from our 3D calculation and from the 1D calculation with quadratic and arithmetic average scale parameters and the angle averaged values of the 3D calculation. In the spherical case, on the left-hand plot, everything was as we would expect; however, on the right-hand plot, when we had a non-spherical source for the 3D calculation, we can see that there was a large difference between the correlation functions, both in the Coulomb interacting and in the free case.

However, we were interested in the question of whether we could use the 1D calculation for the purposes of Coulomb correction only, viz., the ratio of the full and free BEC functions ($K = C_2^{(C)}/C_2^{(0)}$). One can see the comparison of Coulomb corrections in Figure 2 with two sets of non-spherical parameters. The full BEC functions are here the Coulomb-corrected three-dimensional correlation functions (full BEC $= K \cdot C_{2,3D}^{(0)}$). The one-dimensional Coulomb corrections were evaluated at $|\vec{q}|$ in the PCMS, i.e., at q_{inv} and at an average R for R_1, R_2 and R_3. Although the correlation functions were quite different, we can see that the Coulomb corrections were very much the same. Now, we would like to point out the fact that one-dimensional and three-dimensional Coulomb corrections are very similar; therefore, in an experimental analysis, it is sufficient to use a one-dimensional Coulomb correction with the right parameter values. The error caused by the spherical Coulomb correction could be estimated, but it was not in the scope of this paper to give a quantitative limit for this uncertainty.

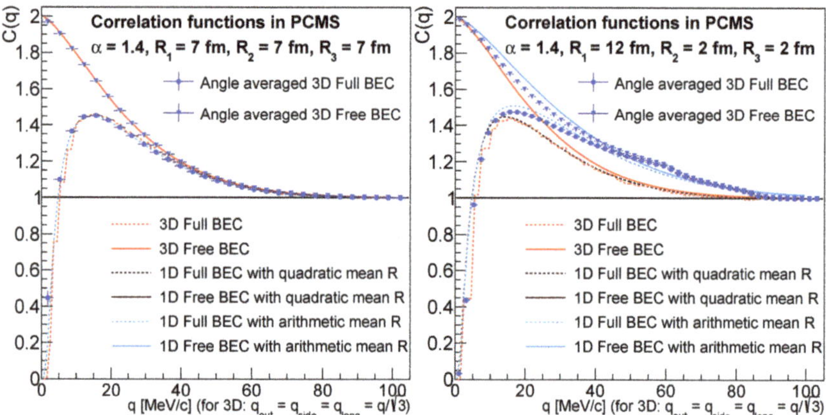

Figure 1. On the left-hand side, the two-particle correlation functions are shown in a spherical case for the three-dimensional calculation in comparison with one-dimensional calculations in the presence of Coulomb interactions in final-state interactions. On the right-hand side, a non-spherical three-dimensional calculation is shown alongside one-dimensional calculations with quadratic and arithmetic mean scale parameters.

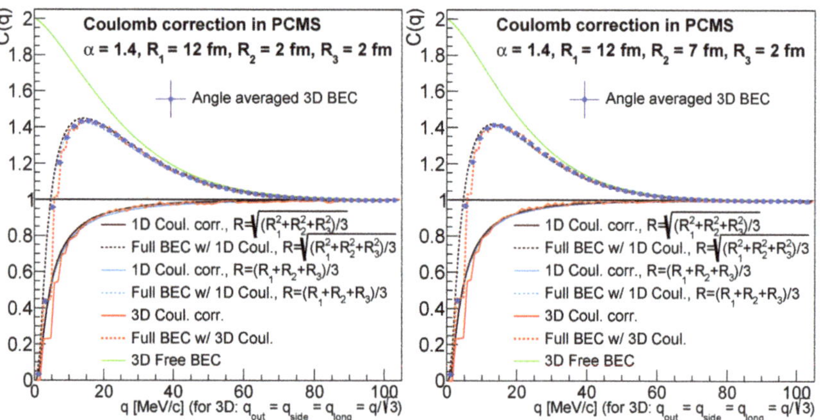

Figure 2. The Coulomb corrections and the Coulomb-corrected three-dimensional two-particle correlation function is shown in two non-spherical cases.

The application of the Coulomb correction in three-dimensional analyses is quite straightforward. If the measurement is in the LCMS and one has the momenta $q^L = q_{out}^L, q_{side}^L, q_{long}^L$ and the Lévy scale parameters $R_{out}, R_{side}, R_{long}$ for particles with an average transverse momentum of K_T, which gives β_T, then one proceeds as follows. We used the assumption that the Coulomb correction transformed as a scalar. We evaluated the Coulomb correction (which was calculated in the PCMS) at momenta $q^P = (\sqrt{1 - \beta_T^2}q_{out}^L, q_{side}^L, q_{long}^L)$ and scale parameters $R_1 = R_{\text{out}}/\sqrt{1 - \beta_T^2}$, $R_2 = R_{\text{side}}$, and $R_3 = R_{\text{long}}$. Accordingly, we used $q_{\text{inv}} = \sqrt{(1 - \beta_T^2)q_{\text{out}}^{L2} + q_{\text{side}}^{L2} + q_{\text{long}}^{L2}}$ and an average of R_1, R_2, and R_3 when we used a 1D Coulomb correction. For example, we could use the quadratic average:

$$R_{\text{PCMS}} = \sqrt{\frac{R_{\text{out}}^2}{1 - \beta_T^2} + R_{\text{side}}^2 + R_{\text{long}}^2}. \tag{30}$$

Therefore, the Coulomb correction that could be applied in a three-dimensional measurement was the following:

$$K_{\text{3D}} = \frac{C_{2,1D}^{(C)}(q_{\text{inv}}, R_{\text{PCMS}}, \alpha)}{1 + \exp\left(-|q_{\text{inv}} R_{\text{PCMS}}|^\alpha\right)}, \tag{31}$$

where $C_{2,1D}^{(C)}$ is the result from the integral of Equation (22) in a spherical case with a radius of R_{PCMS} according to Equation (30) and at momentum q_{inv}, which can be calculated for every point in a three-dimensional measurement in the LCMS.

3.2. Spherical (One-Dimensional) HBT Measurements

Below, we investigate the implications of our calculations for one-dimensional HBT measurements. When we performed a one-dimensional measurement in the LCMS, we assumed that the source was spherical in this frame, i.e., $R = R_{\text{out}} = R_{\text{side}} = R_{\text{long}}$, and we had a single momentum variable $q_{\text{LCMS}} = \sqrt{q_{\text{out}}^{L2} + q_{\text{side}}^{L2} + q_{\text{long}}^{L2}}$. But the Coulomb correction was calculated in the PCMS with R_1, R_2, R_3. This meant that a spherical source in the LCMS would imply a non-spherical ($R_1 = R/\sqrt{1 - \beta_T^2}$, $R_2 = R_3 = R$) source in the PCMS and the need for a three-dimensional Coulomb correction. However, we saw above that the non-spherical Coulomb correction could be well approximated with a spherical Coulomb correction if we used the right average R, viz., instead of $R_{\text{LCMS}} = R$, we had to use

$$R_{\text{PCMS}} = \sqrt{\frac{1 - \frac{2}{3}\beta_T^2}{1 - \beta_T^2}} R, \tag{32}$$

if we used a quadratic average R. Another problem stemmed from the fact that we could not reconstruct q_{inv} from q_{LCMS}. An obvious solution would be to measure all momentum variables instead of just the length of the momentum difference, but then the advantage of the 1D measurement over the 3D measurement (the possibility of a measurement with higher statistical significance) would be lost. We could try to overcome this obstacle in some other ways. One solid approximation could be the following: measure an $A(q_{\text{LCMS}}, q_{\text{inv}})$ distribution of particle pairs, and then use this to obtain a weighted Coulomb-correction, as shown below.

$$K_{\text{weighted}}(q_{\text{LCMS}}) = \frac{\int A(q_{\text{LCMS}}, q_{\text{inv}}) K(q_{\text{inv}}) \mathrm{d}q_{\text{inv}}}{\int A(q_{\text{LCMS}}, q_{\text{inv}}) \mathrm{d}q_{\text{inv}}}. \tag{33}$$

In Figure 3, we see the Coulomb correction and the corrected three-dimensional two-particle correlation functions for $K_T = 0.8\,\text{GeV}/c$ in the LCMS. The parameters were chosen so that in the LCMS we had an approximately spherically symmetric source ($R_{\text{out}} = 2.06\,\text{fm}$, $R_{\text{side}} = R_{\text{long}} = 2\,\text{fm}$). We can see that there was a clear difference between the two one-dimensional corrections, with one having an LCMS average R and the other having an average in accordance with Equation (32). In the low-q region, there was some difference between the angle-averaged, one-dimensional, and three-dimensional Coulomb corrections. Also, the numerical precision of the three-dimensional calculation made it challenging to decide between the options. However, we can see that from $q > 20\,\text{MeV}/c$, the angle-averaged and the three-dimensional Coulomb correction were in good agreement with the one-dimensional Coulomb correction with the average R of Equation (32), and there was a consistent difference compared to the other one. The fact that the angle-averaged case was most similar to the one-dimensional case with the transformed average R of Equation (32) indicated that using the latter for one-dimensional measurements was best. On the left-hand side, the three-dimensional correlation function was taken at a diagonal line in the LCMS ($q_{\text{out}} = q_{\text{side}} = q_{\text{long}}$), and on the right-hand side along the out axis. We did not rely on a weighted average for the one-dimensional Coulomb correction, as we could calculate q_{inv}.

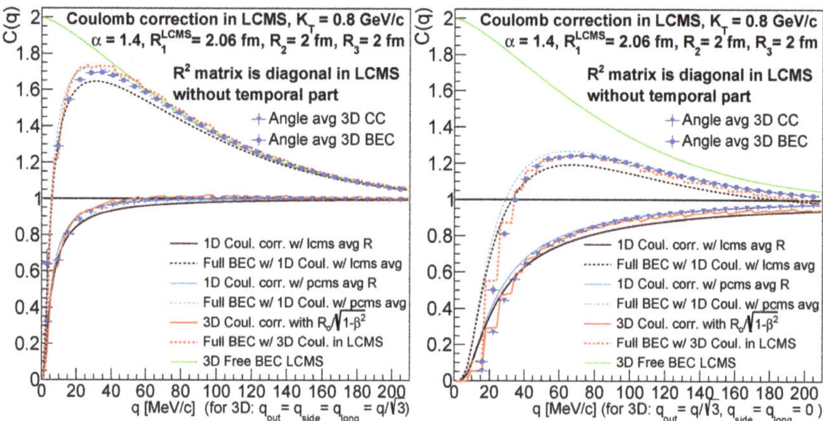

Figure 3. The Coulomb corrections and the Coulomb-corrected three-dimensional two-particle correlation function are shown in the LCMS when the source was spherical in the LCMS but not for the calculation. On the left-hand side, we took the three-dimensional Coulomb correction along a diagonal line, and on the right-hand side along the q_{out} axis.

Let us list the possible approaches to deal with Coulomb interactions in one-dimensional measurements carried out in an LCMS. We only list the options that make use of a one-dimensional calculation for the integral in Equation (22); in these cases, the factor of Ref. [13] can be used. A simpler solution would be to use the Gamow factor, where the source size is neglected. The most sophisticated approach would be to use the angle-averaged Coulomb correction from a three-dimensional calculation, but this would be an overly complex solution. The possibilities for making use of a one-dimensional Coulomb integral calculation are the following, ordered by increasing sophistication:

1. Simply use $C_2^{(C)}(q_{\text{LCMS}}, R_{\text{LCMS}})$, which means that one formally substitutes $q_{\text{LCMS}} = q_{\text{inv}}$ and $R_{\text{PCMS}} = R_{\text{LCMS}}$.
2. Take into account the fact that $q_{inv} \neq q_{\text{LCMS}}$ but neglect the same for the scale parameters, and use the weighting method of Equation (33); however, implement this not for the Coulomb correction, but for the correlation function instead. Thus, use $C_{2,\text{weighted}}(q_{\text{LCMS}}, R_{\text{LCMS}})$ for the fitting:

$$C_{2,\text{weighted}}(q_{\text{LCMS}}, R_{\text{LCMS}}) = \frac{\int A(q_{\text{LCMS}}, q_{\text{inv}}) C_2(q_{\text{inv}}, R_{\text{LCMS}}) \mathrm{d}q_{\text{inv}}}{\int A(q_{\text{LCMS}}, q_{\text{inv}}) \mathrm{d}q_{\text{inv}}}. \tag{34}$$

3. Following the same approach as above, use R_{LCMS} for the Coulomb correction and use a weighted average, though for the Coulomb correction this time. This approach is more sensible if one considers Figure 1, where we saw that the correlation functions could look rather different even if in Figure 2 the Coulomb corrections looked very much the same. Now, one uses $K_{\text{weighted}}(q_{\text{LCMS}}, R_{\text{LCMS}}) \cdot C_2^{(0)}(q_{\text{LCMS}}, R_{\text{LCMS}})$ for fitting.

4. One improvement to the methods mentioned above would be to consider the transformation of scale parameters; thus, use the average as in Equation (32). The simpler version is the same as no. 3 above, i.e., weighing the correlation function and using $C_{2,\text{weighted}}(q_{\text{LCMS}}, R_{\text{PCMS}})$ for fitting. Here, however, one loses the explicit form of $C_2^{(0)}$ in the LCMS, which is known.

5. The most sophisticated option would be to use R_{PCMS} only for the Coulomb correction and use the weighting of Equation (33). The function used for fitting is now $K_{\text{weighted}}(q_{\text{LCMS}}, R_{\text{PCMS}}) \cdot C_2^{(0)}(q_{\text{LCMS}}, R_{\text{LCMS}})$.

6. Finally, an approach that is easier to implement than the previous methods making use of a distribution $A(q_{\text{LCMS}}, q_{\text{inv}})$ is to make an approximation for the q_{LCMS}-q_{inv} relationship that is appropriate for the Coulomb correction. One could be motivated by the left-hand plot of Figure 3, as the one-dimensional Coulomb correction with R_{PCMS} and the angle-averaged three-dimensional calculation were in relatively good agreement. The relationship $q_{\text{inv}} = \sqrt{1 - \beta_T^2/3} q_{\text{LCMS}}$ could be used, as it would hold for the diagonal line $q_{\text{out}} = q_{\text{side}} = q_{\text{long}}$. Therefore, the function that could be used for fitting would be $K(\sqrt{1 - \beta_T^2/3} q_{\text{LCMS}}, R_{\text{PCMS}}) \cdot C_2^{(0)}(q_{\text{LCMS}}, R_{\text{LCMS}})$.

Additionally, either the distribution of particle pairs from same events (usually denoted with A) or some background distribution that has no quantum-statistical effects (B) could be used for weighting C_2 and K [4]. Here, one could argue in favor of the latter; however, it is expected to make a small difference. The soundest approach for one-dimensional analyses is no. 5 in the above list.

4. Conclusions

We investigated Coulomb interactions for HBT measurements in the presence of Lévy sources. Our results can be applied to three-dimensional and one-dimensional measurements alike. The results also hold for Gaussian or Cauchy sources, because these are special cases of the Lévy source ($\alpha = 2$ for Gaussian and $\alpha = 1$ for Cauchy). We learned that a one-dimensional Coulomb correction could be reasonably effectively applied for three-dimensional measurements if we used the appropriately defined average of the three directional scale parameters (as in Equation (30) above) and implemented the q_{inv} invariant momentum difference as the momentum variable for the Coulomb correction. For one-dimensional measurements in the LCMS frame, we saw that one should use the average scale parameter as defined in Equation (32) and evaluate the Coulomb correction at q_{inv} as we calculated this in the PCMS frame, which in practice could be estimated with a weighted Coulomb correction according to option no. 5 in the previous section. The above-detailed treatment of Coulomb interactions in heavy-ion collisions could be readily applied to experimental measurements. To our knowledge this indeed has now been achieved in several analyses from SPS through RHIC to LHC, based on the technique outlined in this paper [8,9,20–24].

Author Contributions: Conceptualization, M.C. and B.K.; methodology, M.C. and M.N.; software, B.K., M.N. and D.K.; formal analysis, B.K.; investigation, B.K.; writing—original draft preparation, B.K. and M.C.; writing—review and editing, D.K. and M.N.; visualization, B.K. and D.K.; supervision, M.C.; project administration, M.C.; funding acquisition, M.C. All authors have read and agreed to the published version of the manuscript.

Funding: NKFIH grant No. K-138136.

References

1. Csörgö, T. *Particle Interferometry from 40 Mev to 40 TeV*; NATO Science Series C: Mathematical and Physical Sciences; Springer: Dordrecht, The Netherlands, 2000; Volume 554, Chapter 8, pp. 203–257. [CrossRef]
2. Bolz, J.; Ornik, U.; Plumer, M.; Schlei, B.; Weiner, R. Resonance decays and partial coherence in Bose-Einstein correlations. *Phys. Rev. D* **1993**, *D47*, 3860–3870. [CrossRef] [PubMed]
3. Adamczyk, L.; Adkins, J.K.; Agakishiev, G.; Aggarwal, M.M.; Ahammed, Z.; Alekseev, I.; Alford, J.; Anson, C.D.; Aparin, A.; Arkhipkin, D.; et al. Beam-energy-dependent two-pion interferometry and the freeze-out eccentricity of pions measured in heavy ion collisions at the STAR detector. *Phys. Rev. C* **2015**, *C92*, 014904. [CrossRef]
4. Adare, A.; Aidala, C.; Ajitanand, N.N.; Akiba, Y.; Akimoto, R.; Alexander, J.; Alfred, M.; Al-Ta'ani, H.; Angerami, A.; Aoki, K.; et al. Lévy-stable two-pion Bose-Einstein correlations in $\sqrt{s_{NN}} = 200$ GeV Au+Au collisions. *Phys. Rev. C* **2018**, *C97*, 064911. [CrossRef]
5. Adler, S.S.; Afanasiev, S.; Aidala, C.; Ajitanand, N.N.; Akiba, Y.; Alexander, J.; Amirikas, R.; Aphecetche, L.; Aronson, S.H.; Averbeck, R.; et al. Evidence for a long-range component in the pion emission source in Au + Au collisions at $\sqrt{s_{NN}} = 200$ GeV. *Phys. Rev. Lett.* **2007**, *98*, 132301. [CrossRef]
6. Afanasiev, S.; Aidala, C.; Ajitanand, N.N.; Akiba, Y.; Alexander, J.; Al-Jamel, A.; Aoki, K.; Aphecetche, L.; Armendariz, R.; Aronson, S.H.; et al. Source breakup dynamics in Au+Au Collisions at $\sqrt{s_{NN}} = 200$ GeV via three-dimensional two-pion source imaging. *Phys. Rev. Lett.* **2008**, *100*, 232301. [CrossRef]
7. Shapoval, V.M.; Sinyukov, Y.M.; Karpenko, I.A. Emission source functions in heavy ion collisions. *Phys. Rev. C* **2013**, *88*, 064904. [CrossRef]
8. Adhikary, H.; Adrich, P.; Allison, K.K.; Amin, N.; Andronov, E.V.; Antićić, T.; Arsene, I.-C.; Bajda, M.; Balkova, Y.; Baszczyk, M.; et al. Measurements of Two-pion HBT Correlations in Be+Be Collisions at 150A GeV/c Beam Momentum, at the NA61/SHINE Experiment at CERN. *arXiv* **2023**, arXiv:2302.04593.
9. Tumasyan, A.; Wolfgang, A.; Andrejkovic, J.W.; Bergauer, T.; Chatterjee, S.; Damanakis, K.; Dragicevic, M.; Escalante Del Valle, A.; Hussain, P.S.; Manfred, J.; et al. Two-particle Bose-Einstein correlations and their Lévy parameters in PbPb collisions at $\sqrt{s_{NN}} = 5.02$ TeV. *arXiv* **2023**, arXiv:2306.11574.
10. Kurgyis, B. Three dimensional Lévy HBT results from PHENIX. *Acta Phys. Polon. Supp.* **2019**, *12*, 477. [CrossRef]
11. Sinyukov, Y.; Lednicky, R.; Akkelin, S.V.; Pluta, J.; Erazmus, B. Coulomb corrections for interferometry analysis of expanding hadron systems. *Phys. Lett.* **1998**, *B432*, 248–257. [CrossRef]
12. Bowler, M.G. Coulomb corrections to Bose-Einstein correlations have been greatly exaggerated. *Phys. Lett. B* **1991**, *B270*, 69–74. [CrossRef]
13. Csanád, M.; Lökös, S.; Nagy, M. Expanded empirical formula for Coulomb final state interaction in the presence of Lévy sources. *Phys. Part. Nucl.* **2020**, *51*, 238–242. [CrossRef]
14. Csörgő, T.; Hegyi, S.; Zajc, W.A. Bose-Einstein correlations for Levy stable source distributions. *Eur. Phys. J.* **2004**, *C36*, 67–78. [CrossRef]
15. Pratt, S.; Csörgő, T.; Zimányi, J. Detailed predictions for two pion correlations in ultrarelativistic heavy ion collisions. *Phys. Rev. C* **1990**, *C42*, 2646–2652. [CrossRef]
16. Bertsch, G.; Gong, M.; Tohyama, M. Pion Interferometry in Ultrarelativistic Heavy Ion Collisions. *Phys. Rev. C* **1988**, *C37*, 1896–1900. [CrossRef]
17. Landau, L.D.; Lifshitz, L.M. *Quantum Mechanics Non-Relativistic Theory*, 3rd ed.; Pergamon Press: Oxford, UK, 1977; Volume 3. [CrossRef]
18. Metropolis, N.; Rosenbluth, A.W.; Rosenbluth, M.N.; Teller, A.H.; Teller, E. Equation of state calculations by fast computing machines. *J. Chem. Phys.* **1953**, *21*, 1087–1092. [CrossRef]
19. Hastings, W.K. Monte Carlo Sampling Methods Using Markov Chains and Their Applications. *Biometrika* **1970**, *57*, 97–109. [CrossRef]
20. Lökös, S. Probing the QCD Phase Diagram with HBT Femtoscopy. *Acta Phys. Polon. Supp.* **2022**, *15*, 30, [CrossRef]
21. Mukherjee, A. Kaon femtoscopy with Lévy-stable sources from $\sqrt{s_{NN}} = 200$ GeV Au + Au collisions at RHIC. *Universe* **2023**, *9*, 300, [CrossRef]
22. Porfy, B. Femtoscopic correlation measurement with symmetric Lévy-type source at NA61/SHINE. *Universe* **2023**, *9*, 298, [CrossRef]
23. Kórodi, B. Centrality dependent Lévy HBT analysis in $\sqrt{s_{NN}} = 5.02$ TeV PbPb collisions at CMS. *Universe* **2023**, *9*, 318, [CrossRef]
24. Kovács, L. Charged kaon femtoscopy with Lévy sources in $\sqrt{s_{NN}} = 200$ GeV Au+Au collisions at PHENIX. Submitted to *Universe*. 2023.

Communication

Femtoscopic Correlation Measurement with Symmetric Lévy-Type Source at NA61/SHINE

Barnabás Pórfy [1,2] on behalf of the NA61/SHINE Collaboration

1   Wigner Research Centre for Physics, Konkoly-Thege Miklós út 29-33, H-1121 Budapest, Hungary; porfy.barnabas@wigner.hu or barnabas.porfy@cern.ch
2   Department of Atomic Physics, Faculty of Science, Eötvös Loránd University, Pázmány Péter sétány 1/A, H-1111 Budapest, Hungary

Abstract: Measuring quantum-statistical, femtoscopic (including final state interactions) momentum correlations with final state interactions in high-energy nucleus-nucleus collisions reveal the space-time structure of the particle-emitting source created. In this paper, we report NA61/SHINE measurements of femtoscopic correlations of identified pion pairs and describe said correlations based on symmetric Lévy-type sources in Ar+Sc collisions at $150A$ GeV/c. We investigate the transverse mass dependence of the Lévy-type source parameters and discuss their possible interpretations.

Keywords: quark-gluon plasma; femtoscopy; critical endpoint; small systems

Citation: Pórfy, B., on behalf of the NA61/SHINE Collaboration. Femtoscopic Correlation Measurement with Symmetric Lévy-Type Source at NA61/SHINE. *Universe* **2023**, *9*, 298. https://doi.org/10.3390/universe9070298

Academic Editor: Roman Pasechnik

Received: 25 May 2023
Revised: 10 June 2023
Accepted: 16 June 2023
Published: 21 June 2023

1. Introduction

The NA61/SHINE is a fixed target experiment using a large acceptance hadron spectrometer located in the North Area H2 beam line of the CERN Super Proton Synchrotron accelerator [1]. Its main goals include the investigation and mapping of the phase diagram of strongly interacting matter, as well as measuring cross sections of processes relevant for cosmic rays and neutrino physics. In this paper, we are focusing on mapping the QCD phase diagram. In order to accomplish this, NA61/SHINE performs measurements of different collision systems at multiple energies. The experiment provides excellent tracking down to $p_T = 0$ GeV/c. This performance is achieved by using four large Time Projection Chambers (TPC's), which cover the full forward hemisphere. The experiment also features a modular calorimeter, called the Projectile Spectator Detector. It is located on the beam axis, after the TPC's, and measures the forward energy which determines the collision centrality of the events. A setup of the NA61/SHINE detector system is shown in Figure 1.

The search for the critical endpoint (CEP) and investigation of the QCD phase diagram requires analysis at different temperatures and baryon-chemical potentials. To study, we need to map the phase diagram using different system sizes at various energies. NA61/SHINE investigations cover several beam momenta ($13A$, $20A$, $30A$, $40A$, $75A$ and $150A$ GeV/c) and collision systems (p+p,p+Pb,Be+Be,Ar+Sc,Xe+La,Pb+Pb). In this paper, we describe the femtoscopic correlations of identical pions emitted from central Ar+Sc collisions at beam momentum of $150A$ GeV/c. This field is often called femtoscopy as it reveals the femtometer scale structure of particle production.

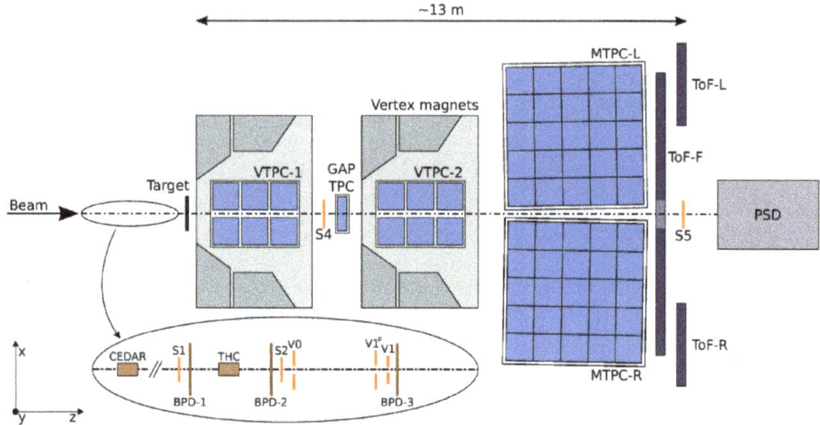

Figure 1. The setup of the NA61/SHINE detector system during the run of Ar+Sc.

2. Femtoscopy with Lévy Shaped Sources

The method of quantum-statistical (Bose-Einstein) correlations is based on the work of R. Hanbury Brown and R. Q. Twiss (HBT) [2], who applied it first in astrophysical intensity correlation measurements. The method was developed to determine the apparent angular diameter of stellar objects. Shortly afterwards, a similar quantum-statistical method was applied in momentum correlation measurements for proton-antiproton collisions [3,4] by Goldhaber and collaborators. Their objective was to understand pion-pion correlations and gain information on the radius, R, of the interaction volume in high-energy particle collisions. The key relationship for measuring Bose-Einstein correlations shows that the spatial momentum correlations, $C(q)$, are related to the properties of the particle emitting source, $S(x)$, that describes the probability density of particle creation for a relative coordinate x as:

$$C_2(q) \cong 1 + |\tilde{S}(q)|^2, \tag{1}$$

where $\tilde{S}(q)$ is the Fourier transform of $S(x)$, and q is the relative momentum of the particle pair (with the dependence on the average momentum, K, of the pair suppressed and described in more detail in [5]). The usual assumption for the shape of the source based on the central limit theorem, is a Gaussian. However, such Gaussian shaped sources lead to Gaussian correlation functions. A more general assumption is the Lévy distribution [6,7]. It exhibits a power-law tail and includes a Gaussian limit, as well. Correlation functions based on this approach have been shown to describe data from different experiments, such as LEP [8], RHIC [9], and LHC [10,11] quite well. Several phenomena could explain the appearance of Lévy shaped sources. The non-Gaussianity of the source could be attributed to critical fluctuations and the emergence of spatial correlations on a large scale, which may indicate the existence of similar sources with power-law tails [12]. Further reasons include the fractal structure of QCD jets [13].

In this paper, the measured femtoscopic correlation (including final state interaction) with spherically symmetric Lévy distributions is defined as:

$$\mathcal{L}(\alpha, R, \vec{r}) = \frac{1}{(2\pi)^3} \int d^3\vec{\zeta} e^{i\vec{\zeta}\vec{r}} e^{-\frac{1}{2}|\vec{\zeta}R|^\alpha}, \tag{2}$$

where R is the Lévy scale parameter and α defined as the Lévy stability index. In addition, $\vec{\zeta}$ is the three-dimensional integration variable with dimensions of MeV/c and $\vec{r}$ is the vector of spatial coordinates. There are two special cases where the distribution can be expressed analytically. One such case is, the already mentioned, Gaussian distribution for $\alpha = 2$. Besides this, the $\alpha = 1$ case leads to a Cauchy distribution. An important difference between Lévy distributions and Gaussians is the presence of a power-law tail $\sim r^{-(d-2+\alpha)}$ in case of $\alpha < 2$, where d represents the number of spatial dimensions. With the assumption of Lévy sources, the femtoscopic correlation functions can be expressed in the following way:

$$C_2(q) = 1 + \lambda \cdot e^{-(qR)^{\alpha}}. \tag{3}$$

$C_2(q)$ is a stretched type of exponential, where the λ intercept parameter is defined as:

$$C_2(q \to 0) = 1 + \lambda. \tag{4}$$

At vanishing relative momentum, the correlation function has a value of $1 + \lambda$. This value is not accessible in the measurements and extrapolation from the region, when two tracks are experimentally resolved, is needed. However, it is commonly observed that the intercept parameter λ is less than 1. The core-halo model, explained in Refs. [14,15], can provide some insights into this parameter.

The model assumes that the source S is made up of two parts, the core and the halo (S_{core} and S_{halo}), respectively. The core contains pions created directly from hadronic freeze-out or from extremely short lived (strongly decaying) resonances. The halo consists of pions created from longer-lived resonances and the general background. It may extend to thousands of femtometers, while core part has a size of around a few femtometers. In this picture, the λ parameter turns out to be connected to the ratio of the core and the halo as:

$$\lambda = \left(\frac{N_{\text{core}}}{N_{\text{core}} + N_{\text{halo}}} \right)^2. \tag{5}$$

Then, one can modify the correlation function to take the effect of the halo into account, by utilizing the Bowler-Sinyukov method [16,17] as:

$$C_2(q) = 1 - \lambda + \lambda \cdot (1 + e^{-|qR|^{\alpha}}). \tag{6}$$

The halo part contributes at very small values of relative momenta, q. Therefore, it does not affect the source radii of the core part [18].

It is well known that critical points are characterized by critical exponents. One, in particular, is related to spatial correlations by the exponent denoted as η. The appearance of the parameter can be explained by the second-order phase transition at the CEP, where fluctuations will appear at all scales causing the spatial correlation function to exhibit a power-law tail with an exponent of $-(d - 2 + \eta)$, with d denoting the dimension. The Lévy exponent, α, is the exponent in the case of, previously defined, Lévy distributed sources and will also exhibit a power-law tail, with an exponent of $-d + 2 - \alpha$ [19]. Hence, α was suggested to be directly related to or being explicitly equal to, the critical exponent, η [12], in absence of other phenomena affecting the source shape. This is the basis of the idea connecting α and η. However, the Lévy-shape of the source can be attributed to several different factors besides critical phenomena, including QCD jets, anomalous diffusion, critical phenomena, and others [6,7,13,20,21]. Hence, while a non-monotonic behavior of α is expected near the critical point, a detailed understanding of the collision energy and system size dependence of α is needed to draw conclusions about the critical point.

It has been suggested that the universality class of QCD is the same as that of the 3D Ising model [22,23]. The value of η in the 3D Ising model is 0.03631 ± 0.00003 [24]. An alternative solution is to use the universality class of the 3D Ising model with a random external field, which yields an η value of 0.50 ± 0.05 [25]. The statements mentioned suggest that α would decrease to 0.50, or below, at the vicinity of the CEP. To confirm this, measurements of α are needed in different collision systems at various energies.

In this analysis, we are dealing with like-charged particles that are influenced by Coulomb repulsion. The final state Coulomb effect has been neglected in the previously defined correlation function. Thus, the correlation function in Equation (6) that lacks this effect will be denoted as $C_2^0(q)$ from now on. The correction necessitated by this effect can be done by simply taking the ratio of C_2^{Coul} and $C_2^0(q)$:

$$K_{\text{Coulomb}}(q) = \frac{C_2^{\text{Coul}}(q)}{C_2^0(q)}, \tag{7}$$

where $C_2^{\text{Coul}}(q)$ is the interference of solutions of the two-particle Schrödinger equation; with a Coulomb-potential [26,27]. The numerator in Equation (7) cannot be calculated analytically and requires a large numerical effort to estimate.

An approximate formula for $C_2^{\text{Coul}}(q)$ was obtained in Ref. [10] for the case of Cauchy-shaped sources. However, a more precise treatment is required due to our assumption of Lévy-shaped sources. We are utilizing a new method in our analysis for estimating the effect of Coulomb repulsion. The treatment includes the numerical calculation presented in Refs. [27,28], the parametrization of its results, and, finally; the parametrization of the dependence of the physical parameters R, λ, and α. Thus Equation (6) is modified as:

$$C_2(q) = N \cdot \left(1 - \lambda + \lambda \cdot (1 + e^{-|qR|^\alpha}) \cdot K_{\text{Coulomb}}(q) \right), \tag{8}$$

where N is introduced as normalization parameter and $K_{\text{Coulomb}}(q)$ denotes the Coulomb correction.

It is important to highlight that the Coulomb correction is calculated in the pair-center-of-mass (PCMS) system, while the measurement is often done in the longitudinally co-moving system (LCMS). The assumption of Coulomb correction in the one-dimensional HBT in LCMS picture is that the shape of the source is spherical, i.e. $R_{\text{out}} = R_{\text{side}} = R_{\text{long}} = R \equiv R_{\text{LCMS}}$. The shape of the source, however, is spherical in the LCMS and not in the PCMS. Therefore, an approximate one-dimensional PCMS size parameter is needed. A study was done, where an average PCMS radius of

$$\overline{R}_{\text{PCMS}} = \sqrt{\frac{1 - \frac{2}{3}\beta_{\text{T}}^2}{1 - \beta_{\text{T}}^2}} \cdot R \tag{9}$$

was calculated [29], with $\beta_{\text{T}} = \frac{K_{\text{T}}}{m_T}$.

3. Measurement Details

For the measurements this paper is based on, we analyzed Ar+Sc collisions at $150A$ GeV/c beam momentum in the 0–10% most central events. The available data set contains around 2.7 million events, which was reduced to around 700,000 events in the analysis. The following paragraph describes the various event, track and pair selection performed on the data. As mentioned, we have selected 0–10% of the most central events by measuring the energy contained in the projectile remnants, with the Projectile Spectator Detector (PSD). We have also selected events where no off-time beam particle was detected. These are particles which come within the drift time of the chambers which are not emptied out. Furthermore, all the events chosen have between ± 10 cm as that is the maximal distance between the main vertex z position and the center of the scandium target. Particle identification was handled by using dE/dx energy deposit in the TPC gas. The tracks were extrapolated to the interaction plane and matched with the distance against the interaction point. If the distance was ≤ 4 cm in the horizontal plane and ≤ 2 cm in the vertical plane, the track was kept. Moreover, track splitting was handled by selecting the ratio of the total number of reconstructed points on the track to the potential number of points to be between 0.5 and 1.0. Finally, to counter track merging, we have used a selection on the momentum space distance between the two tracks. It uses non-standard momentum coordinates, $s_x = p_x/p_{xz}$, $s_y = p_y/p_{xz}$ and $\rho = 1/p_{xz}$.

We then analyzed the combination of negative pion pairs and positive pion pairs. It is important to note that this analysis was done with a one-dimensional relative momentum variable q, calculated in LCMS. These pairs were sorted into eight K_T (average transverse momentum of the pair) bins in the range of 0–450 MeV/c. In each momentum bin, the relative momentum distribution of coincident pion pairs were obtained. Let us call this the actual pair momentum difference distribution, $A(q)$. $A(q)$ contains quantum-statistical correlations, as well as many other residual effects related to kinematics and acceptance. The effects can be removed by constructing a combinatorial background pair distribution, denoted as $B(q)$, which is measured in the same K_T or m_T intervals as the $A(q)$ distribution. The method we use involves randomly selecting the same number of particles as the multiplicity of the actual event. The selected particles are from other events of similar parameters and each is selected from a different event. Let us call the pair momentum difference distribution made from this method as, the background distribution, $B(q)$. This, by construction, enables us to have an uncorrelated pool of events. Then the correlation function is calculated as

$$C_2(q) = \frac{A(q)}{B(q)} \cdot \frac{\int_{q_1}^{q_2} B(q)dq}{\int_{q_1}^{q_2} A(q)dq} \, , \tag{10}$$

in a $[q_1, q_2]$ range where quantum-statistical correlations are not expected. Upon adding the contribution from the background, the previously defined Equation (8) is modified as follows:

$$C(q) = N \cdot (1 + \varepsilon \cdot q) \cdot \left(1 - \lambda + \lambda \cdot \left(1 + e^{-(qR)^\alpha}\right) \cdot K_{\text{Coulomb}}(q)\right), \tag{11}$$

with N being a normalization parameter responsible for the proper normalization of the $A(q)/B(q)$ ratio, ε describing the linearity of background, and $K_{\text{Coulomb}}(q)$ being the Coulomb correction. We then use this formula to our measured $C_2(q)$ as shown in Figure 2. We then use this formula to our measured $C_2(q)$ as shown in Figure 2. To determine the fitting range where the effects of detector resolution do not play a significant role, we have used EPOS simulation [30] with GEANT3 for particle propagation [31]. Note that in the low-q region, the fit does not describe the data. This can be explained, according to Monte Carlo simulations of the detector response, by the limited resolution of pairs with small relative momentum.

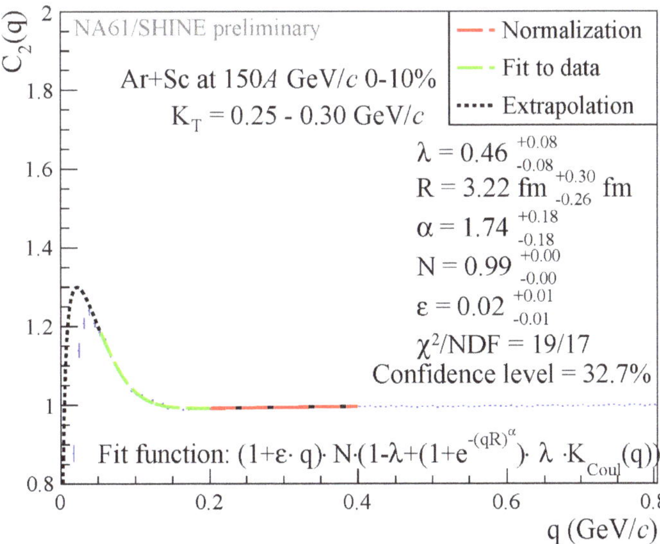

Figure 2. Example fit with Bose–Einstein correlation function at $K_T = 0.25$–0.30 GeV/c for the sum $(\pi^+ + \pi^+) + (\pi^- + \pi^-)$. Blue points with error bars represent the data, the green dash-dotted line

shows the fitted function with Coulomb correction given by Equation (11) within the range of $0.0525\ \mathrm{GeV}/c$ to $0.2\ \mathrm{GeV}/c$ and red dashed line represent normalization to background $0.2\ \mathrm{GeV}/c$ to $0.4\ \mathrm{GeV}/c$. In the low-q region, the black dotted line indicates the extrapolated function outside of the fit range due to prominent detector resolution effects, mentioned in the text.

4. Results

The three physical parameters $(\alpha, R,$ and $\lambda)$ were measured in eight bins of pair transverse momentum, K_{T}. The three mentioned parameters were obtained through fitting the measured correlation functions with the formula shown in Equation (11). The results were investigated regarding their transverse momentum dependence. In the following, we report on the transverse mass dependence of α, λ, and R; where transverse mass is expressed as $m_T = \sqrt{m_\pi^2 c^4 + K_{\mathrm{T}}^2 c^2}$, with m_π being the pion mass.

As explained above, the shape of the source is often assumed to be Gaussian. The Lévy stability exponent, α, can be used to extract the shape of the tail of the source. Our results, shown in Figure 3, yield values for α between 1.5 and 2.0, which imply a source closer to the Gaussian shape than the one in Be+Be collisions [5], but are still significantly lower than the $\alpha = 2$ (Gaussian) case. The observed α parameter is also significantly higher than the conjectured value at the critical point ($\alpha = 0.5$). Altogether, these results suggest that measured correlation functions align with the assumption of a Lévy source, indicating that it is more advantageous over the Gaussian assumption. Further studies are ongoing at NA61/SHINE, using different collision energies and system sizes, in order to map the evolution of the Lévy stability index, α, as a function of collision energy and system size.

As a second parameter, let us look at the Lévy scale parameter, R, visible in Figure 4. It determines the length of homogeneity of the pion emitting source. The parameter R depends on the transverse-mass as $1/\sqrt{m_T}$. This can be derived using simple, hydrodynamical predictions for Gaussian sources [15,32]:

$$\frac{1}{R_{\mathrm{HBT}}^2} = \frac{1}{R_{\mathrm{geom}}^2} + u_{\mathrm{T}}^2 \cdot \frac{m_T}{T_0}, \tag{12}$$

where u_T^2 is the average, transverse expansion and T_0 is the hadronisation temperature. In our case, rather surprisingly, despite the non-Gaussian nature of the source, this formula works in describing the measured femtoscopic radii. More precisely, as mentioned above, observing an $R \sim 1/\sqrt{m_T}$ is particularly interesting as this type of m_T dependence should rise in case of Gaussian sources ($\alpha = 2$) [33]. It is not entirely clear why this happens, the indicated m_T dependence could form in the QGP or at a later stage. This phenomenon was also observed at RHIC [9] and in simulations at RHIC and LHC energies [20,21].

The final parameter being investigated is the intercept (also known as the correlation strength) parameter, λ, defined in Equation (5). The dependence of λ on m_T is shown in Figure 5. One may observe a slight dependence on m_T, but this can still can be considered constant in the investigated range. When compared to measurements from RHIC Au+Au collisions [9,34,35] and from SPS Pb+Pb interactions [36,37], an interesting phenomenon is observed. At the energies of SPS, there is no visible "hole" at lower m_T values, but at RHIC energies, the "hole" appears at m_T values of around a few hundred MeV. This "hole" was interpreted in Refs. [9,38] to be a sign of in-medium mass modification. The results presented in Figure 5, at the given statistical precision, do not indicate the presence of such a low-m_T hole. This trend might imply that this phenomenon can be turned off at SPS. Furthermore, it can be highlighted that the λ values we obtained are significantly below unity. A possible answer can be given by the halo part of the core-halo model. It may indicate that a significant fraction of pions are the decay products of long-lived resonances.

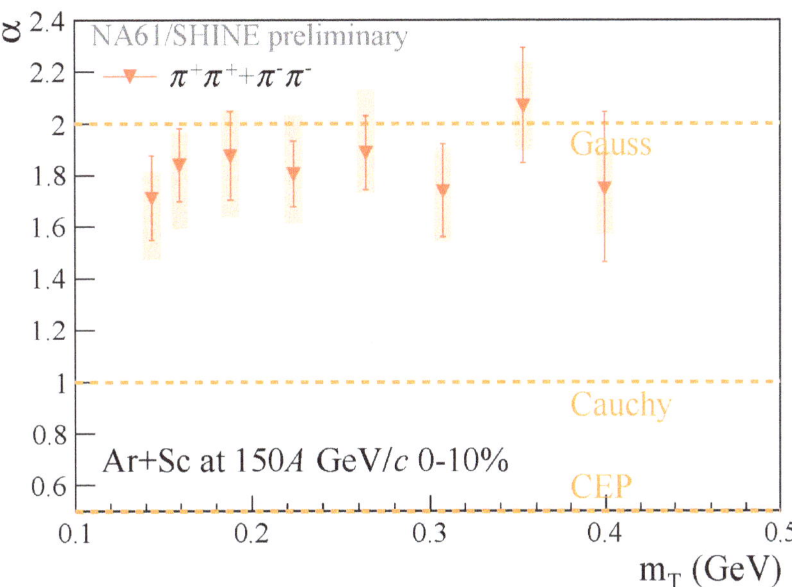

Figure 3. The Lévy stability index, α, for 0–10% central Ar+Sc at 150A GeV/c, as a function of m_T. Special cases corresponding to a Gaussian ($\alpha = 2$) or a Cauchy ($\alpha = 1$) source are shown, as well as $\alpha = 0.5$, the conjectured value corresponding to the critical endpoint. Boxes denote systematic uncertainties, bars represent statistical uncertainties.

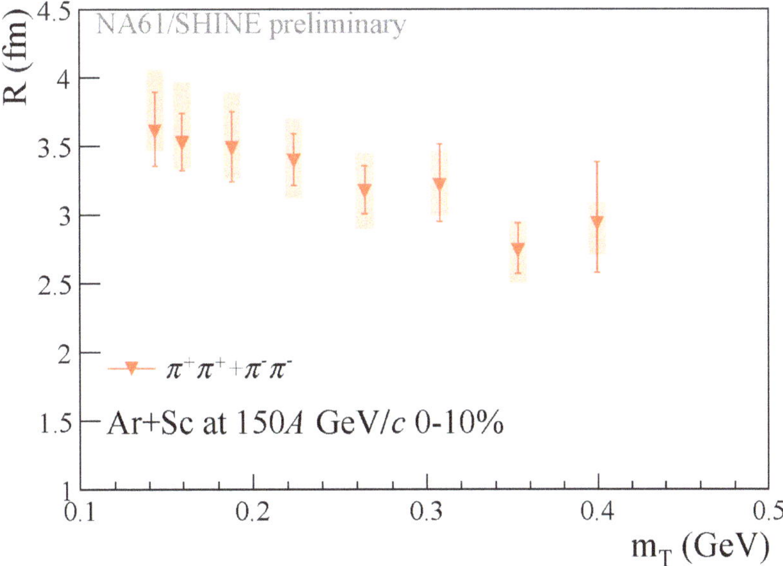

Figure 4. The radial scale parameter, R, for 0–10% central Ar+Sc at 150A GeV/c, as a function of m_T. Boxes denote systematic uncertainties, bars represent statistical uncertainties.

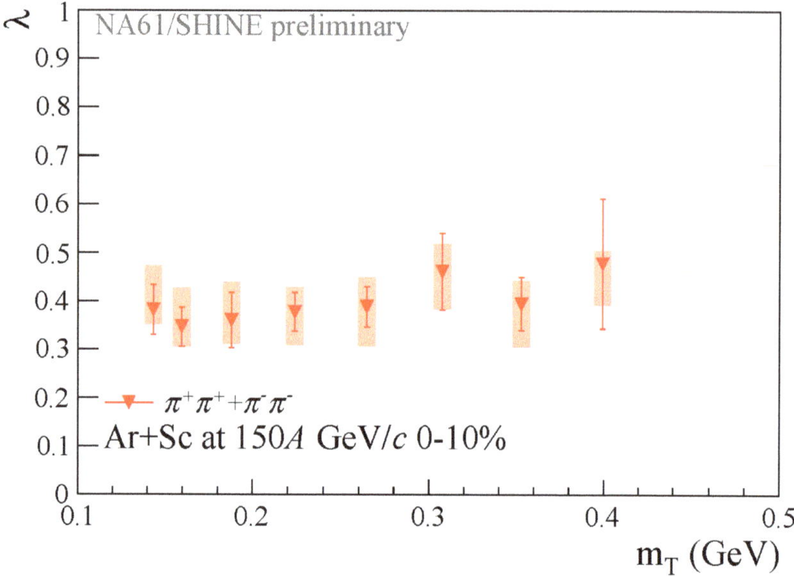

Figure 5. The correlation strength parameter, λ, for 0–10% central Ar+Sc at 150A GeV/c, as a function of m_T. Boxes denote systematic uncertainties, bars represent statistical uncertainties.

5. Conclusions

In the report above, we discussed the NA61/SHINE measurement of one-dimensional, identified, two-pion, femtoscopic correlation functions; in the 0–10% most central Ar+Sc collisions at 150A GeV/c. We discussed the transverse mass dependencies of the Lévy source parameters. Results on the Lévy scale parameter, α, showed a significant deviation from Gaussian sources and are not in the vicinity of the conjectured value at the critical point. The Lévy scale parameter, R, shows a visible decrease with m_T. The correlation strength parameter, λ, does not show any significant m_T dependence, but, maps different patterns at RHIC and similar trends at SPS energies. With these results at hand, we plan to measure Bose-Einstein correlations in larger systems, as well as at smaller energies, to continue mapping the phase diagram of the strongly interacting matter.

Funding: This research was supported by the ÚNKP-22-3 New National Excellence Program of the Ministry for Culture and Innovation from the source of the National Research, Development and Innovation Fund, and the NKFIH OTKA K-138136 grant.

Data Availability Statement: The data presented in this study are available on request from the corresponding author. The data are not publicly available.

Acknowledgments: The author would like to thank the NA61/SHINE collaboration.

Conflicts of Interest: The author declares no conflict of interest.

Abbreviations

The following abbreviations are used in this manuscript:

QCD	quantum chromodynamics
CERN	Conseil européen pour la recherche nucléaire
SPS	Super Proton Synchrotron

HBT	Hanbury Brown and Twiss
BE	Bose-Einstein
CEP	critical endpoint
NA61/SHINE	North Area 61 / SPS Heavy Ion and Neutrino Experiment
LCMS	Longitudinally Co-Moving System

References

1. Abgrall, N. et al. [NA61/SHINE Collaboration]. NA61/SHINE facility at the CERN SPS: Beams and detector system. *JINST* **2014**, *9*, P06005. [CrossRef]
2. Hanbury Brown, R.; Twiss, R.Q. A Test of a new type of stellar interferometer on Sirius. *Nature* **1956**, *178*, 1046. [CrossRef]
3. Goldhaber, G.; Fowler, W.B.; Goldhaber, S.; Hoang, T.F. Pion-pion correlations in antiproton annihilation events. *Phys. Rev. Lett.* **1959**, *3*, 181. [CrossRef]
4. Goldhaber, G.; Goldhaber, S.; Lee, W.Y.; Pais, A. Influence of Bose-Einstein statistics on the anti-proton proton annihilation process. *Phys. Rev.* **1960**, *120*, 300. [CrossRef]
5. Adhikary, H. et al. [NA61/SHINE Collaboration]. Measurements of two-pion HBT correlations in Be + Be collisions at 150A GeV/c beam momentum, at the NA61/SHINE experiment at CERN. *arXiv* **2023**, arXiv:2302.04593.
6. Csörgő, T.; Hegyi, S.; Zajc, W.A. Bose-Einstein correlations for Levy stable source distributions. *Eur. Phys. J. C* **2004**, *36*, 67–78. [CrossRef]
7. Metzler, R.; Barkai, E.; Klafter, J. Anomalous Diffusion and Relaxation Close to Thermal Equilibrium: A Fractional Fokker-Planck Equation Approach. *Phys. Rev. Lett.* **1999**, *82*, 3563–3567. [CrossRef]
8. Achard, P. et al. [L3 Collaboration]. Test of the τ-Model of Bose-Einstein Correlations and Reconstruction of the Source Function in Hadronic Z-boson Decay at LEP. *Eur. Phys. J. C* **2011**, *71*, 1648. [CrossRef]
9. Adare, A. et al. [PHENIX Collaboration]. Lévy-stable two-pion Bose-Einstein correlations in $\sqrt{s_{NN}} = 200$ GeV Au + Au collisions. *Phys. Rev. C* **2018**, *97*, 064911. [CrossRef]
10. Sirunyan, A.M. et al. [CMS Collaboration]. Bose-Einstein correlations in pp, pPb, and PbPb collisions at $\sqrt{s_{NN}} = 0.9 - 7$ TeV. *Phys. Rev. C* **2018**, *97*, 064912. [CrossRef]
11. *Measurement of Two-Particle Bose-Einstein Momentum Correlations and Their Levy Parameters at* $\sqrt{s_{NN}} = 5.02$ *TeV PbPb Collisions*; Technical Report CMS-PAS-HIN-21-011; CERN: Geneva, Switzerland, 2022.
12. Csörgő, T.; Hegyi, S.; Novák, T.; Zajc, W.A. Bose-Einstein or HBT correlation signature of a second order QCD phase transition. *AIP Conf. Proc.* **2006**, *828*, 525–532. [CrossRef]
13. Csorgo, T.; Hegyi, S.; Novak, T.; Zajc, W.A. Bose-Einstein or HBT correlations and the anomalous dimension of QCD. *Acta Phys. Polon. B* **2005**, *36*, 329–337.
14. Csörgő, T. Particle interferometry from 40-MeV to 40-TeV. *Acta Phys. Hung. A* **2002**, *15*, 1–80. . [CrossRef]
15. Csörgő, T.; Lörstad, B. Bose-Einstein correlations for three-dimensionally expanding, cylindrically symmetric, finite systems. *Phys. Rev. C* **1996**, *54*, 1390–1403. [CrossRef] [PubMed]
16. Sinyukov, Y.; Lednicky, R.; Akkelin, S.V.; Pluta, J.; Erazmus, B. Coulomb corrections for interferometry analysis of expanding hadron systems. *Phys. Lett. B* **1998**, *432*, 248–257. [CrossRef]
17. Bowler, M.G. Coulomb corrections to Bose-Einstein correlations have been greatly exaggerated. *Phys. Lett. B* **1991**, *270*, 69–74. [CrossRef]
18. Maj, R.; Mrowczynski, S. Coulomb Effects in Femtoscopy. *Phys. Rev. C* **2009**, *80*, 034907. [CrossRef]
19. Csörgő, T. Correlation Probes of a QCD Critical Point. *arXiv* **2009**, arXiv:0903.0669.
20. Kincses, D.; Stefaniak, M.; Csanád, M. Event-by-Event Investigation of the Two-Particle Source Function in Heavy-Ion Collisions with EPOS. *Entropy* **2022**, *24*, 308. [CrossRef]
21. Kórodi, B.; Kincses, D.; Csanád, M. Event-by-event investigation of the two-particle source function in $\sqrt{s_{NN}} = 2.76$ TeV PbPb collisions with EPOS. *arXiv* **2022**, arXiv:2212.02980.
22. Halasz, A.M.; Jackson, A.D.; Shrock, R.E.; Stephanov, M.A.; Verbaarschot, J.J.M. On the phase diagram of QCD. *Phys. Rev. D* **1998**, *58*, 096007. [CrossRef]
23. Stephanov, M.A.; Rajagopal, K.; Shuryak, E.V. Signatures of the tricritical point in QCD. *Phys. Rev. Lett.* **1998**, *81*, 4816–4819. [CrossRef]
24. El-Showk, S.; Paulos, M.F.; Poland, D.; Rychkov, S.; Simmons-Duffin, D.; Vichi, A. Solving the 3d Ising Model with the Conformal Bootstrap II. c-Minimization and Precise Critical Exponents. *J. Stat. Phys.* **2014**, *157*, 869. [CrossRef]
25. Rieger, H. Critical behavior of the three-dimensional random-field Ising model: Two-exponent scaling and discontinuous transition. *Phys. Rev. B* **1995**, *52*, 6659. [CrossRef] [PubMed]
26. Kincses, D.; Nagy, M.I.; Csanád, M. Coulomb and strong interactions in the final state of Hanbury-Brown–Twiss correlations for Lévy-type source functions. *Phys. Rev. C* **2020**, *102*, 064912. [CrossRef]
27. Csanád, M.; Lökös, S.; Nagy, M. Expanded empirical formula for Coulomb final state interaction in the presence of Lévy sources. *Phys. Part. Nucl.* **2020**, *51*, 238–242. [CrossRef]
28. Csanád, M.; Lökös, S.; Nagy, M. Coulomb final state interaction in heavy ion collisions for Lévy sources. *Universe* **2019**, *5*, 133. [CrossRef]

29. Kurgyis, B.; Kincses, D.; Nagy, M.; Csanád, M. Coulomb interaction for Lévy sources. *arXiv* **2020**, arXiv:2007.10173.
30. Pierog, T.; Werner, K. EPOS Model and Ultra High Energy Cosmic Rays. *Nucl. Phys. Proc. Suppl.* **2009**, *196*, 102–105. [CrossRef]
31. Brun, R.C.F. GEANT Detector Description and Simulation Tool, CERN Program Library Long Writeup W5013. 1993. Available online: http://wwwasdoc.web.cern.ch/wwwasdoc/geant/geantall.html (accessed on 1 May 2023) .
32. Csanád, M.; Vargyas, M. Observables from a solution of 1+3 dimensional relativistic hydrodynamics. *Eur. Phys. J. A* **2010**, *44*, 473–478. [CrossRef]
33. Sinyukov, Y.M. Spectra and correlations in locally equilibrium hadron and quark-gluon systems. *Nucl. Phys. A* **1994**, *566*, 589C–592C. [CrossRef]
34. Vértesi, R.; Csörgő, T.; Sziklai, J. Significant in-medium η' mass reduction in $\sqrt{s_{NN}}$ = 200 GeV Au+Au collisions at the BNL Relativistic Heavy Ion Collider. *Phys. Rev. C* **2011**, *83*, 054903. [CrossRef]
35. Abelev, B.I. et al. [STAR Collaboration]. Pion Interferometry in Au+Au and Cu+Cu Collisions at RHIC. *Phys. Rev. C* **2009**, *80*, 024905. [CrossRef]
36. Beker, H. et al. [NA44 Collaboration]. m(T) dependence of boson interferometry in heavy ion collisions at the CERN SPS. *Phys. Rev. Lett.* **1995**, *74*, 3340–3343. [CrossRef] [PubMed]
37. Alt, C. et al. [NA49 Collaboration]. Bose-Einstein correlations of pi-pi-pairs in central Pb+Pb collisions at A-20, A-30, A-40, A-80, and A-158 GeV. *Phys. Rev. C* **2008**, *77*, 064908. [CrossRef]
38. Vance, S.E.; Csörgő, T.; Kharzeev, D. Partial UA1 restoration from Bose-Einstein correlations. *Phys. Rev. Lett.* **1998**, *81*, 2205–2208. [CrossRef]

Article

Centrality-Dependent Lévy HBT Analysis in $\sqrt{s_{NN}} = 5.02$ TeV PbPb Collisions at CMS

Balázs Kórodi on behalf of the CMS Collaboration

Department of Atomic Physics, Faculty of Science, Eötvös Loránd University, Pázmány Péter Sétány 1/A, H-1111 Budapest, Hungary; balazs.korodi@cern.ch

Abstract: The measurement of two-particle Bose–Einstein momentum correlation functions are presented using $\sqrt{s_{NN}} = 5.02$ TeV PbPb collision data, recorded by the CMS experiment in 2018. The measured correlation functions are discussed in terms of Lévy-type source distributions. The Lévy source parameters are extracted as functions of transverse mass and collision centrality. These source parameters include the correlation strength λ, the Lévy stability index α, and the Lévy scale parameter R. The source shape, characterized by α, is found to be neither Gaussian nor Cauchy. A hydrodynamic-like scaling of R is also observed.

Keywords: heavy ions; quark–gluon plasma; femtoscopy; Lévy HBT

Citation: Kórodi, B., on behalf of the CMS Collaboration. Centrality-Dependent Lévy HBT Analysis in $\sqrt{s_{NN}} = 5.02$ TeV PbPb Collisions at CMS. *Universe* **2023**, *9*, 318. https://doi.org/10.3390/universe9070318

Academic Editor: Jun Xu

Received: 5 June 2023
Revised: 23 June 2023
Accepted: 29 June 2023
Published: 1 July 2023

1. Introduction

The investigation of the femtometer-scale space–time geometry of high-energy heavy-ion collisions has been an important area, called femtoscopy, of high-energy physics for several decades [1]. The main idea of this field originates from astronomy, since it is analogous with the well-known Hanbury Brown and Twiss (HBT) effect that describes the intensity correlation of photons [2,3]. In high-energy physics, however, the observable is the quantum-statistical momentum correlation of hadrons, which carries information about the femtometer-scale structure of the particle-emitting source [4,5]. The measurements of such momentum correlations are partially responsible for establishing the fluid nature of the quark–gluon plasma (QGP) created in heavy-ion collisions [6,7]. Furthermore, the measured source radii provide information about the transition from the QGP to the hadronic phase [8,9], as well as about the phase space of quantum chromodynamics [10].

Recent high-precision femtoscopic measurements [11,12] have shown that the previously widely assumed Gaussian [6,13,14] or Cauchy [15,16] source distributions do not provide an adequate description of the measured correlation functions. Instead, a generalization of these distribution, the Lévy alpha-stable distribution [17], is needed for a statistically acceptable description [11,12]. The shape of the Lévy distribution is characterized by the Lévy stability index α, and can be influenced by various physical phenomena, e.g., anomalous diffusion [18–20], resonance decays [21,22], jet fragmentation [23], and critical phenomena [24]. Until now, the α parameter had not been measured at the largest energies accessible at the LHC. The question of how α changes compared to lower energies signifies the need for a Lévy HBT analysis at LHC energy.

In this paper, the Lévy HBT analysis of two-particle Bose–Einstein momentum correlations is presented using $\sqrt{s_{NN}} = 5.02$ TeV PbPb collision data recorded by the CMS experiment. The source parameters, extracted from the correlations functions, are studied as functions of transverse mass and collision centrality.

2. Femtoscopy with Lévy Sources

The quantum-statistical momentum correlation of identical bosons is called Bose–Einstein correlation. This correlation is in connection with the source function $S(x, p)$ [4,5], which is the phase-space probability density of particle production at space–time point x and four-momentum p. After some approximations detailed in Refs. [4,5], the following formula is obtained:

$$C^{(0)}(Q,K) \approx 1 + \frac{|\widetilde{S}(Q,K)|^2}{|\widetilde{S}(0,K)|^2}, \tag{1}$$

where $C^{(0)}(Q,K)$ is the two-particle momentum correlation function, Q is the pair relative four-momentum, K is the pair average four-momentum, the superscript (0) denotes the neglection of final-state interactions, and $\widetilde{S}(Q,K)$ is the Fourier transform of the source with

$$\widetilde{S}(Q,K) = \int S(x,K)e^{iQx}d^4x. \tag{2}$$

Equation (1) implies that $C^{(0)}(Q=0,K)=2$. In previous measurements, it was found, however, that $C^{(0)}(Q\to 0,K)<2$. This result can be understood via the core–halo model [25,26], wherein the source is divided into two parts, a core of primordial hadrons and a halo of long-lived resonances. The halo is experimentally unresolvable due to its large size, which leads to small momentum in Fourier space. If S represents only the core part of the source, its connection to the correlation function becomes

$$C^{(0)}(Q,K) \approx 1 + \lambda\frac{|\widetilde{S}(Q,K)|^2}{|\widetilde{S}(0,K)|^2}, \tag{3}$$

where λ is the square of the core fraction, and it is often called the correlation strength parameter.

Using Equation (3), a theoretical formula for $C^{(0)}(Q,K)$ can be calculated by assuming a given source distribution. In this analysis, a generalization of the Gaussian distribution, the so-called spherically symmetric Lévy alpha-stable distribution [17], was assumed for the spatial part of the source. This distribution is defined by the following Fourier transform in three dimensions:

$$\mathcal{L}(r;\alpha,R) = \frac{1}{(2\pi)^3}\int d^3q\, e^{iqr}e^{-\frac{1}{2}|qR|^\alpha}, \tag{4}$$

where q is an integration variable, r is the variable of the distribution, α and R are parameters; the Lévy stability index and the Lévy scale parameter, respectively. The α parameter describes the shape of the distribution, with $\alpha=2$ corresponding to the Gaussian and $\alpha=1$ to the Cauchy case. The R parameter describes the spatial scale of the source, as it is proportional to the full width at half maximum. There are many possible reasons [18–24] behind the appearance of the Lévy distribution in heavy-ion collisions, but these possibilities are still under investigation by the community. In case of a spherically symmetric Lévy source, the two-particle correlation function has the form [19]

$$C^{(0)}(q) = 1 + \lambda e^{-(qR)^\alpha}, \tag{5}$$

where $q=|Q|$ is the magnitude of the spatial part of Q.

In the above formulas, the presence of final-state interactions was neglected. In the case of charged particles, the most important final-state interaction is the Coulomb interaction, which is usually taken into account in the form of a Coulomb correction $K_C(q;R,\alpha)$ [27–29]. Using the Bowler–Sinyukov method [30], one obtains

$$C(q) = 1 - \lambda + \lambda(1 + e^{-(qR)^\alpha})K_C(q;R,\alpha). \tag{6}$$

In this analysis, the R and α-dependent Coulomb correction, calculated in Ref. [31], was utilized. A formula based on Equation (6) was used for fitting to the measured correlation functions.

3. Measurement Details

The used data sample contains 4.27×10^9 PbPb events at a center-of-mass energy per nucleon pair of $\sqrt{s_{\mathrm{NN}}} = 5.02$ TeV, recorded by the CMS experiment in 2018. The detailed description of the CMS detector system can be found in Ref. [32]. For the analysis, only events with precisely one nucleus–nucleus collision were used, where the longitudinal distance of the interaction point from the center of the detector was also less than 15 cm. Further event selections were applied to reject events from beam–gas interactions and nonhadronic collisions [33]. The individual tracks were filtered based on their transverse momentum, pseudorapidity, distance to the vertex, the goodness of the track fit, and the number of hits in the tracking detectors.

Particle identification in central PbPb collisions is not possible with the CMS detector; therefore, all charged tracks passing the other selection criteria were used. The majority of these charged particles are pions [34], so the pion mass was assumed for all of them. The largest contamination is caused by kaons and protons [34], and this effect is discussed in Section 4.

Measuring two-particle Bose–Einstein correlation functions means measuring pair distributions. Besides the quantum-statistical effects, these pair distributions are influenced by detector acceptance, kinematics, and other phenomena. In order to remove these unwanted effects, the correlation function is calculated as the normalized ratio of two distributions, the actual (signal) distribution $A(q)$, and the background distribution $B(q)$, with

$$C(q) = \frac{A(q)}{B(q)} \frac{\int B(q)dq}{\int A(q)dq},$$

(7)

where the integrals are calculated over a range where the quantum-statistical effects are not present. The $A(q)$ distribution contains all same charged pairs of a given event, while the $B(q)$ distribution contains all same charged pairs of a mixed event. This mixed event is obtained by randomly selecting particles from different events, as detailed in Refs. [11,35]. For the validity of Equation (7), it was assumed that the produced particles had a uniform rapidity distribution [36].

In the measurement of $C(q)$, the q variable is taken as the magnitude of the relative momentum in the longitudinally comoving system (LCMS), where the longitudinal component of the average momentum is zero. This coordinate system was chosen because, in earlier measurements, the source was found to be approximately spherically symmetric in this frame [6]. The measurement is carried out up to $q = 8$ GeV/c in 6 centrality (0–60%) and 24 average transverse momentum K_{T} (0.5–1.9 GeV/c) classes, separately for positively and negatively charged pairs. In order to remove the merging and splitting effects caused by the finite resolution of the tracking detectors, a pair selection was applied. These artifacts were limited to a region with small $\Delta\eta$ and $\Delta\phi$; therefore, each pair had to satisfy the following condition:

$$\left(\frac{|\Delta\eta|}{0.014}\right)^2 + \left(\frac{|\Delta\phi|}{0.022}\right)^2 > 1,$$

(8)

where $\Delta\eta$ is the pseudorapidity difference and $\Delta\phi$ is the azimuthal angle difference. Tracking efficiency correction factors were also utilized when measuring the $A(q)$ and $B(q)$ distributions.

Even after removing most of the non-quantum-statistical effects by taking the ratio of $A(q)$ and $B(q)$, a structure was observed in $C(q)$ at large q values, where the quantum-statistical effects were not present. This long-range background can be the result of phenomena such as energy and momentum conservation, resonance decays, bulk flow [15], and minijets [15]. To remove any potential influence of the long-range background on the low q region where the Bose–Einstein peak is present, $C(q)$ was divided by a background function $BG(q)$, resulting in the double-ratio correlation function $DR(q)$:

$$DR(q) = \frac{C(q)}{BG(q)}.\tag{9}$$

The explicit form of $BG(q)$ was determined by fitting the following empirically determined formula [15,37,38] to the large q part of $C(q)$:

$$BG(q) = N\left(1 + \alpha_1 e^{-(qR_1)^2}\right)\left(1 - \alpha_2 e^{-(qR_2)^2}\right),\tag{10}$$

where $N, \alpha_1, \alpha_2, R_1, R_2$ are fit parameters with no physical meaning.

The $DR(q)$ distributions were fitted with the following formula based on Equation (6):

$$DR(q) = N(1 + \epsilon q)\left[1 - \lambda + \lambda(1 + e^{-(qR)^\alpha})K_C(q; R, \alpha)\right],\tag{11}$$

where N is a normalization parameter and a possible residual linear background is allowed through the ϵ parameter. The fits were performed using the MINUIT2 package [39,40] and the statistical uncertainties were calculated with the MINOS algorithm [39,40]. The lower and upper fit limits were determined individually in each centrality and K_T class by selecting the limits resulting in the best fit. The goodness of fit was measured by the confidence level, calculated from the χ^2 and the number of degrees of freedom of the fit. This confidence level was in the statistically acceptable range ($>0.1\%$) for each fit. An example fit is shown in Figure 1. In the region below approximately $q = 0.05\,\text{GeV}/c$, the measured data are not reliable due to the finite momentum resolution and pair reconstruction efficiency of the detectors; consequently, that region was not used for fitting.

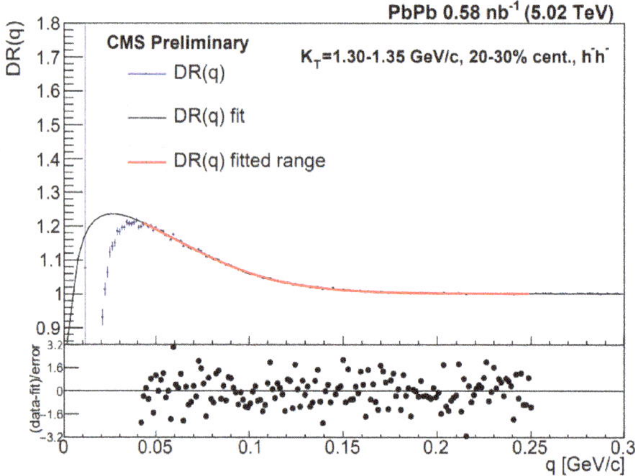

Figure 1. An example fit to the double-ratio correlation function $DR(q)$ of negatively charged hadrons [41]. The fitted function is shown in black, while the red overlay indicates the range used for the fit. The K_T and centrality class is shown in the legend. The lower panel indicates the deviation of the fit from the data.

The systematic uncertainties of R, α, and λ were determined by individually changing each of the analysis settings to slightly larger and smaller values, and conducting the whole analysis procedure again. The deviations from the nominal results were then added in quadrature, resulting in the full systematic uncertainty. The considered analysis settings were the centrality calibration, the vertex selection, the different track selection criteria, the pair selection, and the fit limits. Out of these, the dominant sources of systematic uncertainty were the fit limits. The full systematic uncertainty was separated into correlated and uncorrelated parts, so that the latter could be taken into account when fitting to the parameters.

4. Results and Discussion

As mentioned before, the parameters α, R, and λ were measured separately for positively and negatively charged hadron pairs. As not much difference was observed between the two cases, some of the results for negatively charged pairs are shown only in Appendix A.

The measurement was carried out in K_T classes, but in order to facilitate the comparison with previous measurements and with theory, the parameters are presented as functions of the transverse mass m_T, defined as

$$m_T = \sqrt{\frac{K_T^2}{c^2} + m^2},\qquad(12)$$

where m is the mass of the investigated particle species. Although all charged tracks were used in the analysis, the pion mass was used for m, since above 90% of the identical particle pairs were pion pairs.

The measured α values are shown in Figure 2 as a function of m_T, for positively charged pairs. Within uncertainties, most of the values are between 1.6 and 2.0, meaning that the source follows the general Lévy distribution, instead of the Gaussian. However, the deviation from the Gaussian case is not as large as it was found for 0–30% centrality AuAu collisions at $\sqrt{s_{NN}} = 200$ GeV [11], where a mean value for α of 1.207 was obtained for pion pairs with $|\eta| < 0.35$ and $228 < m_T < 871$ MeV/c^2. For a given centrality class, α is almost constant with m_T. The average of α ($\langle\alpha\rangle$) is indicated in Figure 2 for each centrality class, and it is shown in Figure 3 as a function of the average number of participating nucleons in the collision ($\langle N_{part}\rangle$), for both positively and negatively charged pairs. The $\langle N_{part}\rangle$ values were calculated for each centrality class [42], with a larger value corresponding to a more central case. The $\langle\alpha\rangle$ values show a monotonic increasing trend with $\langle N_{part}\rangle$, which means that the shape of the source is $\langle N_{part}\rangle$ (or equivalently, centrality) -dependent. The shape is closer to the Gaussian distribution in case of more central events. The $\langle\alpha\rangle$ values are slightly higher for positively charged pairs, although the deviations are within systematic uncertainties.

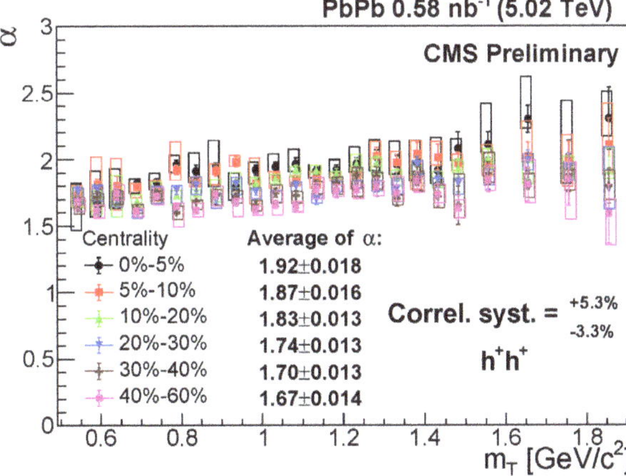

Figure 2. The Lévy stability index α versus the transverse mass m_T in different centrality classes for positively charged hadron pairs [41]. The error bars are the statistical uncertainties, while the boxes indicate the uncorrelated systematic uncertainties. The correlated systematic uncertainty is shown in the legend.

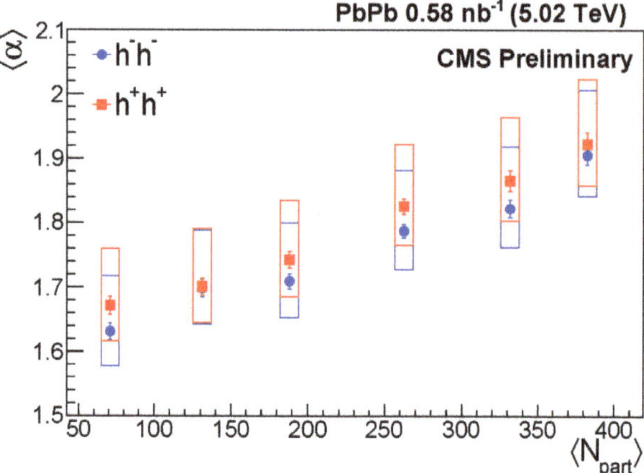

Figure 3. The average Lévy stability index $\langle \alpha \rangle$ versus $\langle N_{\text{part}} \rangle$ in different centrality classes for positively and negatively charged hadron pairs [41]. The error bars are the statistical uncertainties, while the boxes indicate the systematic uncertainties.

The measured R values are shown in Figure 4 as a function of m_{T} for positively charged pairs. A decreasing trend with m_{T} and as the collisions become more peripheral is observed, with the values ranging between 1.6 and 5.8 fm. The centrality dependence confirms the geometrical interpretation of the R parameter, because a smaller source size is expected in case of more peripheral collisions. To further investigate the m_{T} dependence of R, $1/R^2$ was plotted as a function of m_{T}, as shown in Figure 5. In case of a Gaussian source, hydrodynamic models [7,43] predict the linear scaling

$$\frac{1}{R^2} = A m_{\text{T}} + B, \tag{13}$$

where A and B are parameters with physical meaning. The slope A is connected to the Hubble constant (H) of the QGP with [7,44]

$$A = \frac{H^2}{T_{\text{f}}}, \tag{14}$$

where T_{f} is the freeze-out temperature. The intercept B is connected to the size of the source (R_{f}) at freeze-out with [7,44]

$$B = \frac{1}{R_{\text{f}}^2}. \tag{15}$$

In order to verify whether the linear scaling also holds in the Lévy case, a linear fit was performed for each centrality class using Equation (13). The statistical uncertainty and the uncorrelated systematic uncertainty of $1/R^2$ was added in quadrature and used for determining the χ^2 of the fits. In this way, the confidence levels were statistically acceptable for each centrality class, showing that a hydrodynamic-like scaling holds for a Lévy source as well. The fitted lines are shown in Figure 5, and the fit parameters (A and B) are shown in Figure 6 as functions of $\langle N_{\text{part}} \rangle$, for both positively and negatively charged pairs. By assuming a constant freeze-out temperature of $T_{\text{f}} = 156$ MeV [45], the Hubble constant falls between 0.12 c/fm and 0.18 c/fm. Due to the fact that the A parameter decreases toward more central collisions (larger $\langle N_{\text{part}} \rangle$), the Hubble constant also decreases, making the speed of the expansion lower in central collisions. The B parameter has a negative value in each case, which makes it impossible to calculate a freeze-out size using Equation (15). The reasons behind a negative intercept and the interpretation of this result are currently unknown. This may be connected to fluctuations in the initial state [46] which were not taken into account in the hydrodynamic models.

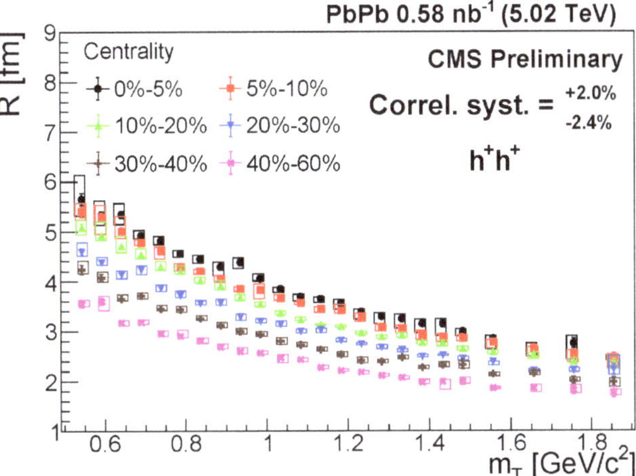

Figure 4. The Lévy scale parameter R versus m_T in different centrality classes for positively charged hadron pairs [41]. The error bars are the statistical uncertainties, while the boxes indicate the uncorrelated systematic uncertainties. The correlated systematic uncertainty is shown in the legend.

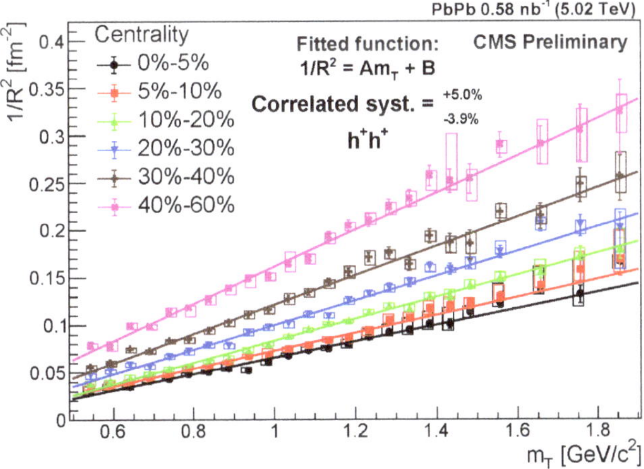

Figure 5. The inverse square of the Lévy scale parameter R versus m_T in different centrality classes for positively charged hadron pairs [41]. The error bars are the statistical uncertainties, while the boxes indicate the uncorrelated systematic uncertainties. The correlated systematic uncertainty is shown in the legend. A line is fitted to the data for each centrality.

The measured λ values are shown in the upper panel of Figure 7 as a function of m_T, for positively charged pairs. A decreasing trend with m_T as the collisions became more central is observed. In case of identified particles, λ is the square of the ratio of core particles. Due to the lack of particle identification, our sample contained particles other than pions, mostly kaons and protons. As a result of this contamination, λ was suppressed by a factor of the square of the pion fraction. The pion fraction was measured by the ALICE Collaboration [34], and it decreased with m_T, resulting in the decreasing trend of λ in the upper panel of Figure 7. For the α and the R parameters, a characteristic m_T dependence was observed; thus, these parameters could not have been influenced by the m_T-dependent effect of the lack of particle identification. To remove the effect of the contamination from λ, the λ^* parameter was introduced by rescaling λ with the square of the pion fraction:

$$\lambda^* = \frac{\lambda}{(N_\mathrm{pion}/N_\mathrm{hadron})^2}. \tag{16}$$

The rescaled correlation strength λ^* is shown in the lower panel of Figure 7. Compared to λ, the decreasing trend with m_T is no longer shown in the data, suggesting that it was caused purely by the lack of particle identification. The centrality dependence, on the other hand, remained the same, which means that the fraction of core pions is smaller in more central collisions.

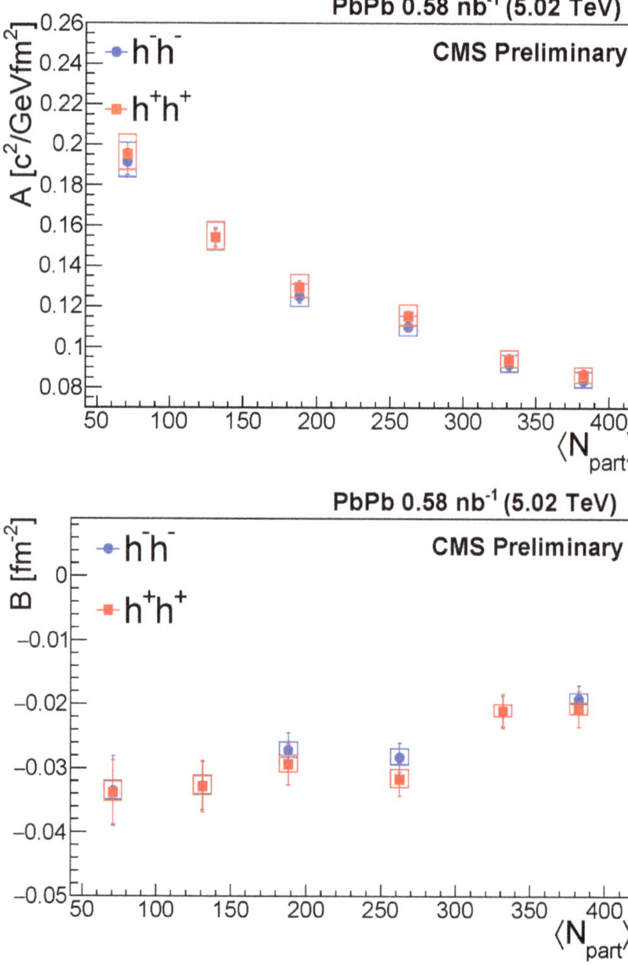

Figure 6. The two fit parameters from the linear fit: the slope A (**upper**) and the intercept B (**lower**) versus $\langle N_\mathrm{part}\rangle$ for negatively and positively charged hadron pairs [41]. The error bars are the statistical uncertainties, while the boxes indicate the systematic uncertainties.

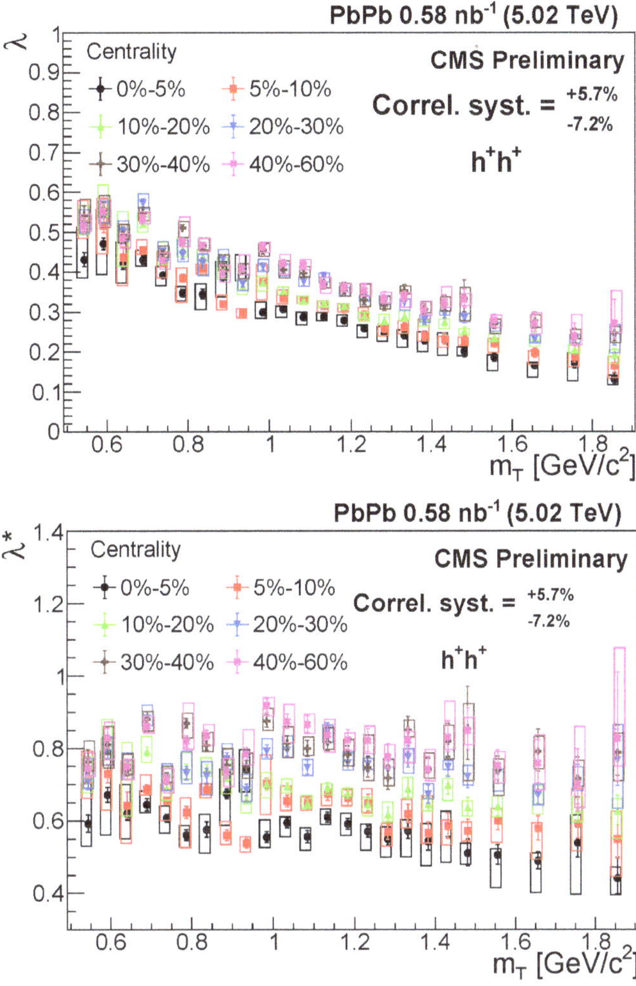

Figure 7. The correlation strength λ and the rescaled correlation strength λ^* versus m_T in different centrality classes for positively charged hadron pairs [41]. The error bars are the statistical uncertainties, while the boxes indicate the uncorrelated systematic uncertainties. The correlated systematic uncertainty is shown in the legend.

5. Conclusions

In this paper, a centrality-dependent Lévy HBT analysis of two-particle Bose–Einstein correlations was presented, using $\sqrt{s_{\mathrm{NN}}} = 5.02$ TeV PbPb collision data recorded by the CMS experiment. The measured correlation functions were described by the assumption of a Lévy alpha-stable source distribution. Three source parameters, the Lévy stability index α, the Lévy scale parameter R, and the correlation strength λ were determined, and their centrality and transverse mass (m_T) dependence was investigated.

The α parameter was found to be centrality-dependent, but constant in m_T, with the average values ranging between 1.6 and 2.0. A decreasing trend with m_T and as the collisions become more peripheral was observed for the R parameter, which could be explained by the hydrodynamic-like scaling and the geometrical interpretation, respectively. The λ parameter showed a decreasing trend with m_T, but after removing the effects of the lack of particle identification, a constant behavior was obtained. A decrease toward more central collisions was also observed for λ.

Funding: B. Kórodi was supported by the ÚNKP-21-2 New National Excellence Program of the Ministry for Innovation and Technology from the source of the National Research, Development and Innovation Fund. This research was supported by the NKFIH OTKA K-138136 and K-128713 grants.

Data Availability Statement: The data presented in this study are available on request from the corresponding author. The data are not publicly available.

Conflicts of Interest: The author declares no conflict of interest.

Abbreviations

The following abbreviations are used in this manuscript:

QGP	quark–gluon plasma
LHC	Large Hadron Collider
HBT	Hanbury Brown and Twiss
PbPb	lead–lead
CMS	Compact Muon Solenoid
AuAu	gold–gold

Appendix A. Results for Negatively Charged Pairs

The results for negatively charged hadron pairs are presented. Due to the fact that they are very similar to the results for positively charged pairs presented in Section 4, the interpretations of these results are the same. The Lévy stability index α is shown as a function of m_T in Figure A1. The Lévy scale parameter R and its inverse square $1/R^2$ are shown as functions of m_T in Figures A2 and A3, respectively. The correlation strength λ and the rescaled correlation strength λ^* are shown as functions of m_T in Figure A4.

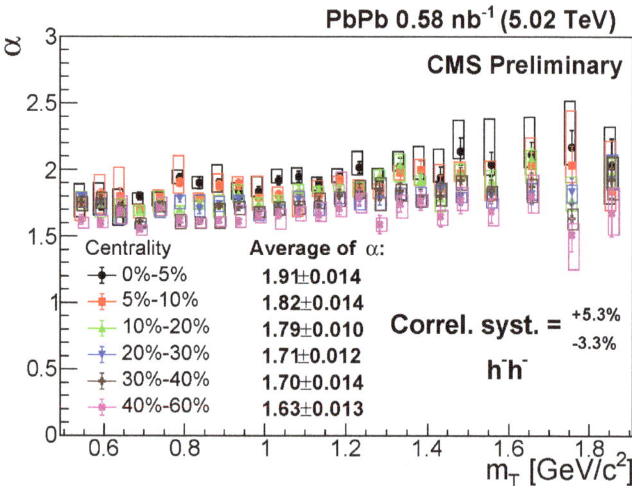

Figure A1. The Lévy stability index α versus the transverse mass m_T in different centrality classes for negatively charged hadron pairs [41]. The error bars are the statistical uncertainties, while the boxes indicate the uncorrelated systematic uncertainties. The correlated systematic uncertainty is shown in the legend.

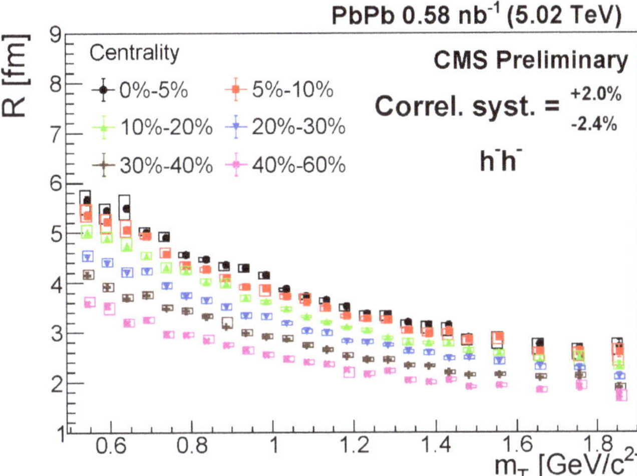

Figure A2. The Lévy scale parameter R versus m_T in different centrality classes for negatively charged hadron pairs [41]. The error bars are the statistical uncertainties, while the boxes indicate the uncorrelated systematic uncertainties. The correlated systematic uncertainty is shown in the legend.

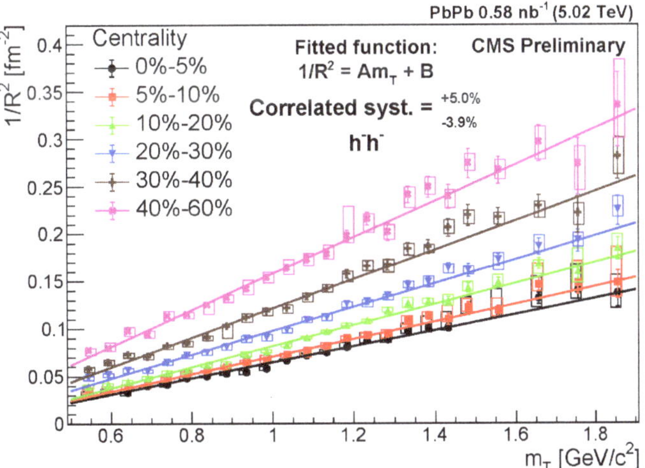

Figure A3. The inverse square of the Lévy scale parameter R versus m_T in different centrality classes for negatively charged hadron pairs [41]. The error bars are the statistical uncertainties, while the boxes indicate the uncorrelated systematic uncertainties. The correlated systematic uncertainty is shown in the legend. A line is fitted to the data for each centrality.

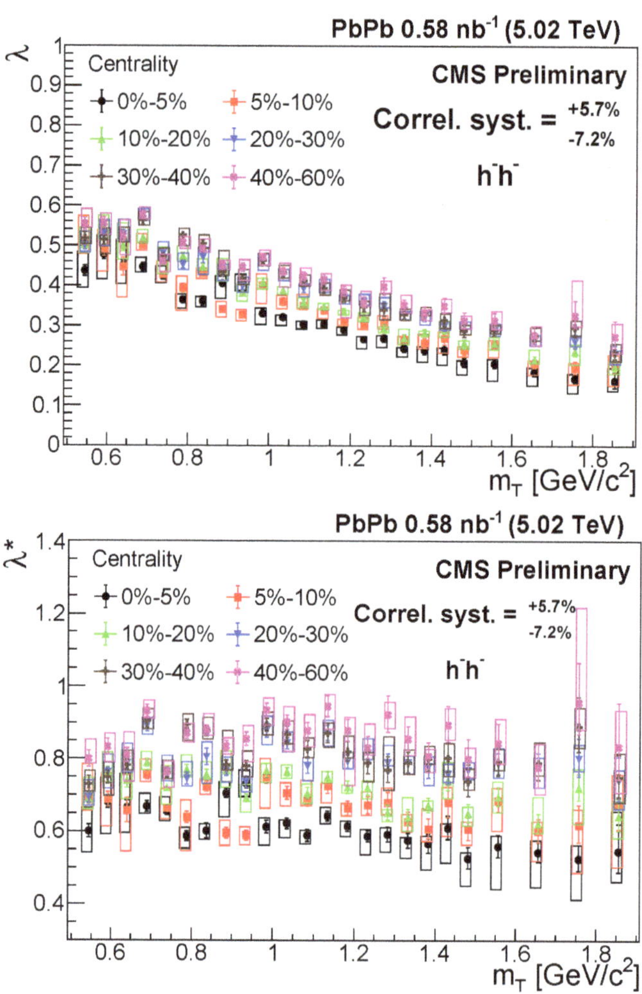

Figure A4. The correlation strength λ and the rescaled correlation strength λ^* versus m_T in different centrality classes for negatively charged hadron pairs [41]. The error bars are the statistical uncertainties, while the boxes indicate the uncorrelated systematic uncertainties. The correlated systematic uncertainty is shown in the legend.

References

1. Lednicky, R. Femtoscopy with unlike particles. In Proceedings of the International Workshop on the Physics of the Quark Gluon Plasma, Palaiseau, France, 4–7 September 2001.
2. Hanbury Brown, R.; Twiss, R.Q. A Test of a new type of stellar interferometer on Sirius. *Nature* **1956**, *178*, 1046–1048. [CrossRef]
3. Glauber, R.J. Photon correlations. *Phys. Rev. Lett.* **1963**, *10*, 84. [CrossRef]
4. Csörgő, T. Particle interferometry from 40 MeV to 40 TeV. *Acta Phys. Hung. A* **2002**, *15*, 1–80. [CrossRef]
5. Wiedemann, U.A.; Heinz, U.W. Particle interferometry for relativistic heavy ion collisions. *Phys. Rep.* **1999**, *319*, 145–230. [CrossRef]
6. PHENIX Collaboration. Bose-Einstein correlations of charged pion pairs in Au+Au collisions at $\sqrt{s_{NN}}$ = 200 GeV. *Phys. Rev. Lett.* **2004**, *93*, 152302. [CrossRef]
7. Csörgő, T.; Lörstad, B. Bose-Einstein correlations for three-dimensionally expanding, cylindrically symmetric, finite systems. *Phys. Rev. C* **1996**, *54*, 1390. [CrossRef]
8. Csanád, M.; Csörgő, T.; Lörstad, B.; Ster, A. Indication of quark deconfinement and evidence for a Hubble flow in 130 GeV and 200 GeV Au+Au collisions. *J. Phys. G* **2004**, *30*, S1079. [CrossRef]

9.   Pratt, S. Resolving the HBT Puzzle in Relativistic Heavy Ion Collision. *Phys. Rev. Lett.* **2009**, *102*, 232301. [CrossRef]
10. Lacey, R.A. Indications for a Critical End Point in the Phase Diagram for Hot and Dense Nuclear Matter. *Phys. Rev. Lett.* **2015**, *114*, 142301. [CrossRef]
11. PHENIX Collaboration. Lévy-stable two-pion Bose–Einstein correlations in $\sqrt{s_{NN}} = 200$ GeV Au+Au collisions. *Phys. Rev. C* **2018**, *97*, 064911. [CrossRef]
12. NA61/SHINE Collaboration. Measurements of two-pion HBT correlations in Be+Be collisions at 150A GeV/c beam momentum, at the NA61/SHINE experiment at CERN. *arXiv* **2023**, arXiv:2302.04593.
13. STAR Collaboration. Pion interferometry in Au+Au collisions at $\sqrt{s_{NN}} = 200$ GeV. *Phys. Rev. C* **2005**, *71*, 044906. [CrossRef]
14. ALICE Collaboration. One-dimensional pion, kaon, and proton femtoscopy in Pb-Pb collisions at $\sqrt{s_{NN}} = 2.76$ TeV. *Phys. Rev. C* **2015**, *92*, 054908. [CrossRef]
15. CMS Collaboration. Bose-Einstein correlations in pp, pPb, and PbPb collisions at $\sqrt{s_{NN}} = 0.9 - 7$ TeV. *Phys. Rev. C* **2018**, *97*, 064912. [CrossRef]
16. ATLAS Collaboration. Femtoscopy with identified charged pions in proton-lead collisions at $\sqrt{s_{NN}} = 5.02$ TeV with ATLAS. *Phys. Rev. C* **2017**, *96*, 064908. [CrossRef]
17. Uchaikin, V.V.; Zolotarev, V.M. *Chance and Stability: Stable Distributions and Their Applications*; De Gruyter: Berlin, Germany, 2011. [CrossRef]
18. Metzler, R.; Barkai, E.; Klafter, J. Anomalous Diffusion and Relaxation Close to Thermal Equilibrium: A Fractional Fokker-Planck Equation Approach. *Phys. Rev. Lett.* **1999**, *82*, 3563. [CrossRef]
19. Csörgő, T.; Hegyi, S.; Zajc, W.A. Bose-Einstein correlations for Lévy stable source distributions. *Eur. Phys. J. C* **2004**, *36*, 67–78. [CrossRef]
20. Csanád, M.; Csörgő, T.; Nagy, M. Anomalous diffusion of pions at RHIC. *Braz. J. Phys.* **2007**, *37*, 1002–1013. [CrossRef]
21. Kincses, D.; Stefaniak, M.; Csanád, M. Event-by-event investigation of the two-particle source function in heavy-ion collisions with EPOS. *Entropy* **2022**, *24*, 308. [CrossRef]
22. Kórodi, B.; Kincses, D.; Csanád, M. Event-by-event investigation of the two-particle source function in $\sqrt{s_{NN}} = 2.76$ TeV PbPb collisions with EPOS. *arXiv* **2022**, arXiv:2212.02980.
23. Csörgő, T.; Hegyi, S.; Novák, T.; Zajc, W.A. Bose-Einstein or HBT correlations and the anomalous dimension of QCD. *Acta Phys. Pol. B* **2005**, *36*, 329–337. [CrossRef]
24. Csörgő, T.; Hegyi, S.; Novák, T.; Zajc, W.A. Bose-Einstein or HBT correlation signature of a second order QCD phase transition. *AIP Conf. Proc.* **2006**, *828*, 525–532. [CrossRef]
25. Bolz, J.; Ornik, U.; Plumer, M.; Schlei, B.R.; Weiner, R.M. Resonance decays and partial coherence in Bose-Einstein correlations. *Phys. Rev. D* **1993**, *47*, 3860. [CrossRef] [PubMed]
26. Csörgő, T.; Lörstad, B.; Zimányi, J. Bose-Einstein correlations for systems with large halo. *Z. Phys. C* **1996**, *71*, 491–497. [CrossRef]
27. Kurgyis, B.; Kincses, D.; Csanád, M. Coulomb interaction for Lévy sources. *arXiv* **2020**, arXiv:2007.10173.
28. Kincses, D.; Nagy, M.; Csanád, M. Coulomb and strong interactions in the final state of Hanbury-Brown–Twiss correlations for Lévy-type source functions. *Phys. Rev. B* **2020**, *102*, 064912. [CrossRef]
29. Csanád, M.; Lökös, S.; Nagy, M. Coulomb final state interaction in heavy ion collisions for Lévy sources. *Universe* **2019**, *5*, 133. [CrossRef]
30. Sinyukov, Y.; Lednicky, R.; Akkelin, S.V.; Pluta, J.; Erazmus, B. Coulomb corrections for interferometry analysis of expanding hadron systems. *Phys. Lett. B* **1998**, *432*, 248–257. [CrossRef]
31. Csanád, M.; Lökös, S.; Nagy, M. Expanded empirical formula for Coulomb final state interaction in the presence of Lévy sources. *Phys. Part. Nucl.* **2020**, *51*, 238. [CrossRef]
32. CMS Collaboration. The CMS experiment at the CERN LHC. *J. Instrum.* **2008**, *3*, S08004. [CrossRef]
33. CMS Collaboration. Charged-particle nuclear modification factors in PbPb and pPb collisions at $\sqrt{s_{NN}} = 5.02$ TeV. *J. High Energy Phys.* **2017**, *04*, 039. [CrossRef]
34. ALICE Collaboration. Production of charged pions, kaons, and (anti-)protons in Pb-Pb and inelastic pp collisions at $\sqrt{s_{NN}} = 5.02$ TeV. *Phys. Rev. C* **2020**, *101*, 044907. [CrossRef]
35. L3 Collaboration. Test of the τ-Model of Bose-Einstein Correlations and Reconstruction of the Source Function in Hadronic Z-boson Decay at LEP. *Eur. Phys. J. C* **2011**, *71*, 1648. [CrossRef]
36. Vechernin, V.; Andronov, E. Strongly intensive observables in the model with string fusion. *Universe* **2019**, *5*, 15. [CrossRef]
37. CMS Collaboration. Bose-Einstein correlations of charged hadrons in proton-proton collisions at $\sqrt{s_{NN}} = 13$ TeV. *J. High Energy Phys.* **2020**, *3*, 14. [CrossRef]
38. CMS Collaboration. K_S^0 and $\Lambda(\overline{\Lambda})$ two-particle femtoscopic correlations in PbPb collisions at $\sqrt{s_{NN}} = 5.02$ TeV. *arXiv* **2023**, arXiv:2301.05290.
39. James, F.; Roos, M. Minuit: A System for Function Minimization and Analysis of the Parameter Errors and Correlations. *Comput. Phys. Commun.* **1975**, *10*, 343–367. [CrossRef]
40. Brun, R.; Rademakers, F. ROOT: An object oriented data analysis framework. *Nucl. Instrum. Meth. A* **1997**, *389*, 81–86. [CrossRef]
41. CMS Collaboration. Measurement of Two-Particle Bose–Einstein Momentum Correlations and Their Lévy Parameters at $\sqrt{s_{NN}} = 5.02$ TeV PbPb Collisions. CMS Physics Analysis Summary CMS-PAS-HIN-21-011. 2022. Available online: https://cds.cern.ch/record/2806150 (accessed on 20 April 2022).

42. Loizides, C.; Kamin, J.; d'Enterria, D. Improved Monte Carlo Glauber predictions at present and future nuclear colliders. *Phys. Rev. C* **2018**, *97*, 054910. [CrossRef]
43. Makhlin, A.N.; Sinyukov, Y.M. Hydrodynamics of Hadron Matter Under Pion Interferometric Microscope. *Z. Phys. C* **1988**, *39*, 69–73. [CrossRef]
44. Chojnacki, M.; Florkowski, W.; Csörgő, T. On the formation of Hubble flow in little bangs. *Phys. Rev. C* **2005**, *71*, 044902. [CrossRef]
45. Steinbrecher, P. The QCD crossover at zero and non-zero baryon densities from Lattice QCD. *Nucl. Phys. A* **2019**, *982*, 847–850. [CrossRef]
46. Sputowska, I. Forward-backward correlations and multiplicity fluctuations in Pb–Pb collisions at $\sqrt{s_{NN}} = 2.76$ TeV from ALICE at the LHC. *Proceedings* **2019**, *10*, 14. [CrossRef]

Communication

Charged Kaon Femtoscopy with Lévy Sources in $\sqrt{s_{NN}} = 200$ GeV Au+Au Collisions at PHENIX

László Kovács on behalf of the PHENIX Collaboration

Department of Atomic Physics, Faculty of Science, Eötvös Loránd University, Pázmány Péter Sétány 1/A, H-1111 Budapest, Hungary; kovacs.laszlo9201@gmail.com

Abstract: The PHENIX experiment measured two-particle Bose–Einstein quantum-statistical correlations of charged kaons in Au+Au collisions at $\sqrt{s_{NN}} = 200$ GeV. The correlation functions are parametrized assuming that the source emitting the particles has a Lévy shape, characterized by the Lévy exponent α and the Lévy scale R. By introducing the intercept parameter λ, we account for the core–halo fraction. The parameters are investigated as a function of transverse mass. The comparison of the parameters measured for kaon–kaon with those measured from pion–pion correlation may clarify the connection of Lévy parameters to physical processes .

Keywords: heavy ions; quark–gluon plasma; femtoscopy; Lévy HBT

1. Introduction

To study the space-time structure of the quark–gluon plasma, the most commonly used method is femtoscopy. It is a sub-field of high-energy particles and nuclear physics, and it allows us to explore the properties of the matter created in particle collisions on the femtometer scale. Femtoscopy typically investigates correlations of particle pairs. However, it is worth mentioning that prior to the development of femtoscopy, a similar physical phenomenon was discovered and utilized in the field of radio astronomy. R. Hanbury Brown and R. Q. Twiss measured the angular size of stars by analyzing intensity correlations, which became known as the Hanbury Brown and Twiss effect (HBT) [1]. Roy Glauber's work that laid the foundations of quantum optics [2–4] greatly increased our understanding of this effect. G. Goldhaber and his collaborators observed intensity correlations among same-charged pions while searching for ρ mesons in high-energy collisions. These correlations were explained by G. Goldhaber, S. Goldhaber, W-Y. Lee, and A. Pais (GGLP), based on the Bose–Einstein symmetrization of the wave-function of identical pion pairs [5]. This is the reason why these correlations are often called Bose–Einstein correlations. Due to the relationship between the two-particle Bose–Einstein correlation function and the phase-space density of the particle-emitting source, by measuring the correlation function we can obtain information about the source function.

Let us note that while there are conceptual similarities between the HBT effect in radio astronomy and the correlations studied in femtoscopy, there are also fundamental differences. In femtoscopy, we can extract information about the space-time structure of the particle source, often represented by femtoscopic radii. On the other hand, the HBT effect in radio astronomy provides insights into the spectral-angular structure of the source radiation.

The fundamentals of modern correlation femtoscopy were established by Kopylov and Podgoretsky [6,7]. They successfully overcame the drawback of the GGLP technique by utilizing the momentum differences of the particle pairs instead of the opening angles.

Based on the central limit theorem, it is a good approach to assume a Gaussian shape for the phase-space density of the particle-emitting source. However, we can go further and take a more general approach. Anomalous diffusion indicates the appearance of Lévy-stable distributions for the source [8,9]. In Ref. [10], it was found that Lévy-stable

Citation: Kovács, L., on behalf of the PHENIX Collaboration. Charged Kaon Femtoscopy with Lévy Sources in $\sqrt{s_{NN}} = 200$ GeV Au+Au Collisions at PHENIX. *Universe* **2023**, *9*, 336. https://doi.org/10.3390/universe9070336

Academic Editor: Jun Xu

Received: 30 May 2023
Revised: 8 July 2023
Accepted: 11 July 2023
Published: 17 July 2023

source distributions in $\sqrt{s_{NN}} = 200$ GeV Au+Au collisions give a high-quality, statistically acceptable description of the measured correlation functions in the case of pions. In the present paper, we will investigate kaon–kaon correlation functions assuming a Lévy shaped source. By comparing the obtained results with the pion data, more insights can be gained regarding the Lévy parameters.

The dataset used in this analysis is Au+Au collisions at $\sqrt{s_{NN}} = 200$ GeV recorded by the PHENIX (**P**ioneering **H**igh **E**nergy **N**uclear **I**nteraction e**X**periment) detector. It is one of the four experiments that have taken data at the relativistic heavy ion collider (RHIC) in Brookhaven National Laboratory. Its primary mission was to search for a new state of matter called the quark–gluon plasma and to study various different particle types produced in heavy ion collisions, such as photons, electrons, muons, and charged hadrons. A beam view layout of the PHENIX detector can be seen in Figure 1. The detectors can be divided into four main subgroups:

1. Global detectors characterize the nature of heavy ion collision events, i.e., zero degree calorimeters (ZDC) and beam-beam counters (BBC);
2. Mid-rapidity detectors form the "central arm spectrometer", which consists of three sets of pad chambers (PC), drift chambers (DC), electromagnetic calorimeters (Em-Cal), and time-of-flight detectors (ToF), are used for energy, momentum, and mass measurements;
3. Two muon spectrometers at forward rapidity;
4. A triggering and computing system to select and archive events of potential physics interest.

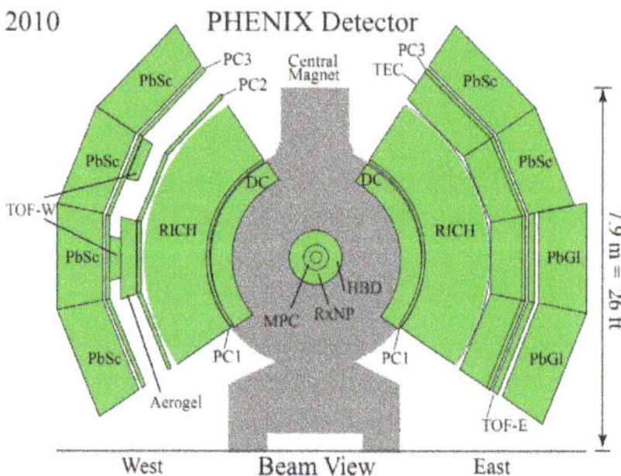

Figure 1. View of the PHENIX central arm spectrometer detector setup in the 2010 data-taking period.

2. Femtoscopy and Lévy Sources

As we mentioned in the previous section, there is a connection between the Bose–Einstein correlation function and the phase-space density of the particle-emitting source. Let us discuss this relationship in more detail. The one- and two-particle momentum distributions can be expressed as [11]

$$N_1(p) = \int d^4 r S(r,p)|\psi_p(r)|^2, \tag{1}$$

$$N_2(p_1,p_2) = \int d^4 r_1 d^4 r_2 S(r_1,p_1)S(r_2,p_2)|\psi^{(2)}_{p_1,p_2}(r_1,r_2)|^2, \tag{2}$$

where $S(r,p)$ is the source function, which describes the probability density of particle creation at space-time point r with four-momentum p; $\psi_p(r)$ denotes the single-particle wave function; and $\psi^{(2)}_{p_1,p_2}$ is the two-particle wave function, which must be symmetric in the spatial variables r_1 and r_2 for bosons. Bose–Einstein correlations arise from this symmetrization effect. Using Equations (1) and (2), we can express the two-particle correlation function as [12,13]

$$C_2(p_1, p_2) = \frac{N_2(p_1, p_2)}{N_1(p_1)N_1(p_2)}.$$

(3)

Let us introduce the average momentum $K = 0.5(p_1 + p_2)$ and relative momentum $q = p_1 - p_2$ as new variables. If $p_1 \approx p_2 \approx K$ and the final state interactions are neglected, the two-particle correlation function can be written as

$$C_2^{(0)}(q, K) \approx 1 + \frac{|\tilde{S}(q, K)|^2}{|\tilde{S}(0, K)|^2},$$

(4)

where the superscript (0) denotes the neglection of final state interactions and $\tilde{S}(q, K)$ is the Fourier transform of the source with

$$\tilde{S}(q, K) = \int S(x, K)e^{iqx}d^4x.$$

(5)

The significance of Equation (4) lies in the fact that by measuring the Bose–Einstein correlation function, we can obtain information about the spatial shape of the source function.

The correlation function depends on the four-momentum difference and the average four-momentum. Since the Lorentz product of q and K is zero in the case of identical particles, the correlation function depends only on the spatial $\mathbf{q}$ instead of the four dimensional q vector:

$$qK = q_0K_0 - \mathbf{q}\mathbf{K} = 0 \quad \Rightarrow \quad q_0 = \frac{\mathbf{q}\mathbf{K}}{K_0}.$$

(6)

In Ref. [7], Kopylov and Podgoretsky showcased the three-dimensional character of the momentum correlation effect due to the particles detected on-mass-shell.

For our correlation function measurements, we use the co-moving system (LCMS) frame, where it was found in earlier measurements [14] that the correlation function is nearly spherically symmetric. Due to the relatively low number of produced kaons and this symmetrical characteristic, we have chosen to perform a one-dimensional (1D) analysis instead of a three-dimensional (3D) one. Based on Ref. [10], we used $Q = |q_{\mathrm{LCMS}}|$ as the 1D variable of the correlation function.

We assumed that the source emitting the particles has a Lévy shape. The symmetric Lévy-stable distribution is defined as

$$\mathcal{L}(\mathbf{r}; R, \alpha) = \frac{1}{(2\pi)^3} \int d^3\varphi\, e^{i\varphi\mathbf{r}} e^{-\frac{1}{2}|\varphi R|^\alpha},$$

(7)

where R is the Lévy scale parameter, α is the Lévy exponent, and φ is a three-dimensional integration variable. The α parameter describes the shape of the distribution, in the case of a Gaussian distribution $\alpha = 2$, while for a Cauchy distribution, the value of α is 1.

In case of such Lévy-stable source functions, the raw (i.e., final-state interaction neglected) correlation function in the observable Q-range will be [8]

$$C_2^{(0)}(Q; \lambda, R, \alpha) = 1 + \lambda e^{-|QR|^\alpha},$$

(8)

where the intercept parameter λ is introduced as the extrapolated $C_2^{(0)}(Q = 0)$ value.

According to the core–halo model [15,16], the source can be divided into two parts: the core, which contains the promptly produced particles, and the halo, which is composed of the products of resonance decays. The ratio of these two parts can be characterized by the correlation strength parameter:

$$\lambda = \left(\frac{N_{\text{core}}}{N_{\text{core}} + N_{\text{halo}}} \right)^2, \tag{9}$$

where N_{core} refers to the number of particles produced in the core, while N_{halo} denotes the number of particles produced in the halo. Considering that the particles from the halo contribute to the correlation function as an unresolvably narrow peak around $Q = 0$, we indeed see that this λ value will be the extrapolated $C_2^{(0)}(Q = 0)$ value.

For charged particles, the most significant final state interaction is the Coulomb interaction. To take care of this effect, we used the Sinyukov–Bowler method [17,18]. Taking an additional possible linear background shape into account, our final assumption for the functional form of the correlation function is

$$C_2(Q; \lambda, R, \alpha) = \left[1 - \lambda + K(q_{\text{inv}}; \alpha, R) \cdot \lambda \cdot \left(1 + e^{-|QR|^{\alpha}} \right) \right] \cdot N \cdot (1 + \epsilon Q), \tag{10}$$

where K is the Coulomb correction, N is the normalization parameter, and ϵ represents a small background long-range correlation effect. Let us note that the Coulomb correction is a function of q_{inv}, which is the Lorentz invariant four-momentum difference[1], while the correlation function has a different variable, denoted by Q. We calculated the Coulomb correction K with the variable Q and analyzed the error coming from this approximation, which was handled as a source of systematic uncertainty the same way as in Ref. [10]. The correction is quite small compared to other sources, so it does not mean a large additional term to the systematics, and this way we have more comparable results to the pions.

3. Motivation

In Ref. [10], the significance of the appearance of the Lévy distribution in the case of pion–pion correlations was investigated. In order to dive into the exploration of the Lévy-shape, we aimed to analyze kaon correlations.

The Lévy source parameters for kaon–kaon two particle correlations have never been measured in PHENIX before.

Anomalous diffusion could be a reason for a Lévy distribution [19]. In such a scenario, the Lévy index for the different particles is different, i.e., $\alpha_{\text{Lévy}}^{\pi} \neq \alpha_{\text{Lévy}}^{K}$. The smaller the cross section is, the longer the mean free path, thus the longer the power-law like "tail" of the source distribution. Kaons have a smaller cross section than pions, so we would expect that $\alpha_{\text{Lévy}}^{\pi} > \alpha_{\text{Lévy}}^{K}$. If this explanation fails to match the reality, other alternatives should be investigated.

The Lévy width (or, equivalently, scale parameter) R has an unclear interpretation. It exhibits similar behavior as the Gaussian source radii, but its precise relation to the geometrical source size is not clear. By measuring this parameter for kaons, we can get closer to clarifying its precise physical interpretation.

4. Measurement Details

In this measurement, we analyzed Au+Au collisions at $\sqrt{s_{\text{NN}}} = 200$ GeV. The dataset consists of about 7.3 billion (minimum bias triggered) events. As the number of produced kaons is relatively low, all minimum bias events (of any centrality) were taken together. We cut out those events whose distance from the nominal collision point was greater than 30 cm along the beam axis.

We also need to take into account the detector inefficiencies and the particularities of the track reconstruction algorithm, which sometimes splits one track into two. On the other hand, when two different tracks are too close to each other, it is possible that they will be detected as a single track. To remove these possible effects, we applied cuts in the

$\Delta\phi$ and Δz variables, where $\Delta\phi$ and Δz stand for the azimuthal angle and longitudinal position difference of track pairs, respectively (as measured in the drift chamber, the main tracking detector).

For particle identification (PID), we calculated the square of the particle mass:

$$m^2 = \frac{\mathbf{p}^2}{c^2}\left[\left(\frac{ct}{L}\right)^2 - 1\right],\tag{11}$$

where t is the time of flight (measured either in the PbSc or the ToF East/West detectors), L is the path length, and $\mathbf{p}$ is the momentum. The distributions of m^2 were fitted mostly using single Gaussians; in cases of merging peaks, a double Gaussian was applied. To identify kaons, we applied a 2.5 standard deviation (σ_p) cut around the nominal kaon m^2 peak position and a $2.5\sigma_\mathrm{p}$ veto cut around the pion and proton m^2 peaks. An example scatter plot of the charge times momentum vs. m^2 before and after the cuts can be seen in Figure 2a,b, respectively.

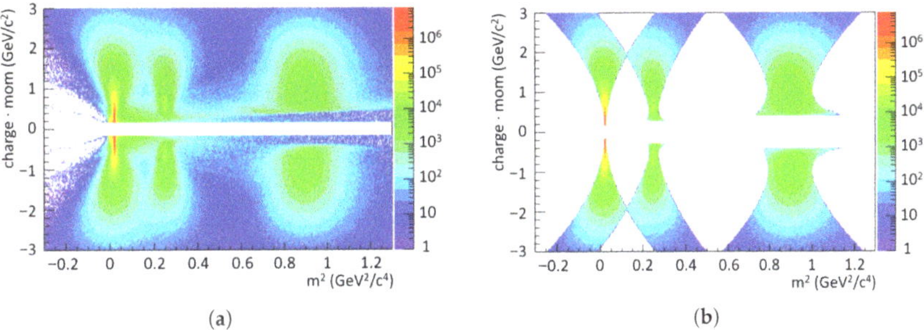

Figure 2. Example plots for PID. (**a**) Scatter plot of the charge times momentum vs. m^2 in TOF West with no cut. (**b**) Scatter plot of the charge times momentum vs. m^2 in TOF West after the applied cuts.

Since the correlation function's dependence on K is smoother than its dependence on Q, it is reasonable to create several K bins, and in each bin, the Q dependence can be investigated. At midrapidity, the transverse mass m_T can be used instead of K:

$$m_\mathrm{T} = \sqrt{m^2 + (K_\mathrm{T}/c)^2},\tag{12}$$

where m is the mass of the particle and

$$K_\mathrm{T} = \sqrt{K_x^2 + K_y^2}\tag{13}$$

is the average transverse momentum. In this analysis, 7 m_T bins were created. In each bin, we analyzed the dependence of the correlation function on Q.

To measure the correlation functions, "actual" (foreground) and "background" distributions of the kaon pairs were created. To construct the actual pair distribution, the momentum differences were calculated of the same-charged particles from the same event and filled into a histogram. Since there are other effects (stemming from acceptance, single-particle distributions, efficiency, etc.) in the actual pair distribution that are not related to the HBT effect, we have to cancel them out with a properly constructed background distribution that contains pairs from different events, where there can be no HBT effect. To create the background pair distribution, we employed the same mixed event method as described in Ref. [10]. The first step is to construct a pool that contains several events. This pool needs to be at least as large as the number of produced kaons in the event with the highest multiplicity. In order to ensure that we meet this condition, a pool with 50 events was used. Every time we process an event to construct the actual pair distribution, we construct a mixed event for the background distribution as well. Since we do not want to introduce any correlations and we would like to avoid the presence of the quantum-statistical correlation between the particles in the background, we have to select the particles as follows. First of all, to ensure that the background event exhibits the same kinematics and acceptance effects, we have to construct the background event from events of similar centrality and with a similar z coordinate of the collision vertex. To accomplish this, we used 5% wide centrality and 2 cm wide z-vertex bins. Secondly, it is essential that the selected particles for the background pair distribution originate from different events. After the particle selection from mixed eventsm we calculate the momentum differences of these particles.

The two-particle correlation function can be calculated from the ratio of the normalized actual and background pair distributions:

$$C_2(m_{\mathrm{T}}, Q) = \frac{A(m_{\mathrm{T}}, Q)}{B(m_{\mathrm{T}}, Q)} \cdot \frac{\int_{Q_{\min}}^{Q_{\max}} B(m_{\mathrm{T}}, Q)}{\int_{Q_{\min}}^{Q_{\max}} A(m_{\mathrm{T}}, Q)}, \tag{14}$$

where A is the actual, B is the background pair distribution, and the integral is performed over a range ($Q_{\min} - Q_{\max}$), where the correlation function does not exhibit quantum statistical features.

We fitted the measured correlation functions with the Coulomb-corrected Lévy-type correlation function and the linear background. Of all the final state interactions, the Coulomb effect has the greatest impact as it causes same-charged pairs to repel each other. As shown in Figure 3, the function drops off sharply at small Q due to the Coulomb effect.

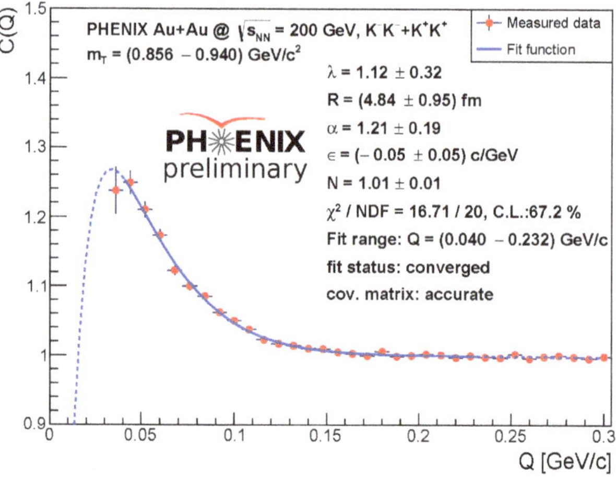

Figure 3. An example fit with the Coulomb-corrected correlation function based on a Lévy source for kaon pairs with transverse mass ranging from $0.856\,\mathrm{GeV}/c^2$ to $0.940\,\mathrm{GeV}/c^2$.

To deal with the Coulomb correction, we applied the same method that was used in Ref. [10] for pions; we did this to obtain comparable results. First, we needed to numerically solve the Coulomb correction; then, the results were loaded into a binary look up table as described in Refs. [20,21]. This numerical table contains the values discretely, so interpolation was needed, which can cause numerical fluctuations. These fluctuations can be handled with a proper iterative fitting procedure. The first round fit was performed with a functional form of the correlation function incorporating the Coulomb correction, which yields a set of parameters λ_0, R_0, and α_0. Based on Equation (10), the second round fit was with

$$C_2^{(0)}(Q; \lambda, R, \alpha) \frac{C_2(Q; \lambda_{\mathrm{n}}, R_{\mathrm{n}}, \alpha_{\mathrm{n}})}{C_2^{(0)}(Q; \lambda_{\mathrm{n}}, R_{\mathrm{n}}, \alpha_{\mathrm{n}})} \cdot N \cdot (1 + \epsilon Q), \tag{15}$$

where the fitted parameters are denoted as λ, R, α, N, and ϵ. In the second round, the values of λ_n, R_n, and α_n were equal to the corresponding values from the first fit. The correlation function without the Coulomb correction is denoted by $C_2^{(0)}(Q; \lambda, R, \alpha)$, while $C_2(Q; \lambda, R, \alpha)$ refers to the Coulomb-corrected one. We continued this iterative procedure until the parameters of the previous fit $(\lambda_n, R_n, \alpha_n)$ and the new ones from the latest fit $(\lambda_{n+1}, R_{n+1}, \alpha_{n+1})$ differed less then 2%. Let us note that usually $N \approx 1$ and $\epsilon \approx 0$, and these parameters converge faster than λ, R and α, so only the latter parameters are used in the test of the convergence criteria.

To determine the systematic uncertainties of the parameters, alternative measurement settings were applied. These considered settings can be seen in Table 1. In case of the PID cut, the default setting was $2.5\sigma_p$, while the lower one was $2.0\sigma_p$ and the upper one was $3.0\sigma_p$. As for the PC3 matching cut, there was no cut in the default setting, but an alternative cut of $2.0\sigma_m$ was applied, where σ_m is the standard deviation of the difference of the projected track position and the closest hit position in the detector, in both the ϕ and z directions. Regarding the EMCal/ToF track matching cut, the default setting was $1.5\sigma_m$, while the lower one was $1.0\sigma_m$ and the upper one was $3.5\sigma_m$. Pair cuts were applied in the $\Delta\phi - \Delta z$ plane by cutting off a two-dimensional region, as described in Ref. [10]. The fit range ($Q_{min} - Q_{max}$) was also varied. We modified the default setting with ± 8 MeV/c. As we mentioned earlier in this paper, the Coulomb correction is a function of q_{inv}, while the correlation function has a different variable, denoted by Q. We calculated the Coulomb correction with the variable Q and treated this approximation as a systematic uncertainty. The parameters were recalculated by individually changing each of the measurement settings; then, we calculated the relative difference of the values of the parameters obtained from the alternative settings and from the default settings. After calculating the relative differences for all the settings, we obtained the final systematics by taking into account the statistical uncertainties of the data points. A similar argument can be found in Ref. [22].

Table 1. The varied settings in order to determine the systematic uncertainties of the results.

Setting Name	Settings
PID cut	3 cut settings
PC3 matching cut	1 cut setting
EMCal/ToF matching cut	3 cut settings
DC pair cut	3 cut settings
ToF East pair cut	3 cut settings
ToF Wast pair cut	3 cut settings
EMCal pair cut	3 cut settings
Fit range (Q_{max})	3 ranges
Fit range (Q_{min})	3 ranges
Coulomb correction variable	2 versions

The method we used is described below. The variance of the difference of two variables is

$$\sigma^2(A_{def} - A_{alt}) = \sigma^2(A_{def}) + \sigma^2(A_{alt}) - 2\text{cov}(A_{def}, A_{alt}), \tag{16}$$

where A_{def} represents the parameter value obtained from the default cut, while A_{alt}, the parameter value, is obtained from the alternative cut; $\text{cov}(A_{def}, A_{alt}) = \rho\sigma(A_{def})\sigma(A_{alt})$ is the covariance matrix; and ρ is the correlation coefficient. The total uncertainties are composed of the systematic and the statistical uncertainties:

$$\sigma_{tot}^2 = \sigma_{stat}^2 + \sigma_{syst}^2 \qquad \text{so} \qquad \sigma_{syst}^2 = \sigma_{tot}^2 - \sigma_{stat}^2. \tag{17}$$

We require that the total uncertainty cover 1 standard deviation (1σ), i.e., $\sigma_{\text{tot}} = |A_{\text{def}} - A_{\text{alt}}|$. Thus,

$$\sigma_{\text{syst}}^2 = (A_{\text{def}} - A_{\text{alt}})^2 - \sigma_{\text{stat}}^2(A_{\text{def}}) - \sigma_{\text{stat}}^2(A_{\text{alt}}) + 2\rho\sigma_{\text{stat}}(A_{\text{def}})\sigma_{\text{stat}}(A_{\text{alt}}). \qquad (18)$$

The advantage of using Equation (18) is that it allows us to consider the impact of the statistical uncertainties. In this analysis, we assumed that A_{def} is completely correlated with A_{alt}, thus $\rho = 1$. The final systematic uncertainties were obtained by taking the squared sum of the calculated σ_{syst} values for each alternative setting.

5. Results

In this section, we present our main results: a comparison of the transverse mass dependence of the Lévy parameters in the case of kaon–kaon and pion–pion correlations.

One of the main reasons why analyzing the kaon–kaon Lévy distribution is interesting is because it could shed light on the physical interpretation of the Lévy exponent. The Lévy exponent α is shown in Figure 4. Within statistical uncertainties, we can draw the conclusion that the value of the parameter is between 1 and 2; however, the systematic uncertainties are quite large. As it is described in Ref. [19], a higher α value is expected for pions than for kaons based on the anomalous diffusion; however, we cannot observe this trend here, which indicates that beside anomalous diffusion of hadrons, there may be other physical processes causing the appearance of Lévy distributions, such as the resonances, as was concluded in Ref. [23]. The violation of the m_{T}-scaling of the two-pion and two-kaon correlations suggested by hydrodynamic models was explained by a rescattering phase in Ref. [24], which was not taken into account in the pure hydrodynamic models. A slight increase in the α values of the kaon measurements can be observed, although the large uncertainties do not allow us to draw any strong conclusions.

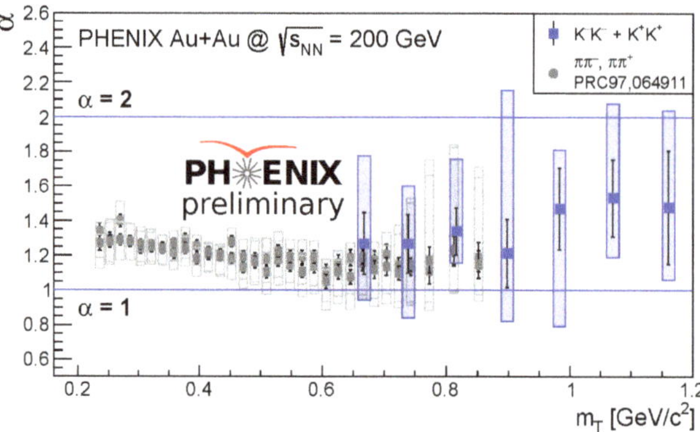

Figure 4. Values of the α parameter in the case of pions and kaons. Boxes indicate the systematic uncertainties, while error bars are used to represent the statistical ones.

The transverse mass dependence of the intercept parameter λ is shown in Figure 5. Compared to the pion data, it is not inconsistent with the given large uncertainties, having approximately matching values at around $m_{\text{T}} = 0.7\,\text{GeV}/c^2$, although their trends appear to be different. No significant m_{T} dependence was observed for this parameter, and it is fairly constant; however, we have to note that a slight decreasing trend is visible. This parameter characterizes the strength of the correlation as it was introduced as the extrapolated $C_2^{(0)}(Q = 0)$ value. Since there is no correlation between particles of different species, a possible worsening of PID efficiency may cause a decrease in the value of this parameter as our dataset may contain particles other than kaons.

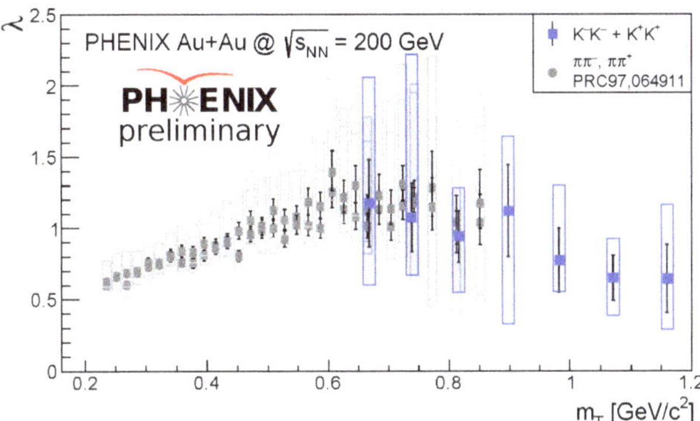

Figure 5. Values of the λ parameter in the case of pions and kaons. Boxes indicate the systematic uncertainties, while error bars are used to represent the statistical ones.

The transverse mass dependence of the Lévy scale parameter R is shown in Figure 6. The m_T scaling of HBT radii across particle species has been predicted in Ref. [25]. From theoretical works, e.g., Refs. [8,26], we know that R is not an RMS so it cannot be related to the source size directly. However, it was clear from previously published analyses (see Refs. [10,27–32]) that the Lévy-scale R exhibits a similar trend as its Gaussian counterpart, namely, it decreases with m_T. In the case of a Gaussian source, hydrodynamic models predict a linear scaling for its inverse square [33–35]:

$$\frac{1}{R^2} = A \cdot m_T + B. \tag{19}$$

As we see it on Figure 7, the linear scaling holds for the Lévy source as well, requiring the need for further theoretical investigations.

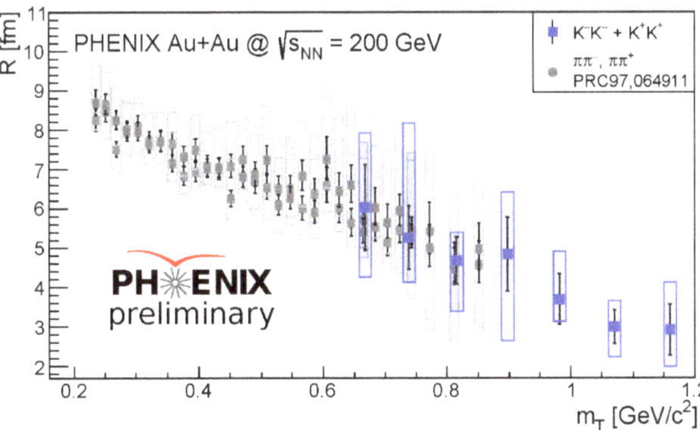

Figure 6. Values of the R parameter in the case of pions and kaons. Boxes indicate the systematic uncertainties, while error bars are used to represent the statistical ones.

It is worthwhile to note that there is a significant amount of point-by-point fluctuation in the systematic uncertainties. Furthermore, the non-fluctuating part of the systematic uncertainty of the pion and kaon data points is also partly correlated.

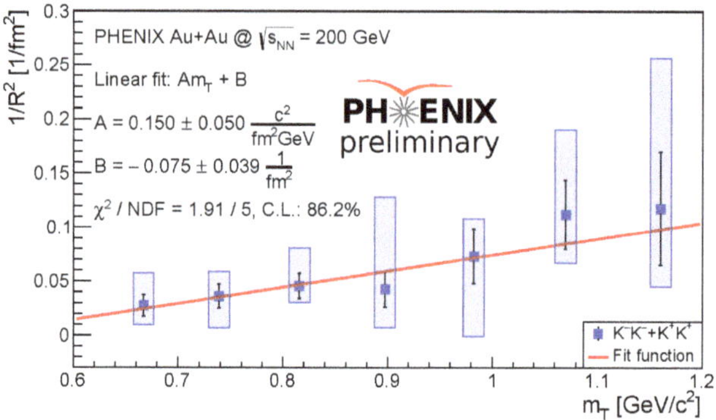

Figure 7. The transverse mass dependence of the $1/R^2$ points. It is worthwhile to note that due to the large uncertainties, one could fit these data points with different powers of m_T as well. A line is fitted to the data points, and the fitted parameters are shown in the legend. Boxes indicate the systematic uncertainties, while error bars are used to represent the statistical ones.

In Ref. [10], a new empirical scaling variable was found:

$$\widehat{R} = \frac{R}{\lambda(1+\alpha)}. \tag{20}$$

The motivation behind this parameter was the fact that the α, R, and λ parameters are strongly correlated, and it is possible to obtain good fits with multiple sets of co-varied parameters. The discovery of $\widehat{R}$ was made without any theoretical motivation, and in Ref. [10] it was observed that $\frac{1}{\widehat{R}}$ scales linearly with m_T. In Figure 8, we can see the same linear behavior for kaons as well.

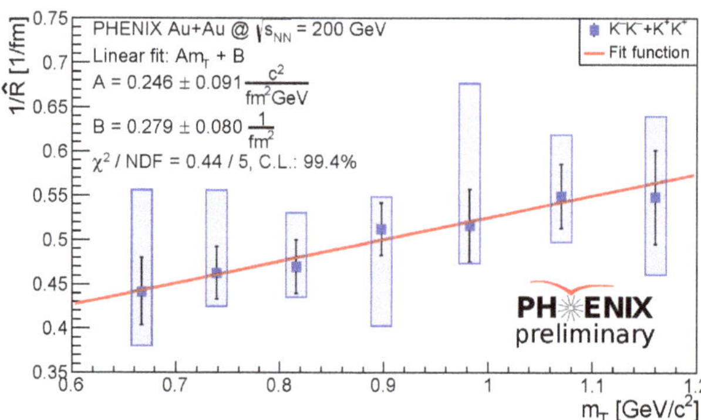

Figure 8. The transverse mass dependence of the $1/\widehat{R}$ points. It is worthwhile to note that due to the large uncertainties, one could fit these data points with different powers of m_T as well. A line is fitted to the data points, and the fitted parameters are shown in the legend. Boxes indicate the systematic uncertainties, while error bars are used to represent the statistical ones.

6. Conclusions

In this paper, we discussed two-kaon Bose–Einstein correlation functions in Au+Au collisions at $\sqrt{s_{\mathrm{NN}}} = 200$ GeV, from the PHENIX experiment. We assumed that the source has a Lévy shape. The Lévy parameters were investigated as functions of m_T and compared to the pion results. In the case of the Lévy stability index α, the large uncertainties prevent us from drawing any strong conclusions. The prediction that was based on anomalous diffusion that $\alpha^\pi_{\text{Lévy}} > \alpha^K_{\text{Lévy}}$ does not seem to be strongly supported. To clarify this question, further measurements and investigations might be necessary. Considering the intercept parameter λ, kaons and pions have matching values at around $m_T = 0.7$ GeV$/c^2$, and a slight decreasing trend is visible for kaons, possibly due to the worsening of the PID efficiency. The Lévy-scale R exhibits a similar trend as its Gaussian counterpart; it decreases with m_T, and its inverse square is linear in m_T, although this was predicted only for the Gaussian width. A new empirical scaling variable, $\widehat{R}$, was found, and it was observed that $\frac{1}{\widehat{R}}$ scales linearly with m_T.

Funding: This research was supported by the NKFIH OTKA K-138136 grant.

Data Availability Statement: The data presented in this study are available on request from the corresponding author. The data are not publicly available.

Acknowledgments: The author would like to thank the PHENIX Collaboration.

Conflicts of Interest: The author declares no conflict of interest.

Abbreviations

The following abbreviations are used in this manuscript:

HBT	R. Hanbury Brown and R. Q. Twiss
GGLP	G. Goldhaber, S. Goldhaber, W-Y. Lee, and A. Pais
RHIC	Relativistic heavy ion collider
LCMS	Longitudinal co-moving system
PCMS	Pair co-moving system
Au+Au	Gold–gold
RMS	Root mean square

Notes

[1]   This variable can be expressed in the PCMS system, which is the pair rest frame: $q_{\text{inv}} = |\mathbf{q}_{\text{PCMS}}|$.

References

1. Hanbury Brown, R.; Twiss, R.Q. A Test of a new type of stellar interferometer on Sirius. *Nature* **1956**, *178*, 1046–1048. [CrossRef]
2. Glauber, R.J. Photon Correlations. *Phys. Rev. Lett.* **1963**, *10*, 84–86. [CrossRef]
3. Glauber, R.J. Nobel Lecture: One hundred years of light quanta. *Rev. Mod. Phys.* **2006**, *78*, 1267–1278. [CrossRef]
4. Glauber, R.J. Quantum Optics and Heavy Ion Physics. *Nucl. Phys. A* **2006**, *774*, 3–13. [CrossRef]
5. Goldhaber, G.; Goldhaber, S.; Lee, W.; Pais, A. Influence of Bose-Einstein Statistics on the Antiproton-Proton Annihilation Process. *Phys. Rev.* **1960**, *120*, 300–312. [CrossRef]
6. Kopylov, G.I. Like particle correlations as a tool to study the multiple production mechanism. *Phys. Lett. B* **1974**, *50*, 472–474. [CrossRef]
7. Podgoretsky, M.I. Interference correlations of identical pions. *Sov. J. Part. Nucl.* **1989**, *20*, 266–282.
8. Csörgő, T.; Hegyi, S.; Zajc, W.A. Bose-Einstein correlations for Levy stable source distributions. *Eur. Phys. J.* **2004**, *C36*, 67–78. [CrossRef]
9. Metzler, R.; Barkai, E.; Klafter, J. Anomalous Diffusion and Relaxation Close to Thermal Equilibrium: A Fractional Fokker-Planck Equation Approach. *Phys. Rev. Lett.* **1999**, *82*, 3563–3567. [CrossRef]
10. Adare, A. et al. [PHENIX Collaboration]. Lévy-stable two-pion Bose-Einstein correlations in $\sqrt{s_{NN}} = 200$ GeV Au+Au collisions. *Phys. Rev. C* **2018**, *97*, 064911. [CrossRef]
11. Yano, F.B.; Koonin, S.E. Determining Pion Source Parameters in Relativistic Heavy Ion Collisions. *Phys. Lett.* **1978**, *78B*, 556–559. [CrossRef]
12. Csörgő, T. Particle interferometry from 40-MeV to 40-TeV. *Acta Phys. Hung. A* **2002**, *15*, 1–80. [CrossRef]
13. Wiedemann, U.A.; Heinz, U.W. Particle interferometry for relativistic heavy ion collisions. *Phys. Rep.* **1999**, *319*, 145–230. [CrossRef]
14. Adler, S.S. et al. [PHENIX Collaboration]. Bose-Einstein Correlations of Charged Pion Pairs in Au+Au Collisions at $\sqrt{s_{NN}} = 200$ GeV. *Phys. Rev. Lett.* **2004**, *93*, 152302. [CrossRef] [PubMed]
15. Lednicky, R.; Podgoretsky, M.I. The Interference of Identical Particles Emitted by Sources of Different Sizes. *Sov. J. Nucl. Phys.* **1979**, *30*, 432.
16. Csörgő, T.; Lörstad, B.; Zimányi, J. Bose-Einstein correlations for systems with large halo. *Z. Phys.* **1996**, *C71*, 491–497. [CrossRef]
17. Sinyukov, Y.; Lednicky, R.; Akkelin, S.V.; Pluta, J.; Erazmus, B. Coulomb corrections for interferometry analysis of expanding hadron systems. *Phys. Lett.* **1998**, *B432*, 248–257. [CrossRef]
18. Bowler, M.G. Coulomb corrections to Bose-Einstein corrections have greatly exaggerated. *Phys. Lett. B* **1991**, *270*, 69–74. [CrossRef]

19. Csanád, M.; Csörgő, T.; Nagy, M. Anomalous diffusion of pions at RHIC. *Braz. J. Phys.* **2007**, *37*, 1002–1013. [CrossRef]
20. Csanád, M.; Lökös, S.; Nagy, M. Expanded empirical formula for Coulomb final state interaction in the presence of Lévy sources. *Phys. Part. Nucl.* **2020**, *51*, 238–242. [CrossRef]
21. Csanád, M.; Lökös, S.; Nagy, M. Coulomb final state interaction in heavy ion collisions for Lévy sources. *Universe* **2019**, *5*, 133. [CrossRef]
22. Barlow, R. Systematic errors: Facts and fictions. In Proceedings of the Conference on Advanced Statistical Techniques in Particle Physics, Durham, UK, 18–22 March 2002; Volume 7, pp. 134–144.
23. Lednicky, R.; Progulova, T. Influence of resonances on Bose-Einstein correlations of identical pions. *Z. Phys. C Part. Fields* **1992**, *55*, 295–305. [CrossRef]
24. Acharya, S. et al. [ALICE Collaboration]. Kaon femtoscopy in Pb-Pb collisions at $\sqrt{s_{NN}}$ = 2.76 TeV. *Phys. Rev. C* **2017**, *96*, 064613. [CrossRef]
25. Csanád, M.; Csörgő, T. Kaon HBT radii from perfect fluid dynamics using the Buda-Lund model. *Acta Phys. Polon. Suppl.* **2008**, *1*, 521–524.
26. Csörgő, T.; Hegyi, S. Model independent shape analysis of correlations in one-dimension, two-dimensions or three-dimensions. *Phys. Lett.* **2000**, *B489*, 15–23. [CrossRef]
27. Lökös, S. Lévy-stable two-pion Bose-Einstein correlation functions measured with PHENIX in $\sqrt{s_{NN}}$ = 200 GeV Au+Au collisions. *Acta Phys. Pol. B Proc. Suppl.* **2019**, *12*, 193. [CrossRef]
28. Lökös, S. Centrality dependent Lévy-stable two-pion Bose-Einstein correlations in $\sqrt{s_{NN}}$ = 200 GeV Au+Au collisions at the PHENIX experiment. *Universe* **2018**, *4*, 31. [CrossRef]
29. Kurgyis, B.; Kincses, D.; Nagy, M.; Csanád, M. Coulomb Corrections for Bose–Einstein Correlations from One- and Three-Dimensional Lévy-Type Source Functions. *Universe* **2023**, *9*, 328. [CrossRef]
30. Kincses, D.; Nagy, M.I.; Csanád, M. Coulomb and strong interactions in the final state of Hanbury-Brown–Twiss correlations for Lévy-type source functions. *Phys. Rev. C* **2020**, *102*, 064912. [CrossRef]
31. Kincses, D. Lévy analysis of HBT correlation functions in $\sqrt{s_{NN}}$ = 62 GeV and 39 GeV Au+Au collisions at PHENIX. *Universe* **2018**, *4*, 11. [CrossRef]
32. Kurgyis, B. Three dimensional Lévy HBT results from PHENIX. *Acta Phys. Pol. B Proc. Suppl.* **2019**, *12*, 477–482. [CrossRef]
33. Chapman, S.; Scotto, P.; Heinz, U. New Cross Term in the Two-Particle Hanbury-Brown–Twiss Correlation Function in Ultrarelativistic Heavy-Ion Collisions. *Phys. Rev. Lett.* **1995**, *74*, 4400–4403. [CrossRef] [PubMed]
34. Makhlin, A.N.; Sinyukov, Y.M. Hydrodynamics of Hadron Matter Under Pion Interferometric Microscope. *Z. Phys. C* **1988**, *39*, 69. [CrossRef]
35. Csörgő, T.; Lörstad, B. Bose-Einstein correlations for three-dimensionally expanding, cylindrically symmetric, finite systems. *Phys. Rev. C* **1996**, *54*, 1390–1403. [CrossRef] [PubMed]

Communication

Kaon Femtoscopy with Lévy-Stable Sources from $\sqrt{s_{\mathrm{NN}}} = 200$ GeV Au + Au Collisions at RHIC

Ayon Mukherjee

Department of Atomic Physics, Eötvös Loránd University (ELTE), Pázmány Péter Stny. 1/A,
H-1117 Budapest, Hungary; ayon.mukherjee@ttk.elte.hu

Abstract: Femtoscopy has the capacity to probe the space-time geometry of the particle-emitting source in heavy-ion collisions. In particular, femtoscopy of like-sign kaon pairs may shed light on the origin of non-Gaussianity of the spatial emission probability density. The momentum correlations between like-sign kaon pairs are measured in data recorded by the STAR experiment, from $\sqrt{s_{\mathrm{NN}}} = 200$ GeV Au + Au collisions at RHIC, BNL. Preliminary results hint at the possible existence of non-Gaussian, Lévy-stable sources, with the likely presence of an anomalous diffusion process in the signal for the identically charged kaon pairs so produced. More statistically significant studies at lower centre-of-mass energies may contribute to the search for the critical end point of QCD.

Keywords: RHIC; STAR; femtoscopy; Bose–Einstein correlations; Lévy distribution; anomalous diffusion; identically charged kaons; heavy-ion collisions; high-energy physics

Citation: Mukherjee, A. Kaon Femtoscopy with Lévy-Stable Sources from $\sqrt{s_{\mathrm{NN}}} = 200$ GeV Au + Au Collisions at RHIC. *Universe* **2023**, *9*, 300. https://doi.org/10.3390/universe9070300

Academic Editor: Jun Xu

Received: 27 May 2023
Revised: 17 June 2023
Accepted: 20 June 2023
Published: 22 June 2023

1. Introduction

Following the discovery of quark–gluon plasma, one of the main thrusts of high-energy nuclear physics has been the understanding and exploration of the space-time geometry of the particle-emitting source created in heavy-ion collisions [1]. The quantity mainly investigated to this end is the two-particle source function, sometimes called the spatial correlation function (CF) or the pair-source distribution. Although this quantity is not easy to reconstruct experimentally, detailed studies of its behaviour are merited by a multitude of reasons, including its connections to hydrodynamic expansion [2,3], critical phenomena [4], light nuclei formation [5], etc. Phenomenological studies and experimental analyses both, emphasise the importance of describing the shape of the source function. Previously conducted hydrodynamical studies seem to suggest a Gaussian source shape [3,6]. Multiple measurements were also taken with the Gaussian assumption [7,8]. However, recent source-imaging studies suggest that the two-particle source function for pions has a long-range component, obeying power-law behaviour [9–14].

Femtoscopy [15] is the sub-field of high-energy heavy-ions physics that allows for the investigation of the space-time structure of femtometre-scale processes encountered in high-energy nuclear and particle physics experiments. Femtoscopic correlations in heavy-ion collisions are currently understood to be caused partly by Bose–Einstein statistics [16–19]. Alternatively, they are called Hanbury-Brown–Twiss (HBT) correlations in recognition of pioneering works by Hanbury-Brown and Twiss [21? ,22] on intensity interferometry in the field of observational astronomy to extract the apparent angular sizes of stars from correlations between the signals of two detectors. Additionally, correlations can arise out of final-state interactions, such as electromagnetic interactions and strong interactions undergone by the investigated particles. These correlations between pairs of identical bosons can be used to explore the properties of the matter created in heavy-ion collisions and to map the geometry of the particle-emitting source [1].

2. Correlations

Femtoscopy works on the principle that the momentum correlation function of a pair of particles is related to the probability density of particle creation at a space-time point (X) for a particle with four-momentum P. This probability density $(S(X, P))$ is also called the source function. Defining $N_1(P)$—obtained by multiplying the particle-creation probability density by $\langle n \rangle$, the average number of particles—as the momentum-invariant distribution and $N_2(P_1, P_2)$—obtained by multiplying the pair-creation probability density by $\langle n(n-1) \rangle$, the average number of pairs—as the pair-momentum distribution, the two-particle correlation function can be written as [23]:

$$C(P_1, P_2) = \frac{N_2(P_1, P_2)}{N_1(P_1)N_1(P_2)} \; ; \tag{1}$$

where

$$N_2(P_1, P_2) = \int S(X_1, P_1)S(X_2, P_2)\left|\psi_{P_1,P_2}(X_1, X_2)\right|^2 d^4X_1 d^4X_2 \, , \tag{2}$$

where $\psi_{P_1,P_2}(X_1, X_2)$ is the two-particle wave function that simplifies to:

$$\left|\psi_{P_1,P_2}(X_1, X_2)\right|^2 = \left|\psi^{(0)}_{P_1,P_2}(X_1, X_2)\right|^2 = 1 + \cos[(P_1 - P_2)(X_1 - X_2)] \tag{3}$$

in the interaction-free case for bosons, when only the quantum-statistical effects are taken into account. Thus, the correlation function can be redefined as [24]:

$$C(Q, K) \simeq 1 + \frac{\left|\tilde{S}(Q, K)\right|^2}{\left|\tilde{S}(0, K)\right|^2} \; ; \tag{4}$$

where

$$Q = P_1 - P_2 \, , \; K = \frac{P_1 + P_2}{2} \tag{5}$$

and

$$\tilde{S}(Q, K) = \int S(X, K)e^{iQX} d^4X \tag{6}$$

denote the pair-momentum difference, the average momentum and the Fourier transform (FT) of the source, respectively, assuming that $Q \ll K$ holds for the kinematic range under investigation. The correlation functions are measured with respect to Q over a range of well-defined K values; then, the properties of the correlation functions are analysed as functions of the average K for each of those ranges.

A significant fraction of the particles created in a heavy-ion collision is secondary, i.e., they come from decay. Whereas the primordial particles are created directly from the hydrodynamic expansion of the collision volume, the decay particles arise from interactions that take place much later. Hence, the source can be assumed to consist of the following two components [25]:

1. a core $(S_\mathrm{C}(X, K))$ consisting of primordial particles created by the hydrodynamically expanding, strongly interacting quark–gluon plasma, along with the decays of resonances with half lives of less than a few fm/c; and
2. a halo $(S_\mathrm{H}(X, K))$ consisting of the products created by the decay of long-lived resonances, including but not limited to η, η', K_S^0 and ω, making it possible to decompose $S(X, K)$ as [25]:

$$S(X, K) = S_\mathrm{C}(X, K) + S_\mathrm{H}(X, K). \tag{7}$$

As explained in detail in Ref. [25], the Fourier-transformed emission function of the halo vanishes for resolvable relative momenta, i.e., the Q values that lie in the experimentally achievable region. Hence, the halo does not contribute to the measured correlation function, which, in turn, is determined by the core component. Hence, the measured correlation function, when extrapolated to $Q = 0$, does not take a value of 2 as expected from Equation (4) but takes the following form:

$$\lim_{Q \to 0} C(Q, K) = 1 + \lambda(K) < 2. \tag{8}$$

This "intercept parameter" of the correlation function, also called the correlation strength ($\lambda(K)$) may depend on the pair-momentum K. It can be understood on the basis of the core-halo model by rewriting the correlation function using Equations (4), (6) and (7); and the fact that $\tilde{S}_H(Q, K) \approx 0$ for experimentally accessible values of Q [9]:

$$C(Q, K) = 1 + \left[\frac{N_C(K)}{N_C(K) + N_H(K)} \right]^2 \frac{\left|\tilde{S}_C(Q, K)\right|^2}{\left|\tilde{S}_C(0, K)\right|^2}$$

$$= 1 + \lambda(K) \frac{\left|\tilde{S}_C(Q, K)\right|^2}{\left|\tilde{S}_C(0, K)\right|^2}, \tag{9}$$

where $N_C(K) = \int S_C(X, K) d^4 X$ and $N_H(K) = \int S_H(X, K) d^4 X$ are contributions of the core and the halo, respectively, and $\lambda(K)$ is:

$$\lambda(K) = \left[\frac{N_C(K)}{N_C(K) + N_H(K)} \right]^2. \tag{10}$$

Realising that $X \equiv X(\vec{r}, t)$, the spatial correlation function:

$$D(\vec{r}, K) = \int S(\vec{r'} + \frac{\vec{r}}{2}, K) S(\vec{r'} - \frac{\vec{r}}{2}, K) d^3 \vec{r'} \tag{11}$$

can be used to rewrite $C(Q, K)$ as [24]:

$$C(Q, K) \approx \frac{\int D(\vec{r}, K) |\psi_Q(\vec{r})|^2 d^3 \vec{r}}{\int D(\vec{r}, K) d^3 \vec{r}}. \tag{12}$$

3. Lévy Distribution

Usually, the shape of the source distribution is assumed to be Gaussian. However, evidence of a non-Gaussian source for correlated pions has been found in multiple studies, necessitating a generalisation of the distribution to a Lévy-stable one [26]:

$$S(x, K) = \mathcal{L}(x; \alpha, \lambda, R) = \frac{1}{2\pi} \int e^{-(Q' R)^\alpha} e^{i Q' x} dQ'; \tag{13}$$

where $\mathcal{L}$ is the one-dimensional Lévy function, R is the Lévy-scale parameter, λ is the correlation strength, $0 < \alpha \leq 2$ is the Lévy exponent and Q' is the integration variable. These parameters are generally understood to depend on K. Salient features of the distribution include moments greater than α being undefined and $D(\vec{r}, K)$ necessarily having a Lévy shape with the same α in cases in which $S(X, K)$ is Lévy-shaped. The distribution exhibits a power-law behaviour for $\alpha < 2$, where $\alpha = 1$ represents a Cauchy distribution and $\alpha = 2$ represents a Gaussian distribution. Multiple factors, such as anomalous diffusion, jet fragmentation and proximity to the critical end point (CEP), can contribute to the appearance of Lévy-stable sources. However, the high-multiplicity, nucleon-on-nucleon nature of the analysed heavy-ion collisions makes it unlikely for jet fragmentation to be the dominant reason for the appearance of Lévy sources in this study—as it has been identified as the cause of Lévy-stable sources in $e^+ e^-$ collisions at LEP [13]. On the other hand, the high centre-of-mass-energy of the collisions explored here rules out the possibility of the system being close to the critical end point [4,27,28].

Interestingly, it is trivial to establish that α is related to one of the critical exponents (η). In the case of a second-order phase transition, the η exponent describes the power-law behaviour of the spatial correlation function at the critical end point with an exponent of $-(d - 2 + \eta)$, where d is the dimension. In a three-dimensional analysis in which $d = 3$, this exponent would compute to $-(1 + \eta)$. However, it is established that the three-dimensional

Lévy distribution describes the power-law tail of the spatial correlation function with an exponent of $-(1 + \alpha)$. Thus, comparing the exponents at the critical end point, it can be easily seen that the Lévy exponent (α) is identical to the critical exponent (η), a conjecture explored in Ref. [29]. The second-order QCD phase transition is expected to be in the same universality class as the three-dimensional Ising model. In that case, the η exponent has a value of 0.03631 ± 0.00003 at the critical end point [30]. However, the universality class of the random-field, three-dimensional Ising model may also be of relevance here, where the value of η is 0.50 ± 0.05 [31]. Thus, extracting α at collision energies lower than those used in this analysis while taking into account finite size and time effects might yield insightful information about the nature of the quark–hadron phase transition and shed light on the location of the critical end point in the QCD phase diagram [4,32–35].

As coordinate-space distributions extracted from experimental data show a heavy tail, the limitations of the hydrodynamical approach—assuming idealised freeze-out with a sudden jump in the mean free path from zero to infinity—become clear. This requires a more realistic approach using hadronic rescattering whereby the system cools as it dilutes with an expanding hadron gas, its mean free path diverges to infinity in a finite time interval and rescattering occurs in the presence of a time-dependent mean free path. This signals the existence of anomalous diffusion—experimentally observed as power-law-shaped tails in coordinate-space distributions of the source—in the system, as opposed to normal diffusion, with the Gaussian source exhibiting a strongly decaying tail caused by the Brownian motion of the particles constituting the system.

The momentum-space diffusion equation of one-dimensional, normal diffusion is expressed as [28]:

$$\frac{\partial W(q,t)}{\partial t} = -k_N q^2 W(q,t); \tag{14}$$

where k_N is the normal-diffusion constant, q is the momentum, t is the time and $W(q,t)$ is the momentum-space probability distribution. The coordinate-space solution to Equation (14) is given by the following Gaussian expression:

$$W(x,t) = \frac{1}{\sqrt{4\pi k_N t}} \exp\left(-\frac{x^2}{4k_N t}\right). \tag{15}$$

For anomalous diffusion, the coordinate-space diffusion equation, in terms of the spatial probability distribution ($W(x,v,t)$), is the generalised Fokker–Planck equation [28]:

$$\frac{\partial W}{\partial t} + v\frac{\partial W}{\partial x} + \frac{F(x)}{m}\frac{\partial W}{\partial v} = \eta_{\alpha'0}D_t^{1-\alpha'}L_{FP}W. \tag{16}$$

where $\eta_{\alpha'}$ is the generalised friction constant of dimension $[\eta_{\alpha'}] = s^{\alpha'-2}$, $_0D_t^{1-\alpha'}$ is the Riemann–Liouville operator:

$$_0D_t^{1-\alpha'}t^p = \left(\frac{\partial}{\partial t}\right)_0D_t^{-\alpha'}t^p = \frac{\Gamma(1+p)}{\Gamma(p+\alpha')}t^{p+\alpha'-1} \tag{17}$$

and L_{FP} is the Fokker–Planck operator:

$$L_{FP} = \frac{\partial}{\partial x}\frac{V'(x)}{m\eta_{\alpha'}} + k_A\frac{\partial^2}{\partial x^2}, \tag{18}$$

where $V'(x)$ is related to the force ($F(x)$) by $F(x) = -\frac{dV(x)}{dx}$, as explained in Refs. [36,37].

The momentum-space solution to Equation (16) is given by:

$$W(q,t) = \exp\left(-tk_A^\alpha |q|^\alpha\right). \tag{19}$$

where $W(q,t)$ happens to be the FT or the characteristic function of Lévy-stable source distributions, where α is the Lévy exponent from Equation (13) and k_A is the anomalous diffusion constant. If a centred, spherically symmetric, Lévy-stable source-distribution is assumed and all final-state interactions are neglected, the one-dimensional, two-particle correlation function takes the following simplified form:

$$C(q) = 1 + \lambda \cdot e^{-(qR)^{\alpha}}, \tag{20}$$

where λ is the correlation strength from Equations (10) and (8), R is the Lévy scale, α is the Lévy exponent and q is the absolute value of the three-momentum difference in the longitudinally co moving system (LCMS) [9]:

$$q = q_{\text{LCMS}} = |\vec{p}_1 - \vec{p}_2|_{\text{LCMS}}. \tag{21}$$

R can be interpreted as the homogeneity length of the particle species, while α represents the extent of the anomalous diffusion occurring in the system. The spherical symmetry in q_{LCMS} is ideal for a one-dimensional analysis of a three-dimensional, spherically symmetric system. Subsequent measurements with higher precision might necessitate a move towards a full, three-dimensional analysis. Until then, the approximations used in Ref. [38] may be utilised for a preliminary analysis.

Momentum correlations of like-sign kaon pairs at $\sqrt{s_{\text{NN}}} = 200$ GeV can be utilised to calculate $C(q)$ and to, consequently, ascertain the shape of the pair-source distribution. If anomalous diffusion is the sole source of non-Gaussianity, then it is expected that the extent of the anomaly will depend on the total cross section or, equivalently, on the mean free path. Since the mean free path is larger for kaons than for pions, the diffusion of the former is expected to be more anomalous in the hadron gas. Hence, the α for kaons (α_K) is expected to be smaller [28]. Thus, measuring α_K may shed light on the role of anomalous diffusion in the hadron gas as the origin of the appearance of Lévy distributions.

4. Measurement

The data used for this analysis were obtained from RHIC's gold-on-gold collisions at a 200 GeV centre-of-mass energy per nucleon pair. The collisions were performed in 2016 and measured by the solenoidal tracker at RHIC (STAR) experiment. The STAR experiment detects multiple particle species emanating from the medium created by the collisions. These different particle species, depending on their mass and charge, produce different shapes when their ionisation energy loss (dE/dx) is plotted as a function of momentum. These shapes can help distinguish the particle species from each other and help isolate the kaons, as observed in Figure 1. For this investigation, the analysis processes about 410 million events in the 0–30% centrality range. They are subjected to strict track and pair selection criteria that, following Ref. [39], include the identification of kaons via energy loss measured by the time-projection chamber (TPC); pair selection based on the fraction-of-merged-hits (FMH) and the splitting-level (SL); and limitations on the track's momentum, rapidity and distance-of-closest-approach (DCA).

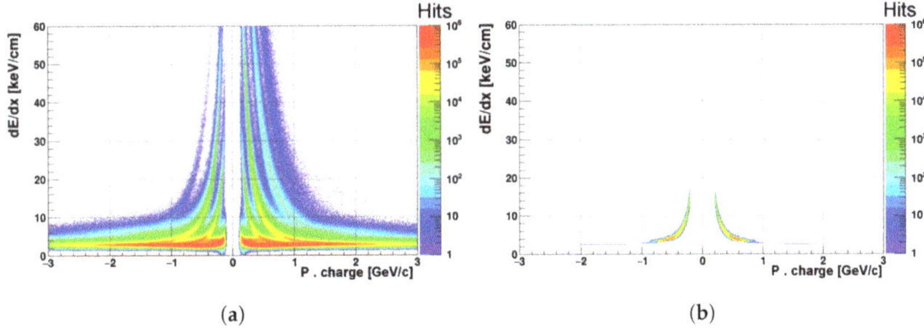

Figure 1. Sample ionisation energy loss as a function of momentum $\times$ charge (**a**) for all available charged particles and (**b**) after being cut only for the charged kaons to be isolated.

The one-dimensional, like-sign, two-kaon, femtoscopic correlation functions are then experimentally constructed using the event-mixing technique [40]. $A(q)$—the actual pair distribution—is formed from kaon pairs, with members of the pair belonging to the same event. This distribution is affected by various effects, such as kinematics and acceptance. $B(q)$—a background distribution—is constructed with the pairs to correct for these effects and the members of this distribution originate from separate events. In this analysis, the method for event mixing described in Refs. [9,13] is used to correct for residual correlations. For a set of event classes based on the centrality and the location of the collision vertex, a background event pool is established. Then, for each real event, a mixed event of the same event class is created from this pool, making sure that each particle in this mixed event belongs to a different real event. Subsequently, pairs within the mixed event contribute to the formation the aforementioned $B(q)$. Finally, the pre-normalised correlation function is calculated as:

$$C(q) = \frac{A(q)}{B(q)} \cdot \frac{\int B(q)\, dq}{\int A(q)\, dq}, \tag{22}$$

for three different ranges of transverse mass (m_T), defined as $m_T = \sqrt{m^2 + (K_T/c)^2}$, where m is the kaon mass and K_T is the average transverse momentum of the pair. The normalisation integrals are performed over a range in which the correlation function is not supposed to exhibit quantum-statistical features. It is to be noted that the method outlined here is applied to pairs belonging to a given range of average momenta. Furthermore, in the event-mixing technique described above, the number of actual and background pairs is the same, aside from the effect of the pair-selection criteria mentioned earlier.

With the momentum correlations, obtained from experimental data (both measured and the empirical values of the calculated correlation function), preparations are made to fit the Lévy function detailed in Equation (20) to $C(q)$. However, the assumptions behind Equation (20) make it unsuitable for a direct fit to experimentally obtained data. As mentioned in Section 1, final-state interactions have considerable effects on the momentum correlations between like-sign kaon pairs; therefore, they need to be imbibed into the analysis as corrections to Equation (20) in order to make it suitable and physically relevant as a fit for the correlation function obtained above. Multiple factors can contribute to the final-state modifications of the momentum correlations, the leading of which are Coulomb interactions, as a gas of charged hadrons can never be entirely devoid of Coulomb repulsion.

The final-state Coulomb interactions are incorporated into the CF by using the Bowler–Sinyukov formula, which includes a correction term for Coulomb repulsion, which is expressed as [41,42]:

$$C(q) = \left[1 - \lambda + \lambda \cdot \mathcal{K}(q; \alpha, R) \cdot \left(1 + e^{-(qR)^\alpha}\right)\right] \cdot N \cdot (1 + q\varepsilon), \tag{23}$$

where N is a normalisation factor; ε is responsible for a linear, residual, long-range background; and $\mathcal{K}$ is the Coulomb correction [24]:

$$\mathcal{K}(q; \alpha, R) = \frac{\int D(\vec{r}) \left|\psi_q^{\mathrm{Coul}}(\vec{r})\right|^2 d^3\vec{r}}{\int D(\vec{r}) \left|\psi_q(\vec{r})\right|^2 d^3\vec{r}}, \tag{24}$$

where $D(\vec{r})$ is the simplified spatial pair distribution, and $\psi_q^{\mathrm{Coul}}(\vec{r})$ is the solution to the two-particle Schrödinger equation in the presence of a Coulomb potential. In this study, $\mathcal{K}(q; \alpha, R)$ for kaons is calculated by numerically employing the procedure used in Refs. [12,24,43]. The inclusion of other final-state contributions, such as the strong interaction, can resolve the possible underestimation regarding R and λ and overestimation regarding α of the Lévy parameters [44]. However, the statistical significance of such precise corrections turns out to be negligible in the context of the current measurement.

5. Results

As illustrated in Figure 2, the Coulomb-corrected Lévy distribution function is in agreement with the measured $C(q)$ over the entire q_{LCMS} range. The femtoscopic peak [18,19] and the Coulomb hole [42] are both observed as expected. The values of the normalisation factor (N) and the linear background factor (ε) are observed to be close to 1 and 0, respectively.

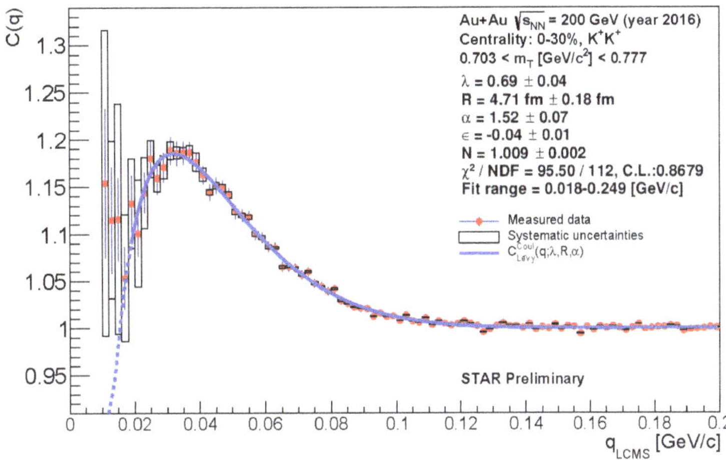

Figure 2. $C(q)$ as a function of q_{LCMS} for positively charged kaon pairs in the m_T range (703–777 MeV/c^2) and the centrality range (0–30%). The red dots denote the measured data, and the blue lines (solid and dotted) denote the fit. The systematic uncertainties are shown as hollow rectangles.

The systematic uncertainties are obtained by combining the uncertainties arising from variations in the event- and pair-selection criteria, denoted by Δ_{cuts}, as mentioned above, and those arising out of variations to the range of the fit are denoted by Δ_{fits}. At this preliminary stage, systematic uncertainties arising out of variations in the track-selection criteria are not included. Thus, the final systematic uncertainties (Δ_{total}) are obtained as:

$$\Delta_{\text{total}} = \sqrt{(\Delta_{\text{fits}})^2 + (\Delta_{\text{cuts}})^2} \,. \tag{25}$$

Figure 3 shows the kaon homogeneity length (R), otherwise known as the Lévy scale, as a function of m_T. It is observed to exhibit large, systematic uncertainties, a very weak dependence on m_T and a possible decrease with respect to it, as reported in previous studies [2,3,6,8,14,45]. However, hydrodynamical studies predicting a decrease in the Lévy scale as a function of m_T are based on the Gaussian source assumption [2,3]. Hence, more investigations on this topic are in order. The extracted values of the Lévy scale in this charged-kaon analysis are also found to be similar to PHENIX's like-sign pion results [9], with $R_\pi \sim 5$–7 fm for the m_T range of 600–700 MeV/c^2. A more detailed comparison of the m_T dependence of Lévy scales of different particle species could shed light on the origin of the appearance of Lévy-stable sources, given that, according to calculations based on hydrodynamics, species-independent m_T scaling was predicted in Ref. [46].

The intercept of the correlation function—the correlation strength, i.e., λ—is shown in Figure 4. Values extracted from the fits show that it is close to unity, as is to be expected based on the small fraction of decay kaons present in the system.

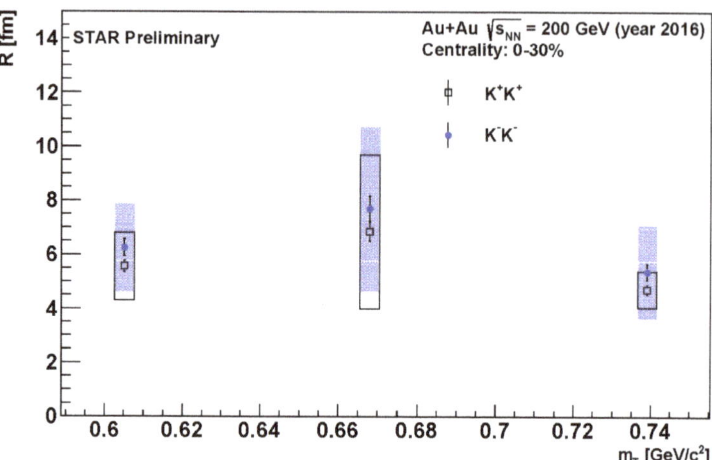

Figure 3. R as a function of m_T for 0–30% centrality. The hollow, blue squares denote positively charged kaon pairs and the solid, and blue circles denote negatively charged kaon pairs, along with their error bars. The systematic uncertainties are shown as hollow (K^+K^+) and shaded (K^-K^-) rectangles.

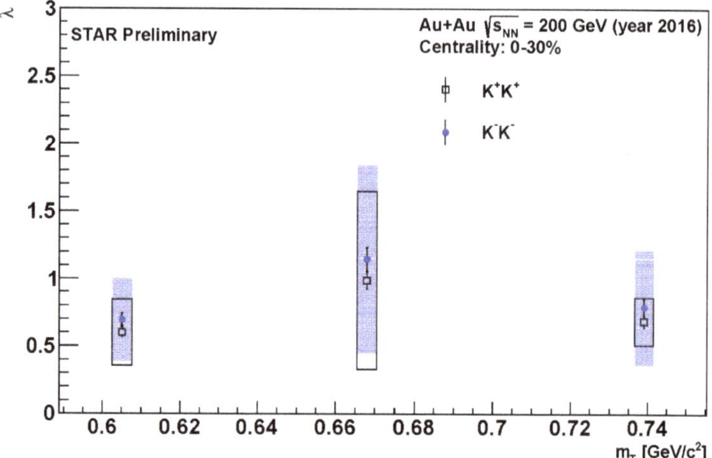

Figure 4. λ as a function of m_T for 0–30% centrality. The hollow, blue squares denote positively charged kaon pairs, and the solid, blue circles denote negatively charged kaon pairs, along with their error bars. The systematic uncertainties are shown as hollow (K^+K^+) and shaded (K^-K^-) rectangles.

The extent of the anomalous diffusion might be gleaned from the Lévy exponent (α) as shown in Figure 5, also illustrating the values corresponding to the Gaussian and Cauchy distributions, with dashed and dotted blue lines, respectively. The Lévy exponent is observed to have values between those two extremes, indicating power-law behaviour and the likely existence of anomalous diffusion. The extracted values of $\alpha \sim 1.0$–1.5 for kaons are similar to PHENIX's pion results, with $\alpha_\pi \sim 1.2$ in the same transverse mass range. α_K not being smaller than α_π hints at the existence of other factors on top of anomalous diffusion, contributing to the appearance of non-Gaussian source shapes. However, the current statistics prevent the drawing of more definitive conclusions.

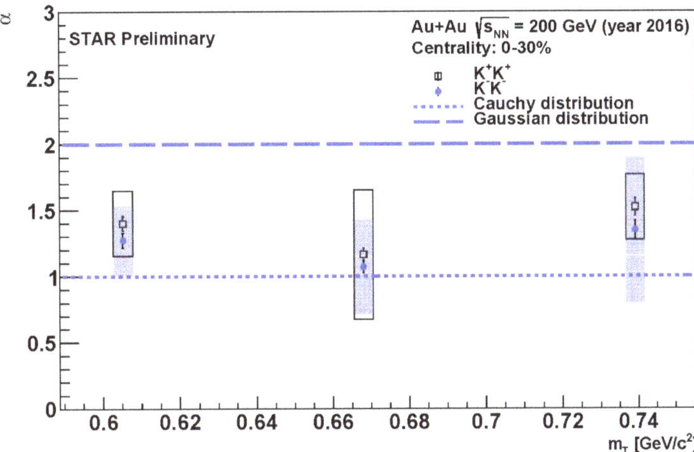

Figure 5. α as a function of m_T for 0–30% centrality. The hollow, blue squares denote positively charged kaon pairs, and the solid, blue circles denote negatively charged kaon pairs, along with their error bars. The systematic uncertainties are shown as hollow (K^+K^+) and shaded (K^-K^-) rectangles.

6. Summary

Preliminary analysis of data collected by STAR from RHIC's 2016 BES $\sqrt{s_{NN}} = 200$ GeV Au + Au collisions suggests a non-Gaussian, Lévy-stable source shape for pairs of the identically charged kaons produced in the collisions. The Lévy-stability exponent α_K is observed to be comparable to that of like-sign pion pairs obtained from PHENIX.

However, anomalous diffusion may not be solely responsible for the heavy tails observed in the source distributions, as suggested by the comparability of α_K to α_π. It is to be noted that, a complete systematic uncertainty analysis, which is currently ongoing, is required to draw definitive conclusions about any and all claims made herein. Because Lévy-stable sources can arise in strongly interacting systems due to their proximity to the QCD critical end point at higher chemical potentials, similar studies at lower beam energies would likely strengthen the search for the QCD critical end point.

Funding: This project was funded, in part, by the TKP2021-NKTA-64 and K-136138 grants from the NKFIH, Hungary, and, in part, by the Department of Energy of the United States of America.

Data Availability Statement: Data publicly unavailable since publication is a communication.

Acknowledgments: This manuscript was prepared for a Special Issue corresponding to the Zimányi School Winter Workshop 2022. The author would like to thank the STAR collaboration and the people behind the RHIC framework, whose incessant efforts have produced every bit of data analysed as a part of this project. The author is also deeply indebted to Koushik Mandal for his invaluable help with discussions about the ROOT framework and the C++ programming language in particular and about the myriad intricacies of navigating the field of experimental heavy-ion physics in general.

Conflicts of Interest: The author declares no conflict of interest.

References

1. Lisa, M.; Pratt, S.; Soltz, R.; Wiedemann, U. Femtoscopy in relativistic heavy ion collisions. *Ann. Rev. Nucl. Part. Sci.* **2005**, *55*, 357–402. [CrossRef]
2. Makhlin, A.; Sinyukov, Y. Hydrodynamics of hadron matter under pion interferometric microscope. *Z. Phys. C* **1988**, *39*, 69. [CrossRef]
3. Csörgő, T.; Lörstad, B. Bose-Einstein correlations for three-dimensionally expanding, cylindrically symmetric, finite systems. *Phys. Rev. C* **1996**, *54*, 1390–1403. [CrossRef] [PubMed]
4. Csörgő, T.; Hegyi, S.; Novák, T.; Zajc, W. Bose-Einstein or HBT correlation signature of a second order QCD phase transition. *AIP Conf. Proc.* **2006**, *828*, 525–532. [CrossRef]

5.	Oliinychenko, D. Overview of light nuclei production in relativistic heavy-ion collisions. *Nucl. Phys. A* **2021**, *1005*, 121754. [CrossRef]

6.	Csanád, M.; Vargyas, M. Observables from a solution of 1 + 3 dimensional relativistic hydrodynamics. *Eur. Phys. J. A* **2010**, *44*, 473–478. [CrossRef]

7.	Adams, J. et al. Pion interferometry in Au + Au collisions at $\sqrt{s_{NN}}$ = 200 GeV. *Phys. Rev. C* **2005**, *71*, 044906. [CrossRef]

8.	Adler, S. et al. Bose-Einstein correlations of charged pion pairs in Au + Au collisions at $\sqrt{s_{NN}}$ = 200 GeV. *Phys. Rev. Lett.* **2004**, *93*, 152302. [CrossRef]

9.	Adare, A. et al. Lévy-stable two-pion Bose-Einstein correlations in $\sqrt{s_{NN}}$ = 200 GeV Au + Au collisions. *Phys. Rev. C* **2018**, *97*, 064911. [CrossRef]

10.	Adler, S. et al. Evidence for a long-range component in the pion emission source in Au + Au collisions at $\sqrt{s_{NN}}$ = 200 GeV. *Phys. Rev. Lett.* **2007**, *98*, 132301. [CrossRef]

11.	Afanasiev, S. et al. Source breakup dynamics in Au + Au Collisions at $\sqrt{s_{NN}}$ = 200 GeV via three-dimensional two-pion source imaging. *Phys. Rev. Lett.* **2008**, *100*, 232301. [CrossRef] [PubMed]

12.	Sirunyan, A. et al. Bose-Einstein correlations in pp, pPb, and PbPb collisions at $\sqrt{s_{NN}}$ = 0.9–7 TeV. *Phys. Rev. C* **2018**, *97*, 064912. [CrossRef]

13.	Achard, P. et al. Test of the τ-model of Bose-Einstein correlations and reconstruction of the source function in hadronic Z-boson decay at LEP. *Eur. Phys. J. C* **2011**, *71*, 1648. [CrossRef]

14.	Kincses, D.; Stefaniak, M.; Csanád, M. Event-by-event investigation of the two-particle source function in heavy-ion collisions with EPOS. *Entropy* **2022**, *24*, 308. [CrossRef] [PubMed]

15.	Lednicky, R. Correlation femtoscopy: Origins and achievements. *Phys. Part. Nucl.* **2020**, *51*, 221–226. [CrossRef]

16.	Bose, S. Plancks gesetz und lichtquantenhypothese. *Z. Phys.* **1924**, *26*, 178–181. [CrossRef]

17.	Einstein, A. Quantentheorie des einatomigen idealen gases. *Sitzungsber. Preuss. Akad. Wiss. Phys. Math. Kl.* **1925**, *1*, 3–14.

18.	Goldhaber, G.; Fowler, W.; Goldhaber, S.; Hoang, T.; Kalogeropoulos, T.; Powell, W. Pion-pion correlations in antiproton annihilation events. *Phys. Rev. Lett.* **1959**, *3*, 181–183. [CrossRef]

19.	Goldhaber, G.; Goldhaber, S.; Lee, W.; Pais, A. Influence of Bose-Einstein statistics on the antiproton-proton annihilation process. *Phys. Rev.* **1960**, *120*, 300–312. [CrossRef]

20.	Hanbury-Brown, R.; Twiss, R. A new type of interferometer for use in radio astronomy. *Lond. Edinb. Dublin Philos. Mag. J. Sci.* **1954**, *45*, 663–682. [CrossRef]

21.	Hanbury-Brown, R.; Twiss, R. Correlation between photons in two coherent beams of light. *Nature* **1956**, *177*, 27–29. [CrossRef]

22.	Hanbury-Brown, R.; Twiss, R. A test of a new type of stellar interferometer on Sirius. *Nature* **1956**, *178*, 1046–1048. [CrossRef]

23.	Yano, F.; Koonin, S. Determining pion source parameters in relativistic heavy-ion collisions. *Phys. Lett. B* **1978**, *78*, 556–559. [CrossRef]

24.	Csanád, M.; Lökös, S.; Nagy, M. Expanded empirical formula for Coulomb final state interaction in the presence of Lévy sources. *Phys. Part. Nucl.* **2020**, *51*, 238–242. [CrossRef]

25.	Csörgő, T.; Lörstad, B.; Zimányi, J. Bose-Einstein correlations for systems with large halo. *Z. Phys. C* **1996**, *71*, 491–497. [CrossRef]

26.	Nolan, J. Parameterizations and modes of stable distributions. *Stat. Probab. Lett.* **1998**, *38*, 187–195. [CrossRef]

27.	Csörgő, T.; Hegyi, S.; Novák, T.; Zajc, W. Bose-Einstein or HBT correlations and the anomalous dimension of QCD. *Acta Phys. Polon. B* **2005**, *36*, 329–337.

28.	Csanád, M.; Csörgő, T.; Nagy, M. Anomalous diffusion of pions at RHIC. *Braz. J. Phys.* **2007**, *37*, 1002–1013. [CrossRef]

29.	Csörgő, T. Correlation probes of a QCD critical point. *PoS* **2008**, *HIGH-PTLHC08*, 027. [CrossRef]

30.	El-Showk, S.; Paulos, M.; Poland, D.; Rychkov, S.; Simmons-Duffin, D.; Vichi, A. Solving the 3d Ising model with the conformal bootstrap II. c-minimization and precise critical exponents. *J. Stat. Phys.* **2014**, *157*, 869–914. [CrossRef]

31.	Rieger, H. Critical behaviour of the three-dimensional random-field Ising model: Two-exponent scaling and discontinuous transition. *Phys. Rev. B Condens. Matter* **1995**, *52*, 6659–6667. [CrossRef] [PubMed]

32.	Halasz, M.; Jackson, A.; Shrock, R.; Stephanov, M.; Verbaarschot, J. Phase diagram of QCD. *Phys. Rev. D* **1998**, *58*, 096007. [CrossRef]

33.	Stephanov, M.; Rajagopal, K.; Shuryak, E. Signatures of the tricritical point in QCD. *Phys. Rev. Lett.* **1998**, *81*, 4816–4819. [CrossRef]

34.	Lacey, R. Indications for a critical point in the phase diagram for hot and dense nuclear matter. *Nucl. Phys. A* **2016**, *956*, 348–351. [CrossRef]

35.	Lacey, R.; Liu, P.; Magdy, N.; Schweid, B.; Ajitanand, N. Finite-size scaling of non-Gaussian fluctuations near the QCD critical point. *arXiv* **2016**, arXiv:1606.08071.

36.	Metzler, R.; Klafter, J. The random walk's guide to anomalous diffusion: A fractional dynamics approach. *Phys. Rep.* **2000**, *339*, 1–77. [CrossRef]

37.	Csörgő, T.; Hegyi, S.; Zajc, W. Bose-Einstein correlations for Lévy stable source distributions. *Eur. Phys. J. C* **2004**, *36*, 67–78. [CrossRef]

38.	Kurgyis, B.; Kincses, D.; Nagy, M.; Csanád, M. Coulomb interaction for Lévy sources. *arXiv* **2023**, arXiv:2007.10173.

39.	Adamczyk, L. et al. Beam-energy-dependent two-pion interferometry and the freeze-out eccentricity of pions measured in heavy ion collisions at the STAR detector. *Phys. Rev. C* **2015**, *92*, 014904. [CrossRef]

40.	Kincses, D. Shape analysis of HBT correlations at STAR. *Phys. Part. Nucl.* **2020**, *51*, 267–269. [CrossRef]

41. Bowler, M. Coulomb corrections to Bose-Einstein corrections have greatly exaggerated. *Phys. Lett. B* **1991**, *270*, 69–74. [CrossRef]
42. Sinyukov, Y.; Lednicky, R.; Akkelin, S.; Pluta, J.; Erazmus, B. Coulomb corrections for interferometry analysis of expanding hadron systems. *Phys. Lett. B* **1998**, *432*, 248–257. [CrossRef]
43. Csanád, M.; Lökös, S.; Nagy, M. Coulomb final state interaction in heavy ion collisions for Lévy sources. *Universe* **2019**, *5*, 133. [CrossRef]
44. Kincses, D.; Nagy, M.; Csanád, M. Coulomb and strong interactions in the final state of Hanbury-Brown–Twiss correlations for Lévy-type source functions. *Phys. Rev. C* **2020**, *102*, 064912. [CrossRef]
45. Afanasiev, S. et al. Kaon interferometric probes of space-time evolution in Au + Au collisions at $\sqrt{s_{NN}} = 200$ GeV. *Phys. Rev. Lett.* **2009**, *103*, 142301. [CrossRef]
46. Csanád, M.; Csörgő, T. Kaon HBT radii from perfect fluid dynamics using the Buda-Lund model. *Acta Phys. Polon. Supp.* **2008**, *1*, 521–524.

Communication

Charged Particle Pseudorapidity Distributions Measured with the STAR EPD

Mátyás Molnár for the STAR Collaboration

Department of Atomic Physics, Insitute of Physics and Astronomy, Eötvös Loránd University, 1117 Budapest, Hungary; molnarmatyas@student.elte.hu

Abstract: In 2018, in preparation for the Beam Energy Scan II, the STAR detector was upgraded with the Event Plane Detector (EPD). The instrument enhanced STAR's capabilities in centrality determination for fluctuation measurements, event plane resolution for flow measurements, and in triggering overall. Due to its fine radial granularity, it can also be utilized to measure pseudorapidity distributions of the produced charged primary particles, in EPD's pseudorapidity coverage of $2.15 < |\eta| < 5.09$. As such a measurement cannot be done directly, the response of the detector to the primary particles has to be understood well. The detector response matrix was determined via Monte Carlo simulations, and corrected charged particle pseudorapidity distributions were obtained in Au+Au collisions at the center of mass collision energies $\sqrt{s_{NN}} = 19.6$ and 27.0 GeV using an iterative unfolding procedure. Several systematic checks of the method were also done.

Keywords: high-energy heavy-ion collisions; STAR EPD; pseudorapidity distributions; Bayesian unfolding

Citation: Molnár, M., for the STAR Collaboration. Charged Particle Pseudorapidity Distributions Measured with the STAR EPD. *Universe* **2023**, *9*, 335. https://doi.org/10.3390/universe9070335

Academic Editor: Jun Xu

Received: 31 May 2023
Revised: 7 July 2023
Accepted: 10 July 2023
Published: 15 July 2023

1. Introduction

According to quantum chromodynamics, quarks cannot be observed in their free form—only in hadrons due to the color confinement. This effect also causes the strong interaction to have a finite range of around 10^{-15} m—even though the gluon mass is known to be zero. In the very early Universe with enormous pressure and temperature, it is assumed that these particles could exist in a form of quark–gluon plasma (QGP). To create such a state experimentally, particle accelerators that perform high-energy heavy-ion collisions are utilized. Since the lifetime of the QGP is short, the information about the partonic state has to be deduced from the final-state particles, e.g., hadronic jets.

One of the experimental facilities studying the formation and the evolution of the QGP is the Relativistic Heavy Ion Collider (RHIC) at the Brookhaven National Laboratory, and one of its experiments is the Solenoidal Tracker at RHIC (STAR) [1]. The complex STAR detector system consists of several instruments; one of them is the Event Plane Detector (EPD) [2].

In these proceedings, measurements of charged particle[1] pseudorapidity distributions in Au+Au collision data at $\sqrt{s_{NN}} = 19.6$ and 27.0 GeV utilizing the EPD are presented. Detailed systematic uncertainty checks are also discussed.

1.1. The EPD

The EPD was installed in 2018, as a part of the preparation for the BES-II program. Among motivations behind building the detector were: improving the event plane resolution for flow measurements, independent centrality determination for fluctuation measurements, and using it as a trigger in the high-luminosity environment during the BES-II program.

The EPD is a completely new subdetector that was supposed to improve the event plane resolution—for example, by about a factor of 2 in Au + Au collisions at

$\sqrt{s_{NN}} = 19.6$ GeV [3]. Its predecessor (in event plane determination), the Beam–Beam Counter (BBC) has much less fine granularity than the EPD: only 36 tiles, with the 18 inner smaller tiles used - compared to the 372 tiles of the EPD [2]. It also has smaller acceptance of $3.3 < |\eta| < 5.0$ in pseudorapidity [4].

The detector consists of two "wheels" on either (West and East) side of the STAR detector system, installed ± 375 cm from the nominal interaction point (the detector's center). Each wheel consists of 12 "supersectors" covering $\phi = 30°$ in azimuthal angle, each further segmented to 31 "tiles", thus giving 16 radial segments so-called "rings"[2] covering a relatively large forward pseudorapidity range of $2.15 < |\eta| < 5.09$ (or, range of $0.7° < \theta < 13.5°$ angle to particle beam axis). Each supersector is connected to a bundle of 31 optical cables that transport light to high-efficiency silicone photo-multipliers (SiPM). The signals are then sent to the digital data acquisition systems [2].

Each tile registers hits, mostly Minimum Ionizing Particles (MIPs). Assuming that the probability distribution of the measured signal of a single hit can be described by a Landau distribution, the presence of multiple hits will result in a convolution of multiple Landau distributions. The measured Analog Digital Count (ADC) distributions were fitted with a multi-MIP Landau function, shown in Figure 1. The different Landau distributions corresponding to the ADC contribution caused by n number of MIPs were convolved with different convolution weights (n-MIP weight).

The conclusion drawn was that convolving with less than 5 n-MIP weights are adequate to achieve a good fit, as the contribution of the 5-MIP weight was already zero within uncertainties—under the asssumption that the MIP weights were Poisson-distributed which was validated during data analysis. In view of this result, the systematic uncertainty contribution from this source—that is, fitting only up to 5 n-MIP weights—can be considered negligible.

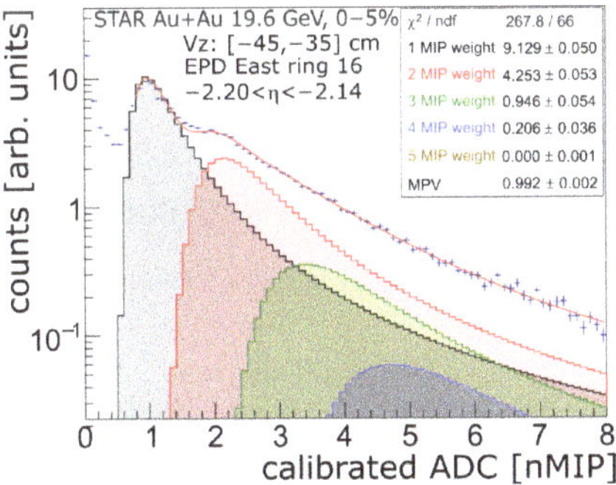

Figure 1. Example multi-MIP Landau fit of ADC count distribution in ring #16, with ADC counts in arbitrary units. Blue points with error bars represent the data, red continuous line shows the fitted function.

2. Methodology

2.1. Charged Particle Pseudorapidity Measurement with the EPD

The aim is to measure the angle θ between the three-momentum **p** of the particle and the beam. Instead, a more convenient[3] quantity, the pseudorapidity η is used, which is defined as:

$$\eta \equiv -\ln\left[\tan\left(\frac{\theta}{2}\right)\right] = \frac{1}{2}\ln\left(\frac{|\mathbf{p}| + p_z}{|\mathbf{p}| - p_z}\right), \tag{1}$$

where p_z is the z component of the momentum, and the z direction is chosen to coincide with the direction of the beam [5].

Beyond the event plane determination, the EPD's fine radial granularity allows for pseudorapidity measurements to be performed. The raw EPD hit numbers could be used to calculate the pseudorapidity distribution of charged particles ($dN_{ch}/d\eta$) by using the corresponding η value of the given ring.

However, this also includes the secondary particles that do not originate from the primary vertex. As the EPD is preceded by the rest of the detector system and is relatively far from the interaction point, multiple factors distort ("blur") the measured distribution.

The factors assumed to cause the most significant distortion effect are as follows. First of all, charged primary particles scatter in detector material (or in rare cases with each other), creating secondary particles contributing to $dN_{ch}/d\eta$ significantly. This is demonstrated in Figure 2a, where the vertices (origins) of particles hitting the EPD in a detector material simulation are depicted. Second, neutral primary particles contribute through decays (e.g., a neutral Λ baryon decaying into proton and pion). In Figure 2b, it is clearly demonstrated that this contribution is non-negligible (based on the same simulation as mentioned above).

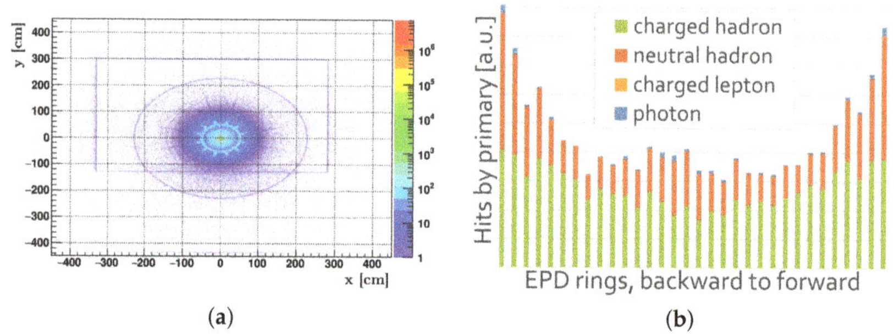

(a) (b)

Figure 2. (a) Vertices of particles registered by the EPD, based on a `HIJING` [6] + `Geant4` [7] MC detector simulation. The plots shows the vertex distribution in the x–y plane, integrated along the z axis, revealing the detector structure and surrounding materials. (b) Distribution of various types of simulated primary particles hitting EPD, ring-by-ring, where rings in the backward direction are in the left hand part of this panel, while rings in the forward direction are in the right hand side—ordered by apparent spatial rapidity of the given ring.

2.2. From Raw EPD Data to Pseudorapidity Distribution [$dN/d\eta$]

Using the previously mentioned multi-MIP Landau fit, one can extract the number of EPD hits for each ring; denoted as $N(i_{Ring})$ in the ith ring. Given the underlying pseudorapidity distribution of the primary particles ($dN/d\eta$), assuming linear dependence from the $dN/d\eta$, the number of hits in a given ring can be calculated formally as a convolution:

$$dN(i_{Ring}) = \int R(\eta, i_{Ring}) \frac{dN}{d\eta} d\eta, \qquad (2)$$

where R denotes the response matrix, which encodes response of the detector, i.e., connects a detector-level distribution with the true distribution to be measured. In this analysis, it contains the number of hits in the given ring number distribution's bin, originating from a particle at given η pseudorapidity distribution's bin.

No probabilistic consideration guarantees this matrix to be invertible; therefore, a simple (or even a regularized) matrix inversion might not be an option even if the exact form of R would be known. Instead, a method called Bayesian iterative unfolding [8] ("deblurring") is used.

Using this approach, the R needs to be extracted from simulations that are as close to the real system as possible. Using a complex event generator, a list of primary particles is obtained, along with a list of EPD hits—preferably all linked to primary tracks causing them.

In this analysis, the events were generated using the STAR's `HIJING` Monte Carlo event generator combined with `Geant4` to simulate the precise geometry of the EPD. In the following, the abbreviation MC will indicate data from these simulations. Such a response matrix can be seen in Figure 3.

It should be noted that no (light) ion fragments can be simulated in HIJING, which are, in reality, inevitable with heavy-ion collisions. However, this shortfall should not change the results significantly, according to PHOBOS results [9]: the contribution from light ion fragments causes at least an order of magnitude smaller contribution to $dN/d\eta$ than the results in this analysis (see Section 4).

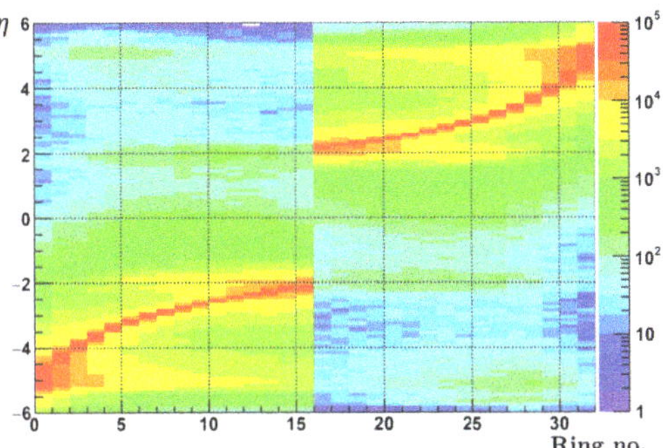

Figure 3. Heatmap visualization of the R response matrix, connecting bins containing numbers of EPD ring hits (caused by either primary or secondary particles) with bins corresponding to primary particles at given η pseudorapidity. The left side corresponds to East EPD wheel, the right side to West EPD wheel. It is worth noting that many primaries create hits even in the opposite side EPD via secondaries, as seen in upper left and bottom right quarters.

In the following step, the unfolding technique is utilized to determine an uncorrected $dN/d\eta$. The software used for this purpose is the RooUnfold [10] framework, implemented in C++, running within the ROOT environment [11]. The package itself defines classes for the different unfolding algorithms—among others, the Bayesian iterative unfolding.

The response matrix class of the software includes functions for populating the response matrix[4] as well as for managing the background (missed hits from real primaries and hits resulting from other sources[5]).

During the unfolding, one can choose to propagate the statistical uncertainty in different ways; in this case, the most appropriate method should be propagating the (mostly badly conditioned, thus noninvertible) covariance matrix [8].

The resulting EPD ring distribution[6] needs to be corrected for the multiple counting (efficiency, ϵ), explained as follows. The unfolding procedure results in one unfolded track for each individual EPD hit. However, it should be noted that one primary track can cause multiple hits. This effect needs to be corrected for—either via a bin-by-bin correction calculated from MC data (via a Number of hits from 1 primary(η) distribution), or by weighing the values filled in response matrix such that it could compensate for the multiple counts during the unfolding. In this analysis, the first method was used.

2.3. Extracting Charged Particle Pseudorapidity Distribution

In order to obtain the charged particle distribution ($dN_{ch}/d\eta$) from $dN/d\eta$, either different bin-by-bin corrections can be used, or neutral particles can be marked as background ("fake") using RooUnfold's Fake() method. In this analysis, the following methods were used as the charged factor correction:

1. Bin-by-bin correction of the already unfolded $dN/d\eta$ using the charged particle fraction $N_{charged}(\eta)/N_{all}(\eta)$ from MC data;

2. Bin-by-bin correction of the raw EPD data via $N_{charged}(i_{Ring})/N_{all}(i_{Ring})$ from MC data; then unfolding of the EPD charged particle distribution.[7]

3. Mark neutral particles as background and fill the response matrix as in the second method, except that the hits from neutral primaries are considered as "fake".

The three different methods can later be used to estimate the systematic uncertainty of the unfolding procedure itself.

2.4. Consistency Check of the Unfolding Methods

Before unfolding the real data, a closure test was done to check whether the unfolding method can recover the "true" training data itself (MC "truth").

It was found that unfolding done on the input training MC sample reproduces well the input η distribution. When some noise ($\pm$1–10%) was added to the training sample, the resulting unfolded distribution was in agreement with the input distribution within <3%. All in all, the unfolding itself was found to work well.

Furthermore, after applying the multiple counting correction and the three different methods of charged factor correction on the unfolded distribution[8], the resulting distributions were compared to the original MC dataset's $dN_{ch}/d\eta$. As it is visible in Figure 4, the maximal relative deviation is up to 2% in certain bins for the first method and less than 0.1% for the other two methods.

It is worth noting that although the third method (marking neutral particles) shows the most precise result here, the systematic checks showed that it is the least reliable in terms of most heavily depending on the MC input provided to the response matrix.

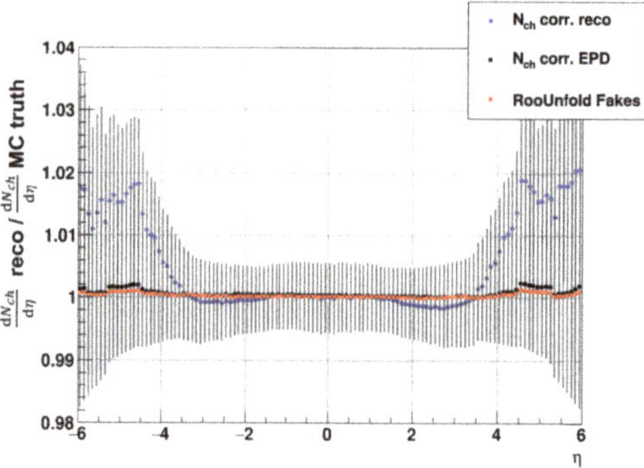

Figure 4. Consistency check of the three different methods to get $dN_{ch}/d\eta$ from MC EPD ring distribution. The difference is shown as unfolded $dN_{ch}/d\eta$ over MC "truth", the distributions divided bin-by-bin. Blue marker represents the first method (η-dependent charged factor correction), black shows the second method (EPD ring number dependent charged factor correction), and red represents the third method (marking neutral particles), relative to MC truth's $dN_{ch}/d\eta$. The errorbars are only plotted for informative purposes: they were calculated using the ROOT's TH1 class' default square root of sum of squares of weights.

Given the result of the closure test, the unfolding and correction methods were considered adequately self-consistent.

3. Systematic Checks

In the following section, the examined systematic uncertainty sources and their contribution to the results are discussed.

3.1. Dependence on Input MC Distribution

The Bayesian iterative unfolding process, via its iterative nature, should mostly overcome differences in response matrix from real response that are not related to distortion effects, such as detector geometry [8]. However, as the exact response matrix cannot be determined even with precise MC simulations and the unfolding process itself is not perfect, some dependencies on the various parameters in the MC simulations can occur. Those are considered as systematic uncertainties of the measurement.

3.1.1. Tightening and Shifting the Input MC $dN/d\eta$

First, the simulated sample's $dN_{ch}/d\eta$ was modified ("suppressed") using a Gaussian shape with width σ and mean η_0. These suppression factors can be seen in Figure 5a. This was done via a random selection based on Gaussian distribution while filling the response matrices.

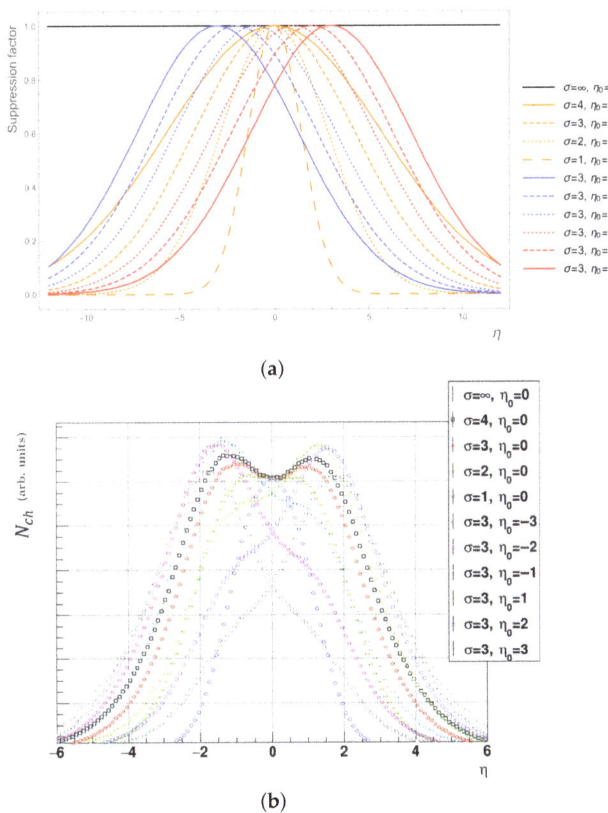

Figure 5. Tightening and shifting the MC input distribution using random selection based on Gaussian distribution of σ width and η_0 curve peak position. (**a**) Demonstration of the Gaussian suppression factors used. (**b**) The $dN_{ch}/d\eta$ of the distorted MC input samples.

Using this approach, all combinations could be analyzed, that is, unfolding the ith MC sample's EPD ring hit distribution via response from jth MC sample. In case of $i = j$, the unfolding was as close to perfect as expected, discussed in Section 2.4.

Unfolding results with the Gaussian width of $\sigma \lesssim 1$ were not considered here as in this case there are almost no particles in the EPD range. Otherwise, there was less than a few percent variation in the EPD's η region.

Overall, in the analysis, the effect of tightening the $dN_{ch}/d\eta$ of the training sample to $\sigma = 2$ and shifting it by ± 3 units of pseudorapidity was investigated.

3.1.2. Broadening the Input MC dN/dη

Similar to modification done in Section 3.1.1, here, the tracks were modified with a factor of

$$\exp\left(\frac{\eta^2 - \eta_{\max}^2}{2\sigma_{\text{broad}}}\right).$$

(3)

There was no suppression utilized for $|\eta| > \eta_{\max}$, with $\eta_{\max} = 6$. The resulting shape of the distributions can be seen in Figure 6.

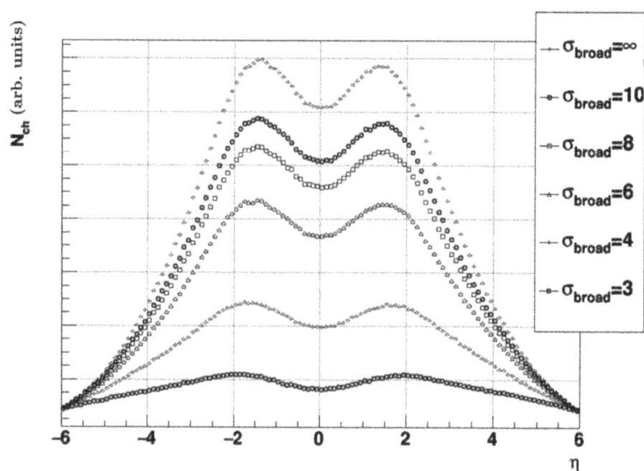

Figure 6. Broadening the MC input distribution using random selection based on Gaussian distribution of σ_{broad} width.

While unfolding the data with these input MC distributions, a significant decrease at midrapidity values was observed. However, this occurred mostly outside the EPD's η region; the unfolding was considered acceptable down to $\sigma_{\text{broad}} \approx 3$.

3.2. Changing the Charged Fraction in the MC Training Dataset

The fraction of the charged particles in the MC input data was changed by $\pm 15\%$. This was achieved by randomly rejecting either the neutral or the charged particles.

3.3. Changing the p_T Slope of the MC Training Dataset

The transverse momentum (p_T) distribution slope of the MC input data was changed by $\pm 10\%$ via randomly rejecting particles of small or large p_T.

3.4. Centrality and z-Vertex Selection

It was investigated, by how much the unfolded distribution would change if either the z-vertex or the centrality selection are modified. For the former investigation, a ± 5 cm calibration uncertainty in the z-vertex measurement of the real EPD data was employed; for the second one, $\pm 5\%$ calibration uncertainty was assumed in centrality determination of the real EPD data.

3.5. z-Vertex Choice

Due to the detector geometry, it is important to also take into account the interaction point's z-vertex position in the calculations, as the resulting pseudorapidity distribution should not depend on it.

The EPD data, as well as the responses, were collected in nine different z-vertex classes, equally distributed from -45 to $+45$ cm. Depending on which range was unfolded, the resulting distribution still may differ and has to be taken into account as systematic uncertainty.

3.6. Unfolding Method Choice

The most significant systematic uncertainty contribution was caused by the difference between the results achieved using different unfolding and correction methods (as listed in Section 2.3). The first method was used as benchmark, from which the differences were calculated.

3.7. EPD Related Uncertainties

As previously stated, the EPD electronics were considered fully efficient (except some "dead areas" in the detector from, e.g., glue and gaps, but these were assumed to be correctly handled in the simulation). The uncertainty from multi-MIP Landau fit was considered negligible compared to other systematic sources.

In conclusion, the systematic uncertainties coming from the detector system itself were considered negligible.

The different systematic uncertainty sources and their contribution with informative percentage values can be seen in Table 1.

Table 1. Summary of systematic uncertainty sources and their contribution.

Source	Systematic Uncertainty
MC input $dN_{ch}/d\eta$ tightening, shifting	6%
MC input $dN_{ch}/d\eta$ broadening	4%
Charged fraction in MC	6%
p_T slope change in MC	1%
Centrality selection	2%
z-vertex selection	negligible
z-vertex choice	1%
Unfolding method choice	8%
EPD related uncertainties, electronics, efficiency	negligible

4. Results

In this manuscript, charged particle pseudorapidity distributions with systematic uncertainties listed in Section 3 were obtained at two RHIC energies in the EPD pseudorapidity range. The results at $\sqrt{s_{NN}} = 19.6$ and 27.0 GeV can be seen in Figures 7 and 8, respectively. The caption #MIP ≤ 5 written on the plot indicates the number of convolution members in the multi-MIP Landau fit, as described in Section 1.1.

Comparison with the PHOBOS Results

Another experiment of the RHIC complex was the PHOBOS experiment, which completed data taking in 2006. The PHOBOS was a large acceptance silicon detector, covering almost 2π in azimuth and $|\eta| < 5.4$ in pseudorapidity [9]. Compared to STAR's EPD, there are differences in both detector topology and granularity: the silicone pad detectors measure the total number of charged particles emitted in the collision, with modules mounted onto a centrally located octagonal frame (Octagon) covering $|eta| \leq 3.2$, as well as three annular frames (Rings) on either side of the collision vertex, extending the coverage out to $|\eta| \leq 5.4$ [12].

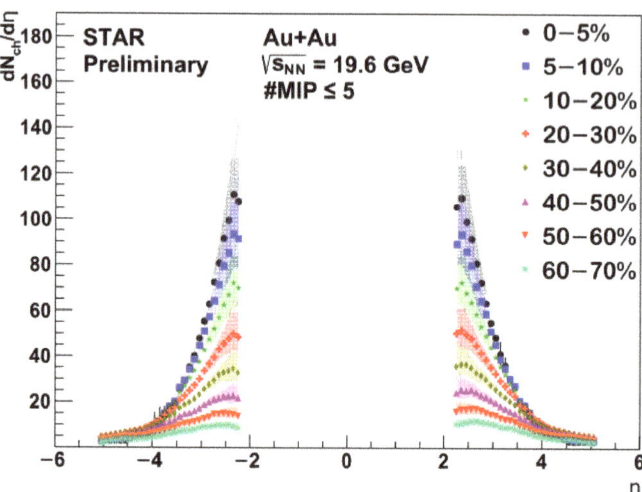

Figure 7. Charged particle pseudorapidity distributions measured with STAR EPD on RHIC energy $\sqrt{s_{NN}} = 19.6$ GeV. The data was processed in eight centrality classes, presented with the different markers. The statistical uncertainties, marked by errorbars, are not visible on this plot, as the markers themselves are larger. The colored area indicates the systematic uncertainties of the measurement.

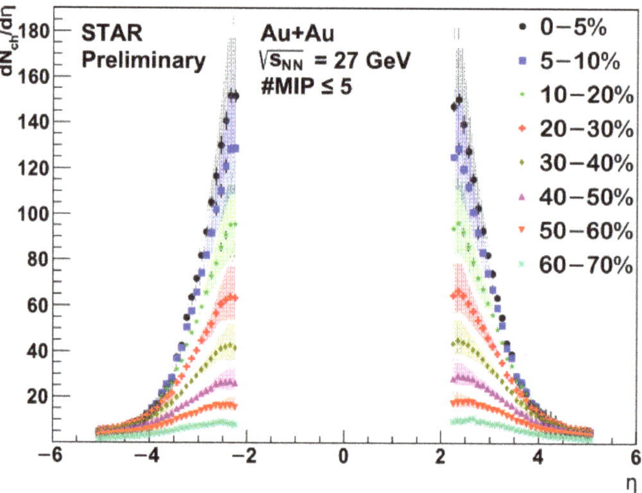

Figure 8. Charged particle pseudorapidity distributions measured with STAR EPD on RHIC energy $\sqrt{s_{NN}} = 27.0$ GeV. The data were processed in eight centrality classes, presented with the different markers. The errorbars represent the statistical uncertainty, and the colored area indicates the systematic uncertainties of the measurement.

The PHOBOS also measured $dN_{ch}/d\eta$ at 19.6, 62.4, 130, 200 GeV energies [13]. Although in that paper a slightly different centrality binning was used (0–3%, 3–6% and 6–10% instead of 0–5% and 5–10%; the other centrality classes were the same), at 19.6 GeV the results can be compared.

In Figure 9, it is apparent that the two measurements show sizeable differences, depending on η: around up to a factor of two, increasing from small $|\eta|$ toward forward/backward rapidities.

The exact reasons behind this discrepancy are not yet known but the difference cannot be explained by the systematic uncertainties described in Section 3.

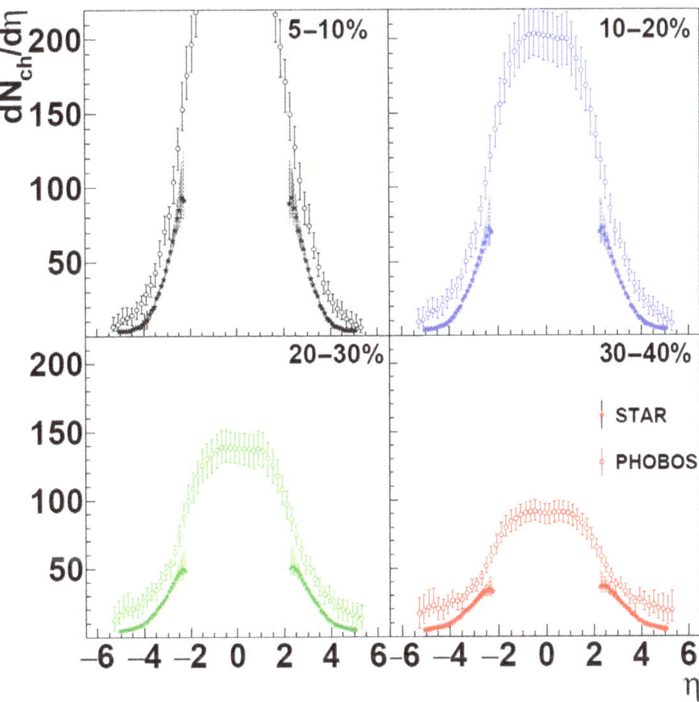

Figure 9. Charged particle pseudorapidity distributions measured in PHOBOS (hollow circles) and STAR (star markers). Note that on the upper left graph, the centrality class of the PHOBOS experiment's result is actually 6–10%.

5. Discussion

In summary, based on EPD ring-by-ring distributions, charged particle pseudorapidity measurements at $\sqrt{s_{NN}} = 19.6$ and 27.0 GeV were performed with detailed systematic investigations regarding simulation data, calibration data, and unfolding methods.

The results at $\sqrt{s_{NN}} = 19.6$ GeV show significant difference compared to the results from PHOBOS. There are four components in this comparison: EPD spectrum measurement, Geant4 simulation, unfolding procedure from the STAR part, and the PHOBOS data itself.

The method presented in this manuscript is to be extended to other $\sqrt{s_{NN}}$ values (as part of the BES-II program) and to fixed target data—mainly at energies where the QCD critical point is expected [14]. Refining this measurement method is also important for the search of the QCD critical point, in order to fine-tune the models used in these analyses.

Measuring pseudorapidity values of charged particles is important due to the possibility of estimating the initial energy density of the quark–gluon plasma created in the collisions, based on them [15,16]. Furthermore, the forward and backward rapidity measurements can provide information about the nuclear-matter effects as well [17].

Funding: This research was supported by the NKFIH OTKA K-138136 grant.

Institutional Review Board Statement: Not applicable.

Informed Consent Statement: Not applicable.

Acknowledgments: The author would like to thank the STAR collaboration and his supervisor, Máté Csanád from Eötvös Loránd University. We acknowledge the support from NKFIH OTKA K-138136 grant.

Conflicts of Interest: The author declares no conflict of interest.

Abbreviations

The following abbreviations are used in this manuscript:

QGP	quark–gluon plasma
RHIC	Relativistic Heavy-Ion Collider
STAR	Solenoidal Tracker
EPD	Event Plane Detector
BES	Beam Energy Scan
MIP	minimum ionizing particle
ADC	analog-to-digital converter
MC	Monte Carlo (simulation)
QCD	quantum chromodynamics

Notes

[1] The EPD is more sensitive to charged particles, as detailed subsequently.

[2] The rings are numbered from 1 to 32 in the following manner: the innermost East EPD ring is the #1 which follows outward until #16; then, the #17 continues on the West EPD side's outermost ring until #32 being the innermost one.

[3] In the ultrarelativistic limit, it approaches to rapidity (in $c = 1$ unit system, c being the speed of light): $\eta \approx y \equiv \frac{1}{2} \ln\left(\frac{E+p_z}{E-p_z}\right)$, with E being the energy of the particle.

[4] $\texttt{Fill}(x_{\text{measured}}, x_{\text{truth}})$; naturally, "measured" and "truth" here stand for the training datasets obtained from MC (simulation).

[5] $\texttt{Miss}(x_{\text{truth}})$ and $\texttt{Fake}(x_{\text{measured}})$.

[6] Caused by both primary and secondary particles.

[7] In this case, another type of response matrix has to be used that was filled only with the charged particles' data.

[8] Note that the mentioned unfolding procedure was at this stage still done on the MC EPD ring distribution and thus on the training sample.

References

1. Ackermann, K.; Adams, N.; Adler, C.; Ahammed, Z.; Ahmad, S.; Allgower, C.; Amonett, J.; Amsbaugh, J.; Anderson, B.; Anderson, M.; et al. STAR detector overview. *Nucl. Instrum. Methods Phys. Res. Sect. A Accel. Spectrometers Detect. Assoc. Equip.* **2003**, *499*, 624–632. The Relativistic Heavy Ion Collider Project: RHIC and its Detectors. [CrossRef]
2. Adams, J.; Ewigleben, A.; Garrett, S.; He, W.; Huang, T.; Jacobs, P.; Ju, X.; Lisa, M.; Lomnitz, M.; Pak, R.; et al. The STAR event plane detector. *Nucl. Instrum. Methods Phys. Res. Sect. A Accel. Spectrometers Detect. Assoc. Equip.* **2020**, *968*, 163970. [CrossRef]
3. Tlusty, D. The RHIC beam energy scan phase II: Physics and upgrades. *arXiv* **2018**, arXiv:1810.04767 .
4. Whitten, C.A., Jr. et al. [STAR Collaboration]. The Beam-Beam Counter: A Local Polarimeter at STAR. *AIP Conf. Proc.* **2008**, *980*, 390–396. [CrossRef]
5. Wong, C.-Y. et al. [Oak Ridge National Laboratory]. *Introduction to High-Energy Heavy-Ion Collisions*; World Scientific: Singapore, 1994; p. 24. [CrossRef]
6. Wang, X.N.; Gyulassy, M. HIJING 1.0: A Monte Carlo Program for Parton and Particle Production in High Energy Hadronic and Nuclear Collisions. *Comp. Phys. Comm.* **1995**, *83*, 307–331. [CrossRef]
7. Agostinelli, S.; Allison, J.; Amako, K.; Apostolakis, J.; Araujo, H.; Arce, P.; Asai, M.; Axen, D.; Banerjee, S.; Barrand, G.; et al. Geant4—A simulation toolkit. *Nucl. Instrum. Methods Phys. Res. Sect. A Accel. Spectrometers Detect. Assoc. Equip.* **2003**, *506*, 250–303. [CrossRef]
8. D'Agostini, G. A multidimensional unfolding method based on Bayes' theorem. *Nucl. Instrum. Methods Phys. Res. Sect. A Accel. Spectrometers Detect. Assoc. Equip.* **1995**, *362*, 487–498. [CrossRef]
9. Alver, B. et al. [PHOBOS Collaboration]. Participant and spectator scaling of spectator fragments in Au + Au and Cu + Cu collisions at $\sqrt{s_{NN}} = 19.6$ and 22.4 GeV. *Phys. Rev. C* **2016**, *94*, 024903. [CrossRef]

10. Adye, T. Unfolding algorithms and tests using RooUnfold. *arXiv* **2011**, arXiv:1105.1160. [CrossRef]
11. Antcheva, I.; Ballintijn, M.; Bellenot, B.; Biskup, M.; Brun, R.; Buncic, N.; Canal, P.; Casadei, D.; Couet, O.; Fine, V.; et al. ROOT—A C++ framework for petabyte data storage, statistical analysis and visualization. *Comput. Phys. Commun.* **2011**, *182*, 1384–1385. [CrossRef]
12. Back, B.; Baker, M.; Ballintijn, M.; Barton, D.; Becker, B.; Betts, R.; Bickley, A.; Bindel, R.; Budzanowski, A.; Busza, W.; et al. The PHOBOS perspective on discoveries at RHIC. *Nucl. Phys. A* **2005**, *757*, 28–101.
13. Alver, B.; Back, B.B.; Baker, M.D.; Ballintijn, M.; Barton, D.S.; Betts, R.R.; Bickley, A.A.; Bindel, R.; Budzanowski, A.; Busza, W.; et al. Charged-particle multiplicity and pseudorapidity distributions measured with the PHOBOS detector in Au + Au, Cu + Cu, d + Au, and p + p collisions at ultrarelativistic energies. *Phys. Rev. C* **2011**, *83*, 024913. [CrossRef]
14. Odyniec, G. RHIC beam energy scan program—Experimental approach to the QCD phase diagram. *J. Phys. G Nucl. Part. Phys.* **2010**, *37*, 094028. [CrossRef]
15. Csanád, M.; Csörgő, T.; Jiang, Z.F.; Yang, C.B. Initial energy density of $\sqrt{s} = 7$ and 8 TeV p+p collisions at the LHC. *Universe* **2017**, *3*, 9. [CrossRef]
16. Ze-Fang, J.; Chun-Bin, Y.; Csanad, M.; Csorgo, T. Accelerating hydrodynamic description of pseudorapidity density and the initial energy density in p + p , Cu + Cu, Au + Au, and Pb + Pb collisions at energies available at the BNL Relativistic Heavy Ion Collider and the CERN Large Hadron Collider. *Phys. Rev. C* **2018**, *97*, 064906. [CrossRef]
17. Adare, A. et al. [PHENIX Collaboration]. Cold-Nuclear-Matter Effects on Heavy-Quark Production at Forward and Backward Rapidity in d + Au Collisions at $\sqrt{s_{NN}} = 200$ GeV. *Phys. Rev. Lett.* **2014**, *112*, 252301. [CrossRef] [PubMed]

Article

Measurements of J/ψ Production vs. Event Multiplicity in Forward Rapidity in $p + p$ Collisions in the PHENIX Experiment

Zhaozhong Shi [†] on behalf of the PHENIX Collaboration

Physics Division, Los Alamos National Laboratory, Los Alamos, NM 87545, USA; zhaozhongshi@lanl.gov; Tel.: +1-415-350-3181

† Current address: Physics Department, Brookhaven National Laboratory, Upton, NY 11973, USA.

Abstract: J/ψ, a charmonium bound state made of a charm and an anti-charm quark, was discovered in the 1970s and confirmed the quark model. Because the mass of charm quarks is significantly above the quantum chromodynamics (QCD) scale Λ_{QCD}, charmonia are considered excellent probes to test perturbative quantum chromodynamics (pQCD) calculations. In recent decades, they have been studied extensively at different high-energy colliders. However, their production mechanisms, which involve multiple scales, are still not very well understood. Recently, in high-multiplicity $p + p$ collisions at RHIC and at the LHC, a significant enhancement of J/ψ production yield has been observed, which suggests a strong contribution of multi-parton interaction (MPI). This is different from the traditional pQCD picture, where charm quark pairs are produced from a single hard scattering between partons in $p + p$ collisions. In this work, we will report the J/ψ normalized production yield as a function of normalized charged particle multiplicity over a board range of rapidity and event multiplicity in the $J/\psi \rightarrow \mu^+\mu^-$ channel with PHENIX Run 15 $p + p$ data at $\sqrt{s} = 200$ GeV. The results are compared with PYTHIA 8 simulations with the MPI option turned on and off. Finally, the outlooks of J/ψ in $p + Au$ and $Au + p$ collisions, along with color glass condensate (CGC) predictions and the multiplicity-dependent $\psi(2S)/J/\psi$ ratio in $p + p$ data, will be briefly discussed.

Keywords: quantum chromodynamics; charmonium; multi-parton interactions; production yield; high event multiplicity; $p + p$ collisions; initial state effect; final state interaction; hadronization; color glass condensate

Citation: Shi, Z., on behalf of the PHENIX Collaboration. Measurements of J/ψ Production vs. Event Multiplicity in Forward Rapidity in $p + p$ Collisions in the PHENIX Experiment. *Universe* **2023**, *9*, 322. https://doi.org/10.3390/universe9070322

Academic Editor: Jun Xu

Received: 30 May 2023
Revised: 22 June 2023
Accepted: 27 June 2023
Published: 4 July 2023

1. Introduction

In 1974, J/ψ, a charmonium-bound state made of charm and anti-charm quark ($c\bar{c}$) was discovered at Brookhaven National Laboratory [1] and the Stanford National Linear Accelerator [2] and confirmed the existence of charm quarks [3] and validated the quark model. Because the charm quark mass is above the QCD scale Λ_{QCD}, the production of a $c\bar{c}$ pair in high-energy collisions is perturbative, which makes J/ψ an excellent probe to test pQCD calculations.

The production of J/ψ at high-energy hadronic colliders involves multiple stages across many different scales. In recent decades, J/ψ has been studied extensively at different colliders [4–6]. Nonetheless, the description of J/ψ production is still not fully developed and cannot reach very high precision. However, thanks to the QCD factorization theorem [7], hard processes, which are perturbatively calculable, are factorized from soft processes, which are non-perturbative but can be modeled phenomenologically and constrained by experiments. This allows us to apply pQCD to calculate the production cross-section of J/ψ. We can test QCD at high-energy colliders through a comparison of the J/ψ production data with model calculations.

In hadronic collision events, both elastic and inelastic scatterings may occur. Experimentally, we are interested in inelastic collision events. There are two types of inelastic hadronic collision events: diffractive and non-diffractive dissociations [8]. In this work, we focus on J/ψ, produced in non-diffractive hadronic collision events, which can be denoted as $pp \to J/\psi + X$.

A simple sketch of J/ψ production in high-energy hadronic collision events can be summarized as below:

Initial Dynamics of Partons: According to the Parton Model, the structure of hadrons can be described by constituent partons [9]. The initial dynamics of partons are non-perturbative but can be parametrized by parton distribution function (PDF) [10]. They can be measured in deep inelastic scattering experiments at different colliders and related with each other via scaling [7]. Alternatively, phenomenological approaches [11], such as String Percolation [12] and Color Glass Condensate [13], can be applied to model the incoming hadrons. These models have been compared with experimental data such as charged particle p_T spectra and rapidity distribution $dN_{ch}/d\eta$ and demonstrate reasonably good agreement [14,15].

Initial State Interactions: They occur among energetic partons before hard scatterings. One example is the soft radiation of partons [16], which is called the initial state radiation (ISR). ISR will influence the initial heavy-quark pair production [17]. Usually, the effect can widen $c\bar{c}$ azimuthal angle correlation [18] and broaden the p_T spectra [19].

Hard Partonic Scattering: Energetic partons scatter off each other with large momentum transfers. In the traditional pQCD picture, it is simply described as a single hard scattering between two partons in each collision. They can be calculated analytically by pQCD with Feynman diagrams to a very high precision [20]. At RHIC, the $c\bar{c}$ pair production is dominated by gluon–gluon fusion: $gg \to c\bar{c}$.

Multiple Parton Interaction (MPI): MPI is an elaborate paradigm to describe the partonic interaction stage at high-energy colliders at RHIC, Tevatron, and the LHC [21]. According to MPI, one hard scattering, accompanied by several semi-hard interactions, takes place in each collision. All of them need to be included in the partonic scattering amplitudes. At present, high-energy hadronic colliders create more phase space for MPI to occur. Many studies at the LHC suggest MPI should be included to better describe the data [22].

Hadronization: In the final state, the $c\bar{c}$ pairs will lose energy via radiation and evolve into the color-neutral J/ψ bound state. Because this process is also soft and non-perturbative, many phenomenological models have been developed to describe the $c\bar{c} \to J/\psi$ process in different collision systems. Selected examples of theoretical models are listed below:

Non-Relativistic QCD (NRQCD): This is an effective field theory approach to describe the hadronization of $c\bar{c}$ pairs thanks to their large mass compared to its internal kinetic energy, which results in a slow speed β [23] within the non-relativistic limit $\beta << 1$. NRQCD includes perturbative short-distance and non-perturbative long-distance effects for a range of strong coupling α_s. To the leading order (LO), there are two mechanisms describing charmonium production.

- Color Singlet (CS): The $c\bar{c}$ pairs are in the color-singlet state with the same quantum number as the $c\bar{c}$ bound state in J/ψ. When the $c\bar{c}$ pair kinematics reach the J/ψ mass, they will bind together [24].
- Color Octet (CO): The $c\bar{c}$ pairs are in the color-octet state carrying net color changes and emit extra gluons [25] to reach the color-neutral state, which results in additional hadron production associated with the J/ψ observed in the J/ψ-hadron correlation studies [26].

NRQCD predicts sizable transverse polarization of J/ψ, which has also been observed experimentally [27]. There are other phenomenological models, such as the Color Evaporation Model [28], Statistical Hadronization Model [29], and Color String Reconnection

Model [30], that describe the J/ψ hadronization. Currently, physicists are testing all these models with experimental data.

Final State Interaction (FSI): In the final state, the newly formed J/ψ mesons may still interact with comoving particles nearby [31]. In the elastic scenario, J/ψ kinematics will be modified. Inelastically, J/ψ may possibly be broken up [32]. Hence, the final state comover effect may affect J/ψ production yield and will become more prominent at high levels of multiplicity. Experimentally, the final state comover effect can be studied by J/ψ-hadron femtoscopic correlation measurements [32]. Theoretically, FSI has also been implemented in the EPOS event generator [33].

Experimental Observables: All the above-mentioned processes will contribute to the final production of J/ψ, which can be reconstructed from its decay particles with detectors in the experiment. The experimental observables used to study J/ψ production may be the production yield as a function of event activity for fully reconstructed J/ψ. Experimentally, the event activity is quantified by charged particle multiplicity. The production yield, as a function of event multiplicity, can probe the processes at the partonic level and will shed light on the interplay between soft and hard particle production [34].

In particular, we can use a relative quantity: the normalized J/ψ yield $R \equiv N^{J/\psi}/\langle N^{J/\psi}\rangle$ as a function of normalized charged particle multiplicity $N_{ch}/\langle N_{ch}\rangle$. In experiments, this observable has an advantage because it can cancel the luminosity and some efficiency corrections, such as J/ψ acceptance and reconstruction efficiency, which ultimately reduces the systematic uncertainties. Theoretically, in the string percolation picture [12], there is a simple scaling of $N^{J/\psi}$ by the number of color strings N_s at the partonic level, which is similar to N_{coll} in heavy-ion collisions at the nucleon level. Moreover, N_{ch} is scaled by N_s, another analog to the N_{part} scaling for soft particle production in heavy-ion collisions. Therefore, this is also inspired by theoretical perspectives. The normalized J/ψ yield as a function of normalized charged particle multiplicity measurements was studied by experiments conducted at RHIC and the LHC over different kinematic regions.

Autocorrelation: The J/ψ itself can contribute to the charged particle multiplicity in many different ways, as listed below [35]:

- The J/ψ decay daughters, such as the dipion, dielectron, and dimuon pairs.
- The extra gluons emitted from the $c\bar{c}$ pair in the color-octet state producing additional charged hadrons [26].
- The J/ψ cluster collapsing into hadrons [36].
- The feed down from b-hadron decays for non-prompt J/ψ.

Generally speaking, the autocorrelation increases N_{ch} in J/ψ events compared to minimum bias (MB) events. Reducing the autocorrelation effects can improve our study for dedicated physics processes.

2. Recent Developments

Today, with the advancement of technologies related to detector instrumentation, high-performance computing, and artificial intelligence, we are moving toward a high-precision QCD era. Many novel studies of J/ψ production have been conducted at RHIC and the LHC.

Recently, the ALICE Collaboration reported results on J/ψ production measured in dielectron channel with the LHC Run 2 pp data at $\sqrt{s} = 5, 7, 13$ TeV [37–39]. The measurements of the J/ψ-normalized yield are performed in both middle- and forward-rapidity regions over a wide range of normalized charged particle multiplicities [40]. The normalized J/ψ yield, as a function of $N^{J/\psi}/\langle N^{J/\psi}\rangle$ at mid-rapidity, generally lies above the forward-rapidity region. A significant enhancement of J/ψ production with respect to linear scaling is observed at high multiplicities for both middle and forward rapidities [39]. Several theoretical models incorporating both initial state effects and MPI attempt to explain the data [39].

At RHIC, the STAR experiment carried out the J/ψ studies in the dielectron channel, which only shows up to about three units of average charged particle multiplicities at a rapidity of $|y| < 1$ [41]. The data are presented in different p_T regions. However, the results also suggest a slight enhancement for J/ψ production and are comparable to ALICE at the LHC energy rate. The increase becomes steeper at a higher p_T and multiplicity region, although this difference is not significant due to the large uncertainties and does not occur at a very high event multiplicity, where the FSI is reduced and MPI effect is more prominent. The STAR result are generally well described by CEM, CGC, and NLO with NRQCD calculations at different p_T regions. However, no conclusion regarding the use of MPI for J/ψ production at RHIC at mid-rapidity and near the average charged particle multiplicity has been drawn.

Phenomenologically, the MPI effect plays a significant role in charm-quark production [42]. In the MPI picture, the average number of heavy-quark pairs in pp collision increases compared to the traditional pQCD picture of single hard scattering [43]. Along with the color reconnection model for J/ψ hadronization treatment, a significant enhancement of the J/ψ production cross-section [30] is predicted. Hence, the linear scaling assumed in the traditional pQCD picture does not hold [44].

From the simulation side, the latest versions of PYTHIA 8 event generator incorporated many physics processes, including ISR, hadronization, and FSR, in addition to MPI, to describe underlying events in high-energy pp collisions [45]. PYTHIA 8 simulations are able to reproduce the charged particle p_T spectra and $dN/d\eta$ with reasonably good agreement at RHIC with Detroit tune [46] and the LHC with Monash tune [47]. PYTHIA users can turn MPI on and off, use different underlying event tunes, and adjust the CSM and COM contribution in $c\bar{c} \to J/\psi$ to compare with the data.

The J/ψ produced from the recombination of the $c\bar{c}$ pair described in the Introduction is traditionally considered the dominant production mechanism of J/ψ [48] and will lead to a substantial amount of transverse polarization. However, recently, at the LHC, unpolarized J/ψ production from jets, an alternative production mechanism in pp [49] and PbPb [50] collisions was observed by the CMS experiment. Moreover, LHCb has shown that unpolarized J/ψ, down to low p_T, is produced from jet fragmentation in pp collisions [51]. J/ψ are observed to be hadrons within the jet cones' radius. The J/ψ produced from jets will have different production processes compared to those described above.

Most J/ψ measurements are carried out in non-diffractive dissociation events at hadronic colliders. There are also some theoretical efforts to study novel QCD with J/ψ production in single diffractive pp collisions via Pomerons exchange ($pp \to pX$) [52]. Measurements on a single diffractive pp cross-section have also been carried out by the ALICE [53] and ATLAS experiments [54] at the LHC.

These latest developments motivate us to investigate J/ψ at high event multiplicities in forward-rapidity at RHIC. The PHENIX detector is capable of carrying out this physics [55]. Thanks to the excellent tracking, vertexing, and muon performance of the PHENIX detector, we can perform charmonium studies in the forward-region up to high multiplicities. Historically, the research on the event multiplicity dependence of J/ψ production in small systems with PHENIX dates back to early 2013, focusing on $p+p$ collisions at $\sqrt{s} = 510$ GeV [56]. We will report our latest studies on J/ψ using PHENIX Run 15 $p+p$ data at $\sqrt{s} = 200$ GeV.

3. Experimental Apparatus and Data Samples

The PHENIX experiment [55] is a general-purpose detector at RHIC at Brookhaven National Laboratory for relativistic heavy-ion physics research [57]. It has broad ϕ and η acceptance coverage [58] and can collect large data samples to perform measurements at middle and forward rapidities. The tracking, particle identification, calorimeter, and muon systems of the PHENIX experiment apply various radiation detection techniques to maximize its physics capabilities.

The forward silicon tracker detector (FVTX) employs advanced silicon strip technologies and is installed as four endcaps in the forward and backward regions covering $1.2 < |\eta| < 2.2$ [59]. Its sensor contains two columns of mini-strips with 75 μm pitches in the radial direction and lengths varying from 3.4 to 11.5 mm in the azimuthal direction. The FVTX is capable of excellent tracklet reconstruction and precise vertex determination. In addition to the FVTX, at mid-rapidity $|\eta| < 1$, the Silicon Vertex Tracker (SVX) is a four-layer barrel detector built to enhance the capabilities of the central arm spectrometers and provides excellent position resolution [60], which enables tracking at mid-rapidity.

Two muon arms are built in the forward and backward regions, far away from the beam spot, with a rapidity coverage of $1.2 < |\eta| < 2.4$ [61]. A stack of absorber/low resolution tracking layers allow for excellent muon detection and identification. Along with the three new resistive plate chambers, the rejection factor for muon from pions and kaons is in the order magnitude of 10^3. Each muon arm is equipped with a radial field magnetic spectrometer to provide precision muon tracking. The muon momentum resolution is $\delta p / p \sim 1.7\%$, allowing for an excellent performance in quarkonia reconstruction and clean separation between J/ψ and ψ' [62].

The PHENIX Electromagnetic Calorimeter (EMCAL) uses Pb as the absorber material and a shashlik design with a block size of 5.5 cm $\times$ 5.5 cm and wavelength shifting fibers to measure the electromagnetic shower energy [63]. The EMCAL can provide an excellent energy linearity and resolution for jet reconstruction.

The PHENIX experiment is also equipped with a ring image Cherenkov detector (RICH) to perform electron identification [64]. It can achieve a great electron selection performance from π, K, p separation at a very high p_T.

The beam–beam counters (BBC) are installed in both far-north and far-south directions with advanced electronics to determine the event vertex and activity [65]. BBC uses the coincidence of both sides along with a minimum ADC threshold to select MB events. The zero-degree calorimeter (ZDC) is used in an identical form for all four experiments at RHIC to characterize global event parameters in the very forward direction [66]. It can achieve the precise determination of event activity, luminosity, and forward-neutron-counting through the measurements of beam-fragment energy deposition in the far-forward direction.

With excellent detector hardware capabilities, PHENIX also designs and deploys a dedicated Level 1 trigger to collect data samples for different physics topics [67], applying high-performance computing and electronic readout technologies [68]. Many collisions occur at RHIC when the collider is running. However, only a small fraction of them are relevant to our physics studies. Thus, the MB trigger was developed for general physics studies. The MB trigger uses both BBC and ZDC to select non-diffractive dissociation processes and determine global event parameters such as the collision vertex, luminosity, and impact parameter. The overall efficiency of the MB trigger is approximately $55 \pm 5\%$.

For charmonium physics studies, we need high statistics J/ψ samples. The dimuon trigger samples enrich J/ψ by requiring an MuID trigger to identify muons and applying quality selections to the muon tracks (MuTr). The overall efficiency of the dimuon trigger is approximately $79 \pm 2\%$.

The PHENIX detector is also equipped with a beam clock trigger utilizing the granule timing module with fast electronics [69]. It can operate at high frequencies with an excellent timing resolution to provide precise timing information for the raw data, which allows for synchronization among subdetectors and event-building. In addition, the EMCal/RICH trigger (ERT) is dedicated to sampling hard scattering events for heavy flavor and jet physics studies.

4. Analysis

In 2015, PHENIX acquired $p + p$, $p + Al$, and $p + Au$ data with transversely polarized protons at $\sqrt{s} = 200$ GeV. Based on the PHENIX $p + p$ data, we can define the J/ψ normalized yield $R(J/\psi)$ from quantities as follows:

$$R(J/\psi) = \left[\frac{N_S^{J/\psi}}{N^{MB}} \frac{\epsilon_{trig}^{MB}}{\epsilon_{trig}^{J/\psi}}\right] \bigg/ \left[\frac{N_S^{J/\psi}(total)}{N^{MB}(total)} \frac{\langle\epsilon_{trig}^{MB}\rangle}{\langle\epsilon_{trig}^{J/\psi}\rangle}\right] \times f_{coll} \tag{1}$$

The quantities are defined below:

- $N_S^{J/\psi}$: the J/ψ signal raw yield extracted from dimuon invariant mass $m_{\mu\mu}$ distribution.
- N^{MB}: the number of minimum biased events recorded.
- ϵ_{trig}^{MB}: minimum biased trigger efficiency.
- $\epsilon_{trig}^{J/\psi}$: J/ψ trigger efficiency.
- f_{coll}: correction factor for multiple collisions obtained from a data-model method.

The quantities in the bracket stand for the average value over the integral event multiplicity and the total in the parentheses means the sum over all multiplicity bins. We reconstructed J/ψ from the dimuon decay channel: $J/\psi \to \mu^+\mu^-$. It should be noted that we assumed that the luminosity, the branching ratio of $J/\psi \to \mu^+\mu^-$, the acceptance, and the reconstruction efficiency would cancel out in the normalization because they do not have significant event multiplicity dependence.

The normalized J/ψ yield $R(J/\psi)$ is plotted as a function of normalized charged particle multiplicity $N_{ch}/\langle N_{ch}\rangle$, defined as the number of tracklets reconstructed by FVTX or SVX hits. Thus, the pseudorapidity ranges of N_{ch} are $1.2 < \eta < 2.4$ for FVTX north, $-2.4 < \eta < -1.2$ for FVTX south, and $-1.0 < \eta < 1.0$ for SVX. Our results are presented in charged particle multiplicity bins of $[0, 1, 2, 3, 4, 5, 6, 8, 10, 12, 19]$. In our analysis, we used the MB data sample to obtain the N^{MB}. The dimuon trigger sample was used to reconstruct J/ψ. Finally, the beam clock trigger sample was used for efficiency correction and systematic uncertainties studies in a data-driven manner.

4.1. Event, Track, and J/ψ Candidate Selections

In order to achieve the best analysis results, we need to apply selections to the data samples. We applied event, track, muon, and J/ψ candidate selections to the data sample to reduce the size and ensure the quality of our analysis results. Specifically, we required the z-component of the reconstructed event vertex (z_vtx) to be within 10 cm, which was used to define the charge particle multiplicity counting.

4.2. MB Event Multiplicity Determination

We used the MB sample to determine the N^{MB}. With the N^{MB} as a function of N_{ch}, we can also obtain the $N_{MB}(total)$ by summing the distribution and $\langle N_{ch}\rangle$, taking the average on the distribution. We can then rescale the x-axis to $N_{ch}/\langle N_{ch}\rangle$ and plot the N^{MB} as a function of $N_{ch}/\langle N_{ch}\rangle$.

4.3. J/ψ Signal Extraction

After applying all selections to the dimuon sample, we were able to observe a very clear J/ψ signal with good resolution and a correct peak near the PDG value. The kinematics of the reconstructed J/ψ has $\langle p_T\rangle \sim 1.7$ GeV/c and $1.2 < |y| < 2.2$. To determine the J/ψ raw yield, we need to extract the signal in the dimuon invariant mass in data. We developed a fitting model using a single asymmetric Crystal Ball function to describe the J/ψ signal component to account for the bremsstrahlung tail and an exponential decay function to describe the background component in the data. The functional form of the signal component is given by

$$f_S(x; \alpha, n, \mu, \sigma) = \begin{cases} N \exp\left[-\frac{(x-\mu)^2}{2\sigma^2}\right], & \text{if } \left(\frac{x-\mu}{\sigma} > \alpha\right) \\ NA\left(B - \frac{x-\mu}{\sigma}\right)^{-n} & \text{if } \left(\frac{x-\mu}{\sigma} \leq \alpha\right) \end{cases} \tag{2}$$

where

$$A = \left(\frac{n}{|\alpha|}\right)^n \exp\left(-\frac{|\alpha|^2}{2}\right) \tag{3}$$

and

$$B = \frac{n}{|\alpha|} - |\alpha| \tag{4}$$

The functional form of the background component is given by

$$f_B(x; D, \lambda) = De^{-\lambda x} \tag{5}$$

Hence, the total fit function is given by

$$f = N_S^{J/\psi} \cdot f_s + N_B \cdot f_B \tag{6}$$

We then used the *RooFit* package [70] to fit the data points and obtain the J/ψ signal raw yield $N_S^{J/\psi}$. The invariant mass distribution of J/ψ from the north and south muon arms for inclusive event multiplicity, along with the fits, are shown below in Figure 1.

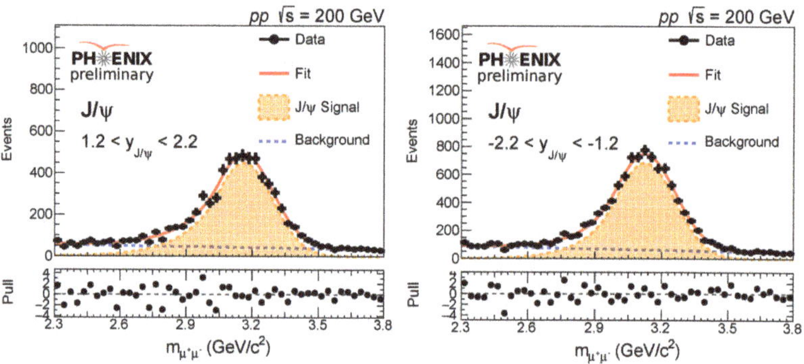

Figure 1. The dimuon invariant mass distributions of the J/ψ at $1.2 < y_{J/\psi} < 2.2$ (**left**) and $-2.2 < y_{J/\psi} < -1.2$ (**right**) with their fits are shown above. It should be noted that the south muon arm on the right has a higher J/ψ yield than the north one with the same selections because it has better performance with a higher muon efficiency in Run 2015 data collection.

The free parameters for the fits are N, A, B, μ, σ, D, and λ. We fixed A and N in the fit on the inclusive north and south muons arm samples to keep the overall shape needed to fit each $N_{ch}/\langle N_{ch}\rangle$ bin. Good statistics, with a reconstruction performance of J/ψ in the dimuon channel, can be observed in Figure 1. We sum $N_S^{J/\psi}$ of all $N_{ch}/\langle N_{ch}\rangle$ bins to obtain $N_S^{J/\psi}(total)$.

4.4. Efficiency Correction

We quoted the $\langle \epsilon_{trig}^{MB} \rangle = 55\%$ and $\langle \epsilon_{trig}^{J/\psi} \rangle = 79\%$, as mentioned in the description for the PHENIX detector. Then, we employed a data-drive method to correct the MB and J/ψ efficiencies.

To determine ϵ_{trig}^{MB} as a function of event multiplicity, we used the the RHIC beam clock trigger data. A collision is declared to have occurred if there is at least one tracklet in the FVTX or SVX. Hence, ϵ_{trig}^{MB} is the ratio of the RHIC beam clock trigger sample with BBC local level 1 trigger, which is also fired, f to the whole sample for BBC rate between 1000 and 1500 kHz. The systematic uncertainties $\sigma(\epsilon_{trig}^{MB})$ are given by the deviation of ϵ_{trig}^{MB} at a BBC rate from 600 to 800 kHz and 2000 to 2500 kHz from the nominal value 1000–1500 kHz as the upper and lower bounds, respectfully.

To determine $\epsilon_{trig}^{J/\psi}$ as a function of event multiplicity, we used the ERT trigger sample. We calculated $\epsilon_{trig}^{J/\psi}$ using the multiplicity distribution of the ERT sample with at least one track as the denominator, and the multiplicity distribution of the ERT sample with at least one track and a valid BBC vertex z-component within 200 cm as the numerator. The statistical uncertainties of the first bin $N_{ch} = 1$ are quoted as global systematic uncertainties $\sigma(\epsilon_{trig}^{J/\psi})$.

4.5. Multiple Collection Factor Correction

Multiple $p + p$ collisions may occur at RHIC. Experimentally, each collision results in a primary vertex. The number of collisions in each event generally obeys the Poisson distribution. According to our studies, the double-collision probability is at the level of a few percent.

Because we focused on J/ψ produced in a single $p + p$ collision, we needed to correct multiple collision effects in our data. We employed a data-model hybrid method to determine f_{coll}. We used a model to calculate N^{MB} as a function of N_{ch}. We divided the normalized N^{MB} distribution for the south FVTX arm near a BBC rate of 830 kHz, which consists of less than 2% of double-collisions, using the single-collision model and the ratio as f_{coll}. We quoted the deviation from the model to the PHENIX data with a BBC rate between 1000 and 1500 kHz as the systematic error on the multiple collision correction factor $\sigma(f_{coll})$, accounting for the disagreement between the model and the data.

4.6. Systematic Uncertainties Estimation

The systematic uncertainties on this measurement consist of the MB trigger efficiency, J/ψ trigger efficiency, multiple-collision correction, and J/ψ reconstruction efficiency. The J/ψ reconstruction efficiency $\epsilon_{reco}^{J/\psi}$ has a weak dependence on N_{ch}. This is treated as a constant but can be assigned using a global systematics of 5% from previous J/ψ measurements in the dimuon channel [71]. Finally, we treated individual uncertainties as uncorrelated, and thus could estimate the total systematic uncertainties as follows:

$$\sigma(total) = \sigma(\epsilon_{trig}^{MB}) \oplus \sigma(\epsilon_{trig}^{J/\psi}) \oplus \sigma(\epsilon_{reco}^{J/\psi}) \oplus \sigma(f_{coll}) \tag{7}$$

5. Results

After finishing the data analysis, we gathered all the ingredients to obtain the final results. Different underlying physics processes can be studied from different rapidity combinations of J/ψ and tracklet multiplicity measurements. Figure 2 illustrates the physics with different measurements.

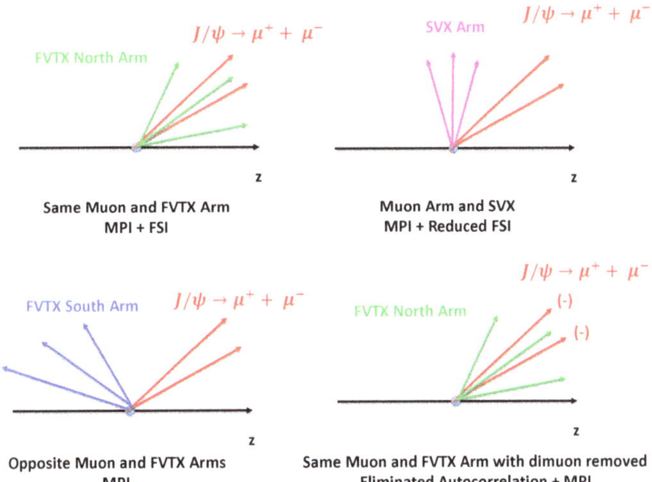

Figure 2. The definitions of four different rapidity measurements of the J/ψ with respect to the silicon tracklet measurements and the physics processes involved are illustrated above.

Phenomenologically, MPI always occurs, regardless of the rapidity of the J/ψ and the charged particles. In the PHENIX experiment, when both the J/ψ and the tracklets point in the same rapidity direction, we expect to find significant FSI contributions to

J/ψ production due to the presence of nearby particles [33]. In the elastic scenario, J/ψ kinematics will be modified. Inelastically, J/ψ may be broken up [32]. As the J/ψ moves away from the charged particles, the comover effect in the final state is expected to diminish. This can be achieved by measuring the SVX and the opposite FVTX arm for the tracklet multiplicity with respect to the muon arms. Finally, muons can also contribute to the event multiplicity. For $J/\psi \to \mu^+\mu^-$, the two muons, on average, increase the N_{ch} by approximately 1.4. After removing this autocorrelation effect from the J/ψ decayed muons, the charged particle multiplicity will become $\tilde{N}_{ch}$. We can also present the normalized J/ψ yield as a function of $\tilde{N}_{ch}/\langle \tilde{N}_{ch} \rangle$ by adjusting the x-axis in our measurement. These cases are all shown in Figure 2.

The final results of J/ψ, reconstructed from the north muon arm located in the forward-rapidity direction $1.2 < y_{J/\psi} < 2.2$ and the south muon arm located at the backward-rapidity $-2.2 < y_{J/\psi} < -1.2$ with respect to FVTX north and south and SVX measurements, are shown in Figure 3.

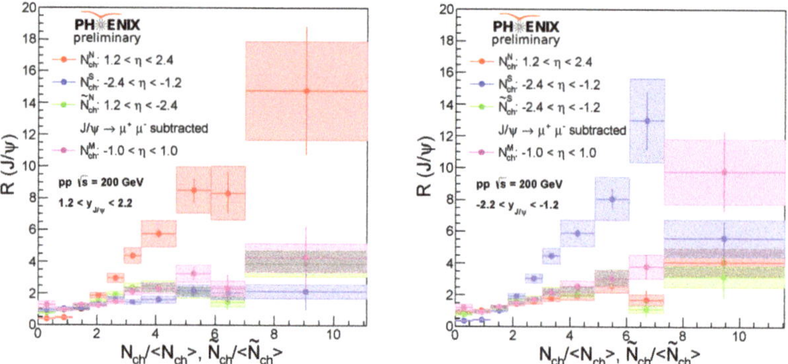

Figure 3. The J/ψ reconstructed from the north muon arm $1.2 < y_{J/\psi} < 2.2$ (**left**) and the south muon arm $-2.2 < y_{J/\psi} < -1.2$ (**right**) are shown above. The J/ψ normalized signal yields $R(J/\psi)$ from the dimuon channel are presented as a function of normalized charged particle multiplicity $N_{ch}/\langle N_{ch} \rangle$ measured by FVTX and SVX. In addition, we show the results when J/ψ and the charged particles are in the same direction, with the dimuon contribution subtracted. In terms of color convention, the FVTX north data are shown in red, FVTX south data in blue, SVX in magenta, and the dimuon-subtracted results are shown in magenta. These four sets of data points are all overlaid with each other in the same figure.

The J/ψ yields up to approximately 10 units of average charged particle multiplicity, which are measured with good precision. A stronger than linear rise is observed at the same rapidity direction between the J/ψ and the charged particles. The enhancement becomes more prominent at high-multiplicity regions. The slope decreases as the rapidity gap between the J/ψ and the charged particles increases when $N_{ch}/\langle N_{ch} \rangle > 1$. Finally, after subtracting the dimuon contributions at the same rapidity directions, the data points drop drastically and become consistent with the opposite rapidity measurements. These results imply that the FSI effect does not have a substantial impact on J/ψ production in $p + p$ collisions. However, MPI effects should be considered in order to the enhancement, particularly in the high-multiplicity region.

We also compare our data with recent measurements from STAR at RHIC [41] and ALICE at the LHC [39], as shown below.

In Figure 4, we can see that PHENIX has broader charged particle multiplicity measurements with better precision than STAR and a comparable reach to ALICE, albeit with lower precision. At a low charged-particle multiplicity, PHENIX data points are systematically below the STAR ones. In a higher-event-multiplicity region, PHENIX data points ($1.2 < |y| < 2.2$) lie in between the ALICE middle ($|y| < 0.9$) and forward-rapidity

(2.5 < y < 4) measurements, filling the missing-rapidity region from ALICE. All data points have slopes significantly above 1 when $N_{ch}/\langle N_{ch}\rangle > 1$. Hence, the comparisons suggest that J/ψ produced in the middle-rapidity is generally above the forward-rapidity at both RHIC and LHC energies, which corresponds to the different phase-space regions of $x_{1,2}$ of the partons during hard interactions.

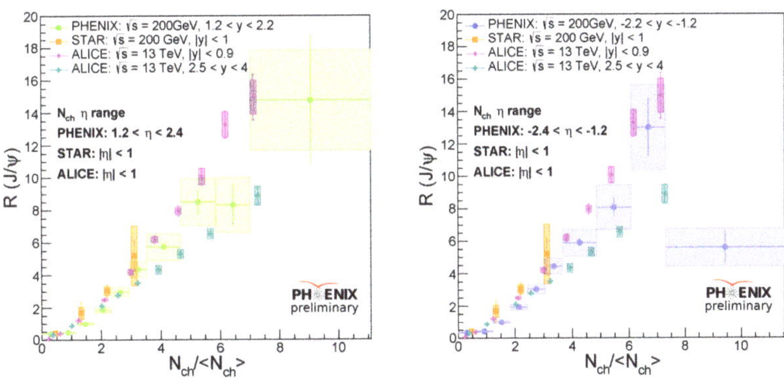

Figure 4. The J/ψ normalized yield as a function of normalized charged particle multiplicity of PHENIX, STAR [41], and ALICE [39] are all shown above. To make our results comparable with STAR and ALICE data, we present the measurements where J/ψ and tracklets are both in the north (green, **left**) and the south (blue, **right**). STAR uses the RHIC $p + p$ data at $\sqrt{s} = 200$ GeV to reconstruct J/ψ from the dielectron channel and the charged particle multiplicity, both measured at the mid-rapidity region $|y| < 1$ (orange). ALICE carries out both middle-rapidity $|y| < 0.9$ with the SPD tracklets and forward-rapidity $2.5 < y < 4$ with the V0 tracklets from $J/\psi \to e^+ e^-$ channel. All these results include daughter lepton tracks from J/ψ decay in the event multiplicity.

Finally, we compare our data with the PYTHIA 8 simulations with Monash and Detroit tunes including, and not including, the MPI effect shown in Figure 5.

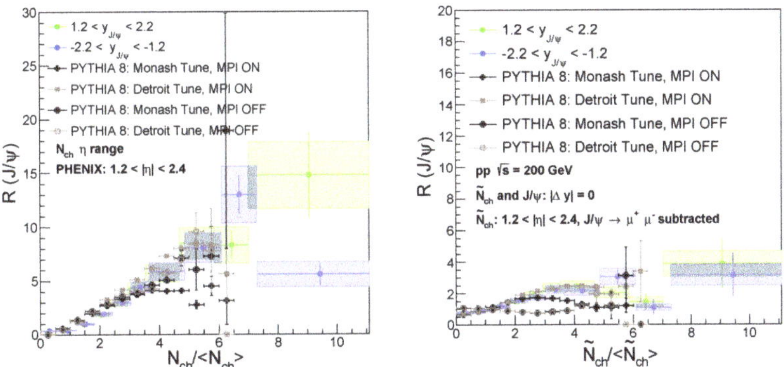

Figure 5. In the figure, the same muon arms and FVTX directions are used for J/ψ and charged particle tracklet reconstructions. J/ψ in the forward (green) and backward (blue) rapidities, and PYTHIA 8 simulations without dimuon subtraction (**left**) and with dimuon subtraction (**right**), are shown above. The PYTHIA 8 simulations present four different combinations, Monash Tune and Detroit Tune, with the MPI option turned on and off, to directly study MPI using the data.

In PYTHIA 8 $p + p$ simulations, we set up the $c\bar{c}$ event with a large $\hat{p}_T$ for J/ψ production and used general inelastic hadronic collisions to model MB events. Because it is unlikely to generate events with high multiplicities, our simulation only covers up to

$N_{ch}/\langle N_{ch}\rangle \simeq 6$ and has large statistical uncertainties at high multiplicities. Nonetheless, PYTHIA 8 simulations with different setups diverge at high multiplicities. PYTHIA 8, when using the Detroit Tune and turning on the MPI effect, can best describe the data. Hence, the MPI effect is significant for J/ψ production in $p+p$ collisions at RHIC, particularly in the high-multiplicity region.

6. Summary

We have reported the measurement of J/ψ normalized yield as a function of normal charged particle multiplicity with PHENIX Run 2015 $p+p$ collisions at $\sqrt{s}=200$ GeV. The J/ψ is reconstructed from the dimuon channel with the PHENIX muon arms in the forward rapidity. The charged particle tracklets are reconstructed with FVTX and SVX detectors. Our results are presented in different combinations of J/ψ with $\langle p_T\rangle \sim 1.7$ GeV/c and $1.2 < |y| < 2.2$ and charged particles at $1.2 < |\eta| < 2.4$ for FVTX and $|y| < 1$ for SVX up to approximately 10 units of normalized event multiplicity. The J/ψ normalized yield beyond linear scaling is observed when the J/ψ and charged particles are both measured at the same rapidity. The enhancement of J/ψ production becomes more pronounced at high event multiplicities, which could possibly be explained by MPI. The J/ψ normalized yield decreases significantly, as the rapidity gap between the J/ψ and the charged particles increases. After subtracting the dimuon contributions from the event multiplicity when the J/ψ and the charged particles point in the same rapidity direction, the results become consistent with the results where J/ψ and charged particles are produced in opposite rapidity directions, which hints at the insignificance of the final-state comover effects for J/ψ production in $p+p$ collisions.

Our forward J/ψ results lie systematically below the STAR measurement in the middle rapidity and in between the ALICE data in the forward and middle rapidities. We notice that J/ψ produced in the middle rapidity is generally below that of the forward rapidity within the same normalized charged-particle multiplicity. This allows for us to probe the parton distribution function in different phase-space regions. Finally, through the comparison of our data with PYTHIA 8 simulation using the Detroit and Monash Tunes with MPI options turned on and off, we found that the Detroit Tune with MPI on best describes our data. Hence, the MPI contribution should be included in order to precisely describe J/ψ production in $p+p$ collisions at RHIC, especially in high-multiplicity events.

To investigate the possibility of J/ψ production from jet fragmentation, we plan to look at our results in different J/ψ p_T regions. We expect J/ψ production from jets to be more likely at a high p_T. This study is currently ongoing. However, because of the limited statistics, particularly for $p_T > 3$ GeV/c, we may not achieve sufficient precision to conclude the possible J/ψ production from jet fragmentation in $p+p$ collisions at RHIC.

We are also carrying out J/ψ production in $p+Au$ collisions to test CGC calculations. In addition, the ongoing measurement of $\psi(2S)/J/\psi$ ratio in $p+p$ collisions will help us to understand charmonium hadronization. Many novel and exciting physics results regarding charmonium production in different collision systems with PHENIX data are coming in the near future.

Funding: This work was supported by Los Alamos National Laboratory Laboratory Directed Research and Development (LDRD) gran number (20220698PRD1). The APC was funded by Knowledge Unlatched (a Wiley brand).

Data Availability Statement: The data in this work come from the PHENIX experiment. It is not available to the public at this time.

Acknowledgments: We would like to thank Cesar da Silva and Ming Liu for their suggestions to improve the analysis and interpret the physics messages. We also appreciate the 2023 Zimányi Winter School organizers for their cordial reception and clarification of all questions related to conference registration and online presentations. Particularly, we are indebted to Máté Csanád for the invitation to contribute this paper as part of the Zimányi School Special Issue.

Conflicts of Interest: The authors declare no conflict of interest.

References

1. Aubert, J.J.; Becker, U.; Biggs, P.J.; Burger, J.; Chen, M.; Everhart, G.; Goldhagen, P.; Leong, J.; McCorriston, T.; Rhoades, T.G.; et al. Experimental Observation of a Heavy Particle. *J. Phys. Rev. Lett.* **1974**, *33*, 1404–1406. [CrossRef]
2. Augustin, J.E.; Boyarski, A.M.; Breidenbach, M.; Bulos, F.; Dakin, J.T.; Feldman, G.J.; Fischer, G.E.; Fryberger, D.; Hanson, G.; Jean-Marie, B.; et al. Discovery of a Narrow Resonance in e^+e^- Annihilation. *Phys. Rev. Lett.* **1974**, *33*, 1406–1408. [CrossRef]
3. Glashow, S.L.; Iliopoulos, J.; Maiani, L. Weak Interactions with Lepton–Hadron Symmetry. *Phys. Rev. D* **1970**, *2*, 1285–1292. [CrossRef]
4. Cacciari, M.; Greco, M.; Mangano, M.L.; Petrelli, A. Charmonium Production at the Tevatron. *Phys. Lett. B* **1995**, *356*, 553–560. [CrossRef]
5. Zhao, X.; Ralf, R. Forward and midrapidity charmonium production at RHIC. *Eur. Phys. J. C* **2009**, *62*, 109–117. [CrossRef]
6. Stahl, A.G. Charmonium production in pp, pPb and PbPb collisions with CMS. *J. Phys. Conf. Ser.* **2017**, *832*, 012031. [CrossRef]
7. Collins, J.C.; Soper, D.E.; Sterman, G. Factorization of Hard Processes in QCD. *Adv. Ser. Direct. High Energy Phys.* **1989**, *5*, 1–91.
8. Wong, C.Y. *Introduction to High-Energy Heavy-Ion Collisions*; World Scientific: Singapore, 1994; 516p.
9. Feynman, R.P. The behavior of hadron collisions at extreme energies. *Conf. Proc. C* **1969**, *33*, 237–258.
10. Soper, D.E. Parton distribution functions. *Nucl. Phys. B Proc. Suppl.* **1997**, *53*, 69–80. [CrossRef]
11. Dias de Deus, J.; Pajares, C. String Percolation and the Glasma. *Phys. Lett. B* **2011**, *695*, 211–213. [CrossRef]
12. Armesto, N.; Braun, M.A.; Ferreiro, E.G.; Pajares, C. Percolation Approach to Quark-Gluon Plasma and J/ψ Suppression. *Phys. Rev. Lett.* **1996**, *77*, 3736 [CrossRef]
13. Gelis, F.; Iancu, E.; Jalilian-Marian, J.; Venugopalan, R. The Color Glass Condensate. *Annu. Rev. Nucl. Part. Sci.* **2010**, *60*, 463–489. [CrossRef]
14. Ma, Y.-Q.; Venugopalan, R. Comprehensive Description of J/ψ Production in Proton-Proton Collisions at Collider Energies. *Phys. Rev. Lett.* **2013**, *113*, 192301. [CrossRef]
15. Braun, M.A.; Del Moral, F.; Pajares, C. Percolation of strings and the first RHIC data on multiplicity and transverse momentum distributions. *Phys. Rev. C* **2002**, *65*, 024907. [CrossRef]
16. Shao, H.S. Initial state radiation effects in inclusive J/ψ production at B factories. *J. High Energ. Phys.* **2014**, *2014*, 182. [CrossRef]
17. Buonocore, L.; Nason, P.; Tramontano, F. Heavy quark radiation in NLO+PS POWHEG generators. *Eur. Phys. J. C* **2018**, *78*, 151. [CrossRef]
18. Catani, S.; Grazzini, M.; Torre, A. Transverse-momentum resummation for heavy-quark hadroproduction. *Nucl. Phys. B* **2014**, *890*, 518–538 [CrossRef]
19. Berger, E.L.; Meng, R. Transverse momentum distributions for heavy quark pairs. *Phys. Rev. D* **1994**, *49*, 3248–3260. [CrossRef] [PubMed]
20. Anastasiou, C.; Dixon, L.; Melnikov, K.; Petriello, F. High precision QCD at hadron colliders: Electroweak gauge boson rapidity distributions at NNLO. *Phys. Rev. D* **2004**, *69*, 094008 [CrossRef]
21. Blok, B.; Dokshitzer, Y.; Frankfurt, L.; Strikman, M. pQCD physics of multiparton interactions. *Eur. Phys. J. C* **2012**, *72*, 1963. [CrossRef]
22. Bartalini, P.; Berger, E.L.; Blok, B.; Calucci, G.; Corke, R.; Diehl, M.; Dokshitzer, Y.; Fano, L.; Frankfurt, L.; Gaunt, J.R.; et al. Multi-Parton Interactions at the LHC. *arXiv* **2011**, arXiv:1111.0469.
23. Soto, J. Overview of Non-Relativistic QCD. *Eur. Phys. J. A* **2007**, *31*, 705–710. [CrossRef]
24. Berger, E.L.; Jones, D. Inelastic photoproduction of J/ψ and Y by gluons. *Phys. Rev. D* **1981**, *23*, 1521. [CrossRef]
25. Bain, R.; Dai, L.; Hornig, A.; Leibovich, A.K.; Makris, Y.; Mehen, T. Analytic and Monte Carlo Studies of Jets with Heavy Mesons and Quarkonia. *JHEP* **2016**, *6*, 121 [CrossRef]
26. Cho, P.L.; Leibovich, A.K. Color octet quarkonia production. *Phys. Rev. D* **1996**, *53*, 150–162. [CrossRef] [PubMed]
27. Adare, A. et al. [PHENIX Collaboration]. Transverse momentum dependence of J/ψ polarization at midrapidity in $p + p$ collisions at $\sqrt{s} = 200$ GeV. *Phys. Rev. D* **2010**, *82*, 012001 [CrossRef]
28. Ma, Y.-Q.; Vogt, R. Quarkonium production in an improved color evaporation model. *Phys. Rev. D* **2016**, *94*, 114029. [CrossRef]
29. Andronic, A.; Braun-Munzinger, P.; Köhler, M.K.; Redlich, K.; Stachel, J. Transverse momentum distributions of charmonium states with the statistical hadronization model. *Phys. Lett. B* **2019**, *797*, 134836 [CrossRef]
30. Kotko, P.; Motyka, L.; Stasto, A. Color Reconnection Effects in J/ψ Hadroproduction. *arXiv* **2023**, arXiv:2303.13128.
31. Crkovska, J. Study of the J/ψ Production in pp Collisions at $\sqrt{s_{NN}} = 5.02$ TeV and of the J/ψ Production Multiplicity Dependence in p-Pb Collisions at $\sqrt{s_{NN}} = 8.16$ TeV with ALICE at the LHC. High Energy Physics-Experiment. Ph.D. Thesis, Université Paris-Saclay, Paris, France, 2018.
32. Bahmani, M.; Kikoła, D.; Kosarzewski, L. A technique to study the elastic and inelastic interaction of quarkonium with hadrons using femtoscopic correlations. *Eur. Phys. J. C* **2021**, *81*, 305. [CrossRef]
33. Werner, K.; Guiot, B.; Karpenko, I.; Pierog, T. Analysing radial flow features in p–Pb and pp collisions at several TeV by studying identified particle production in EPOS3. *Phys. Rev. C* **2014**, *89*, 064903. [CrossRef]
34. Bailung, Y. Measurement of D-meson production as a function of charged-particle multiplicity in proton–proton collisions at $\sqrt{s} = 13$ TeV with ALICE at the LHC. *PoS LHCP* **2021**, *2021*, 190.
35. Weber, S.G.; Dubla, A.; Andronic, A.; Morsch, A. Elucidating the multiplicity dependence of J/ψ production in proton–proton collisions with PYTHIA8. *Eur. Phys. J. C* **2019**, *79*, 36. [CrossRef]

36. Norrbin, E.; Sjöstrand, T. Production mechanisms of charm hadrons in the string model. *Phys. Lett. B* **1998**, *442*, 407–416. [CrossRef]
37. Acharya, S. et al. [ALICE Collaboration]. Inclusive J/ψ production at mid-rapidity in pp collisions at $\sqrt{s}$ = 5.02 TeV. *JHEP* **2019**, *10*, 84.
38. Abelev, B. et al. [ALICE Collaboration]. J/ψ polarization in pp collisions at $\sqrt{s}$ = 7 TeV. *Phys. Rev. Lett.* **2012**, *108*, 082001. [CrossRef]
39. Acharya, S. et al. [ALICE Collaboration]. Multiplicity dependence of J/ψ production at midrapidity in pp collisions at $\sqrt{s}$ = 13 TeV. *Phys. Lett. B* **2020**, *810*, 135758. [CrossRef]
40. Acharya, S. et al. [ALICE Collaboration]. Forward rapidity J/ψ production as a function of charged-particle multiplicity in pp collisions at $\sqrt{s}$ = 5.02 and 13 TeV. *JHEP* **2022**, *6*, 15
41. Adam, J. et al. [STAR Collaboration]. J/ψ production cross section and its dependence on charged-particle multiplicity in $p + p$ collisions at $\sqrt{s}$ = 200 GeV. *Phys. Lett. B* **2018**, *786*, 87–93. [CrossRef]
42. Thakur, D.; De, S.; Sahoo, R.; Dansana, S. Role of multiparton interactions on J/ψ production in $p + p$ collisions at LHC energies. *Phys. Rev. D* **2018**, *97*, 094002. [CrossRef]
43. Egede, U.; Hadavizadeh, T.; Singla, M.; Skands, P.; Vesterinen, M. The role of multi-parton interactions in doubly-heavy hadron production. *Eur. Phys. J. C* **2022**, *82*, 773. [CrossRef]
44. Gotsman, E.; Levin, E. High energy QCD: Multiplicity dependence of quarkonia production. *Eur. Phys. J. C* **2021**, *81*, 99. [CrossRef]
45. Sjöstr, T.; Ask, S.; Christiansen, J.R.; Corke, R.; Desai, N.; Ilten, P.; Mrenna, S.; Prestel, S.; Rasmussen, C.O.; Skands, P.Z. An Introduction to PYTHIA 8.2. *Comput. Phys. Commun.* **2015**, *191*, 159–177. [CrossRef]
46. Aguilar, M.R.; Chang, Z.; Elayavalli, R.K.; Fatemi, R.; He, Y.; Ji, Y.; Kalinkin, D.; Kelsey, M.; Mooney, I.; Verkest, V. PYTHIA 8 underlying event tune For RHIC energies. *Phys. Rev. D* **2022**, *105*, 016011. [CrossRef]
47. Skands, P.; Carrazza, S.; Rojo, J. Tuning PYTHIA 8.1: The Monash 2013 Tune. *Eur. Phys. J. C* **2014**, *74*, 3024. [CrossRef]
48. Butenschoen, M.; Kniehl, B.A. J/ψ polarization at Tevatron and LHC: Nonrelativistic-QCD factorization at the crossroads. *Phys. Rev. Lett.* **2012**, *108*, 172002 [CrossRef] [PubMed]
49. Canelli, F. et al. [CMS Collaboration]. Study of J/ψ production inside jets in pp collisions at $\sqrt{s}$ = 8 TeV. *Phys. Lett. B* **2020**, *804*, 135409. [CrossRef]
50. Bruno, G. et al. [CMS Collaboration]. Fragmentation of jets containing a prompt J/ψ meson in $PbPb$ and pp collisions at $\sqrt{s_{NN}}$ = 5.02 TeV. *Phys. Lett. B* **2022**, *825*, 136842. [CrossRef]
51. Aaij, R. et al. [LHCb Collaboration]. Study of J/ψ Production in Jets. *Phys. Rev. Lett.* **2017**, *118*, 192001. [CrossRef]
52. Yuan, F.; Chao, K.-T. Diffractive J/ψ production as a probe of the gluon component in the Pomeron. *Phys. Rev. D* **1998**, *57*, 5658–5662. [CrossRef]
53. Sahoo, R. et al. [ALICE Collaboration]. Measurement of inelastic, single- and double-diffraction cross sections in proton–proton collisions at the LHC with ALICE. *Eur. Phys. J. C* **2013**, *73*, 2456 [CrossRef] [PubMed]
54. Aad, G. et al. [ATLAS Collaboration]. Measurement of differential cross sections for single diffractive dissociation in $\sqrt{s}$ = 8 TeV pp collisions using the ATLAS ALFA spectrometer. *JHEP* **2020**, *10*, 182.
55. Adcox, K. et al. [PHENIX Collaboration]. PHENIX detector overview. *Nucl. Instrum. Methods Phys. Res. Sec. A* **2003**, *499*, 469. [CrossRef]
56. Adare, A. et al. [PHENIX Collaboration]. Measurements of double-helicity asymmetries in inclusive J/ψ production in longitudinally polarized $p + p$ collisions at $\sqrt{s}$ = 510 GeV. *Phys. Rev. D* **2016**, *94*, 112008 [CrossRef]
57. Adcox, K. et al. [PHENIX Collaboration]. Formation of dense partonic matter in relativistic nucleus–nucleus collisions at RHIC: Experimental evaluation by the PHENIX Collaboration. *Nucl. Phys. A* **2005**, *757*, 184–283 [CrossRef]
58. Brooks, M.L. Physics Potential of and Status Report on the PHENIX Experiment. In Proceedings of the ICPAQGP, Jaipur, India, 17–21 March 1997; pp. 326–334
59. Aidala, C. et al. [PHENIX Collaboration]. The PHENIX Forward Silicon Vertex Detector. *Nucl. Instrum. Meth. A* **2014**, *755*, 44–61. [CrossRef]
60. Akiba, Y. Proposal for a Silicon Vertex Tracker (VTX) for the PHENIX Experiment. DOE Contact Number DE-AC02-98CH10886, United States: N. p. 2004. Available online: https://www.bnl.gov/isd/documents/28627.pdf (accessed on 30 May 2023).
61. Akikawa, H. et al. [PHENIX Collaboration]. PHENIX Muon Arms. *Nucl. Instrum. Meth. A* **2003**, *499*, 537–548. [CrossRef]
62. Newby, J. Single and Dimuon Reconstruction Performance of the PHENIX South Muon Arm. Poster for Quark Matter 2002. Available online: https://www.phenix.bnl.gov/WWW/publish/rjnewby/QM2002Poster.pdf (accessed on 28 May 2023).
63. Aphecetche, L. et al. [PHENIX Collaboration]. PHENIX calorimeter. *Nucl. Instrum. Meth. A* **2003**, *499*, 521–536. [CrossRef]
64. Akiba, Y. Ring imaging Cherenkov detector of PHENIX experiment at RHIC. *Nucl. Instrum. Meth. A* **1999**, *433*, 143–148. [CrossRef]
65. Allen, M. et al. [PHENIX Collaboration]. PHENIX inner detectors. *Nucl. Instrum. Meth. A* **2003**, *499*, 549–559. [CrossRef]
66. Adler, C.; Denisov, A.; Garcia, E.; Murray, M.; Stroebele, H.; White, S. The RHIC zero degree calorimeter. *Nucl. Instrum. Meth. A* **2001**, *470*, 6499. [CrossRef]
67. Belikov, S.; Hill, J.; Lajoie, J.; Skank, H.; Sleege, G. PHENIX trigger system. *Nucl. Instrum. Meth. A* **2002**, *494*, 541–547. [CrossRef]
68. Lajoie, J.G.; Woh, F.K.; Hill, J.C.; Petridis, A.; Wood, L.; Cook, K.; Plagge, T.; Skank, H.D.; Thomas, W.D.; Sleege, G.A. The PHENIX Level-1 trigger system. In Proceedings of the 999 IEEE Conference on Real-Time Computer Applications in Nuclear Particle and Plasma Physics. 11th IEEE NPSS Real Time Conference. Conference Record (Cat. No.99EX295), Santa Fe, NM, USA, 14–18 June 1999; p. 517.

69. Adler, S.S.; Allen, M.; Alley, G.; Amirikas, R.; Arai, Y.; Awes, T.C.; Barish, K.N.; Barta, F.; Batsouli, S.; Belikov, S.; et al. PHENIX on-line systems. *Nucl. Instrum. Meth. A* **2003**, *499*, 560–592. [CrossRef]
70. Verkerke, W.; Kirkby, D. The RooFit toolkit for data modeling. In *Statistical Problems in Particle Physics, Astrophysics and Cosmology*; Imperial College Press: Oxford, UK, 2006; pp. 186–189.
71. Acharya, U.A. et al. [PHENIX Collaboration]. J/ψ and $\psi(2S)$ production at forward rapidity in $p + p$ collisions at $\sqrt{s} = 510$ GeV. *Phys. Rev. D* **2020**, *101*, 052006. [CrossRef]

Communication

Event-Shape-Dependent Analysis of Charm–Anticharm Azimuthal Correlations in Simulations

Anikó Horváth [1,2,*], Eszter Frajna [1,3] and Róbert Vértesi [1,*]

[1] Wigner Research Centre for Physics, P.O. Box 49, H-1525 Budapest, Hungary
[2] Faculty of Science, Eötvös Loránd University, 1/A Pázmány Péter Sétány, H-1111 Budapest, Hungary
[3] Faculty of Natural Sciences, Budapest University of Technology and Economics, 3 Műegyetem rkp., H-1111 Budapest, Hungary
* Correspondence: aniko.horvath@wigner.hu (A.H.); vertesi.robert@wigner.hu (R.V.)

Abstract: In high-energy collisions of small systems, by high-enough final-state multiplicities, a collective behaviour is present that is similar to the flow patterns observed in heavy-ion collisions. Recent studies connect this collectivity to semi-soft vacuum-QCD processes. Here we explore QCD production mechanisms using angular correlations of heavy flavour using simulated proton-proton collisions at $\sqrt{s} = 13$ TeV with the PYTHIA8 Monte Carlo event generator. We demonstrate that the event shape is strongly connected to the production mechanisms. Flattenicity, a novel event descriptor, can be used to separate events containing the final-state radiation from the rest of the events.

Keywords: high-energy collisions; LHC; multiple parton interactions

Citation: Horváth, A.; Frajna, E.; Vértesi, R. Event-Shape-Dependent Analysis of Charm–Anticharm Azimuthal Correlations in Simulations. *Universe* **2023**, *9*, 308. https://doi.org/10.3390/universe9070308

Academic Editors: Máté Csanád, Péter Kovács, Sándor Lökös, Dániel Kincses and Lorenzo Iorio

Received: 19 May 2023
Revised: 15 June 2023
Accepted: 22 June 2023
Published: 27 June 2023

1. Introduction

In high-energy heavy-ion collisions, a strongly interacting quark–gluon plasma (QGP) is created, which was found to behave as an almost-perfect fluid [1,2]. Surprisingly, a similar collective behaviour was observed in small (proton–proton and proton–nucleus) collisions with high final-state multiplicity [3,4]. Whether or not QGP is created in these smaller collision systems is still an open question today. Recent works suggest that vacuum-QCD processes on the soft-hard boundary, such as multiple parton interactions (MPI) with colour reconnection, are able to generate the collective patterns that are observed in such systems [5,6].

Heavy quarks are mostly created in the early stages of the collision, in perturbatively accessible quantum chromodynamics (QCD) processes; a heavy quark can be created from a pair of gluons or light quarks by flavour creation (FLC), a gluon splitting into the quark–antiquark pair (GSP), or through flavour excitation (FLX) [7,8]. Moreover, they may interact in semi-hard processes and participate in the formation of the underlying event [9]. Further insight to the connection of the hard process and the underlying event can be gained by the differential exploration of events with respect to event-shape variables. While, traditionally, the final-state multiplicity is used to categorize events by activity, other recently introduced event-shape variables, such as transverse sphericity and flattenicity, are sensitive to event topology and have a more direct connection to multiple parton interactions and the emerging collective patterns [10,11].

Angular correlation measurements are sensitive probes of parton production and fragmentation down to low momenta where jet reconstruction is problematic in a rich final-state environment. The current experimental precision enables the exploration of heavy-flavour hadron correlations, which provides information about heavy-flavour fragmentation but very little insight to their creation. A recent ALICE measurement did not find event-activity dependence in the angular correlation of D^0 mesons to charged hadrons [12]. The experimental possibilities will be significantly extended with the arrival of LHC Run3 data, where heavy-flavour–heavy-flavour correlations will be possible to reconstruct. The full potential of the forthcoming LHC Run3+Run4 data, however, can be exploited with measurements that are differential in event-activity or event-shape.

In this work, we explore the azimuthal correlations of charm–anticharm quark pairs in Monte Carlo (MC) simulations, in terms of different event-activity variables. We use MC information to explore the connection of the partonic processes with the emerging final state and propose an experimental method to separate them.

2. Methods

We analyzed proton–proton collisions simulated at $\sqrt{s} = 13$ TeV center-of-mass energy with the PYTHIA8 [13] (version 8.308) MC event generator. In PYTHIA, the initial leading-order production process is amended by other partonic processes, initial and final-state radiation (ISR and FSR, respectively), as well as multiple parton interactions. By enabling these processes one by one, we simulated just the initial hard process (all off), the initial hard scattering and multiple parton interactions added (MPI on), with the initial-state gluon radiations included (MPI, ISR on), and with all the previous processes and the final-state radiations also enabled in the events (all on). With each setting, 10 million events were simulated.

We computed the azimuthal correlations of charm quarks with anticharm quarks. Only those quarks that directly hadronised were considered, to avoid multiple counting of the same quark. As an arbitrary choice, charm quarks were used as trigger particles and anticharm quarks as associated particles. Both the charm and anticharm quarks were required to fall within the $|y| < 1.44$ rapidity window. The distribution of the azimuthal angles between each pair ($\Delta\varphi$) was calculated to explore the event structures. In some of our results, we separately analysed the soft and hard production of the c–c̄ pairs by requiring their momenta to fall below or above $p_T = 4$ GeV/c.

We categorized the simulated events by the parton-level production process in which the trigger particles (charm quarks) were created. The flavour creation, flavour excitation, and gluon splitting processes were separated in the simulations by tracing back the trigger charm quark to the first charm quark in the ancestry line and examining the status code of its parents. If this quark only had gluon parents that are not connected to the hardest process, the pair was considered to be a result of gluon splitting. In the case when the charm quark had gluon parents, and it was an incoming particle in the hardest (sub)process, it was categorised as flavour excitation. When both of the parents were incoming light quarks or gluons in the hardest process, and created a charm quark, then the pair was categorised to come from pair creation. In a small number of events, where the charm did not originate from the hardest process, this method was not able to categorize the production process. These charm quarks, coming from subsequent soft processes, were added to the gluon splitting group.

The c–c̄ azimuthal correlations were categorized with respect to charged hadron multiplicity, transverse sphericity and flattenicity. In all three cases, event variable cuts were applied to separate the top and bottom thirds of the sample. Charged hadron multiplicity (N_{ch}) was defined as the number of final charged hadrons with a transverse momentum of $p_T > 0.15$ GeV in the central pseudorapidity range $|\eta| < 1$. The low-activity range was taken as $N_{\mathrm{ch}} \leq 21$, and the high-activity range as $N_{\mathrm{ch}} \geq 38$. Transverse sphericity is calculated by finding the unit vector $\vec{n}$ that minimalises the expression

$$S_0 = \frac{\pi^2}{4} \left(\frac{\Sigma_i |\vec{p_{T_i}} \times \vec{n}|}{\Sigma_i |\vec{p_{T_i}}|} \right)^2,$$

where the sum runs over all final-state charged particles with $p_T > 0.15$ GeV and $|\eta| < 1$. With this definition, transverse sphericity is $0 < S_0 < 1$, where $S_0 \approx 0$ events have a "pencil-like" back-to-back dijet topology, and $S_0 \approx 1$ events are isotropic [11]. For the low-S_0 range $S_0 < 0.53$ was used, and for the high-S_0 range $S_0 > 0.70$ was used. Flattenicity (ρ) is a recently introduced event-shape variable that describes the distribution of transverse momenta over the azimuthal angle–pseudorapidity plane, which is capable of selecting "hedgehog-like" events without any discernible jetty structure in high-multiplicity PP collisions [10]. To calculate this, one has to divide the φ–η plane into equal sections, and take the average transverse momenta of the charged particles in each of them. Flattenicity is the relative standard deviation of the average momentum in a cell:

$$\rho = \frac{\sigma_{p_{\mathrm{T}}^{\mathrm{cell}}}}{\langle p_{\mathrm{T}}^{\mathrm{cell}} \rangle}.$$

Larger ρ implies a greater jetty event, and around $\rho \approx 1$ is where at least one jet can be seen [10]. The low flattenicity range was $\rho < 1.00$, and the high flattenicity range was $\rho > 1.28$.

3. Results

First, we compare c–c̄ azimuthal correlations for the high and low values of the event-shape variables, as well as without selection for the variables. The distributions are normalised with the number of triggers N_{trig}, as well as with the integral of the distributions for the given range divided by the integral of the distribution without selection for the event-shape variable, I_{class}.

Figure 1 shows the azimuthal correlation of c–c̄ pairs in the low and high charged hadron multiplicity ranges, as well as without selection for N_{ch}. We observe that for lower multiplicities the away-side peak of the correlation is sharper than at higher multiplicities. This can be explained by considering that low-multiplicity events are produced more often from simpler back-to-back correlations, while events with more complicated underlying physics tend to have higher multiplicities.

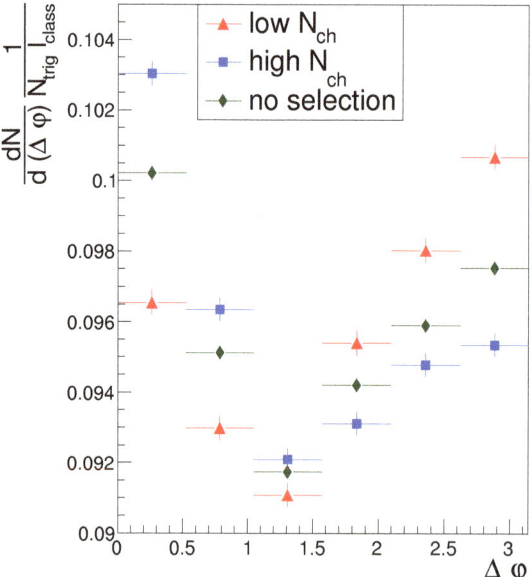

Figure 1. The azimuthal correlation of c–c̄ pairs in the low and high charged hadron multiplicity ranges, as well as without selection for N_{ch}, normalised by the number of triggers and the integral of the interval.

Figure 2 shows the azimuthal correlation of c–c̄ pairs in the low and high transverse sphericity ranges, as well as without selection for S_0. We observe that events with low sphericity, which tend to be more jetty, result in a stronger correlation, and the more isotropic high S_0 range selects more random correlation.

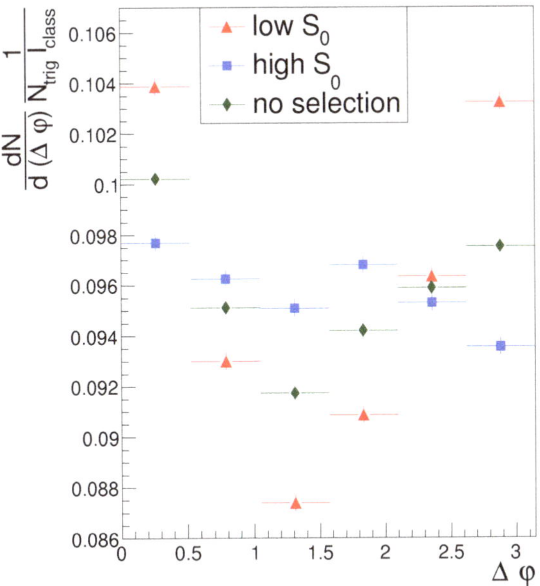

Figure 2. The azimuthal correlation of c–c̄ pairs in the low and high transverse sphericity ranges, as well as without selection for S_0, normalised by the number of triggers and the integral of the interval.

Figure 3 shows the azimuthal correlation of c–c̄ pairs in the low and high flattenicity cuts. We can see that flattenicity highlights the correlation peaks, as high ρ gives both a sharper near-side and away-side peak, similarly to the observations made from the S_0 correlations.

In Figure 4, we see the different trigger quark creation processes in the same flattenicity intervals, normalised by the number of triggers. Both the trigger and associated particle were required to have a transverse momentum $p_T > 4$ GeV/c. The dominant creation process in this high-p_T range is gluon splitting, which gives the largest contribution to the near-side peak, while adding to the away-side peak as well. Flavour creation gave a sharp away-side peak, and, though less visible, flavour excitation also adds mainly to the away-side peak. We can see that the flattenicity cut separates the peaks of gluon splitting (high ρ GSP) from mostly random correlation (low ρ GSP), which can be attributed to flattenicity geometrically separating isotropic events from jetty events. We can also note that above a flat baseline of mostly gluon splitting, the away-side peak in the low ρ range arises mainly due to flavour creation. On the other hand, the near-side peak in the high ρ range is created by gluon splitting. We can see that flattenicity has the ability to geometrically separate these different creation processes via azimuthal correlation of c–c̄ quarks. This could provide an opportunity to experimentally separate different QCD production processes by observing the distribution of final-state particles through correlations of heavy-flavour jets.

Figure 5 shows the contributions of different PYTHIA8 parton-level processes for high and low ρ values (top and bottom rows, respectively), both in the $p_T < 4$ GeV/c and $p_T > 4$ GeV/c momentum ranges separately (left and right panels). The c–c̄ pairs created back-to-back in the initial leading-order production result in an away-side peak. Multi-parton interactions and initial-state radiations also add to the away-side peak, while contributing to the baseline as well. The near-side peak arises from final-state radiations. Contrasting the two rows, we see that the flattenicity cut isolates most of the final-state radiation from multi-parton interaction and initial-state radiation. We also observe that the correlation peaks are stronger in the high ρ events, that on average correspond to more jetty topologies. As expected, higher transverse momenta also results in less baseline.

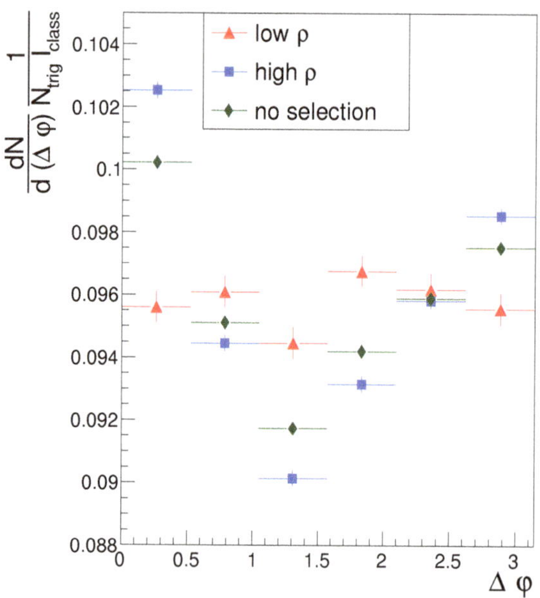

Figure 3. The azimuthal correlation of c–c̄ in the low and high flattenicity ranges, as well as without selection for ρ, normalised by the number of triggers and the integral of the interval.

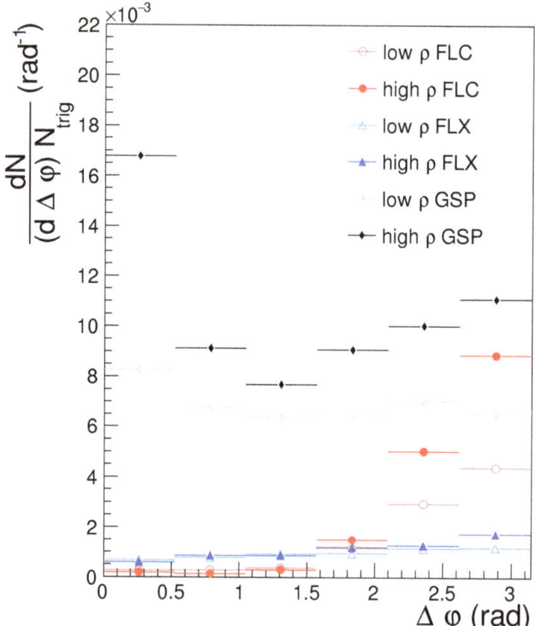

Figure 4. The azimuthal correlation of c–c̄ pairs separated by the different parton-level quark creation processes: flavour creation (in red), flavour excitation (in blue) and gluon splitting (in grey), in the c and c̄ momentum range of $p_T > 4\,\text{GeV}/c$, in the low and high ρ ranges (empty and full markers, respectively).

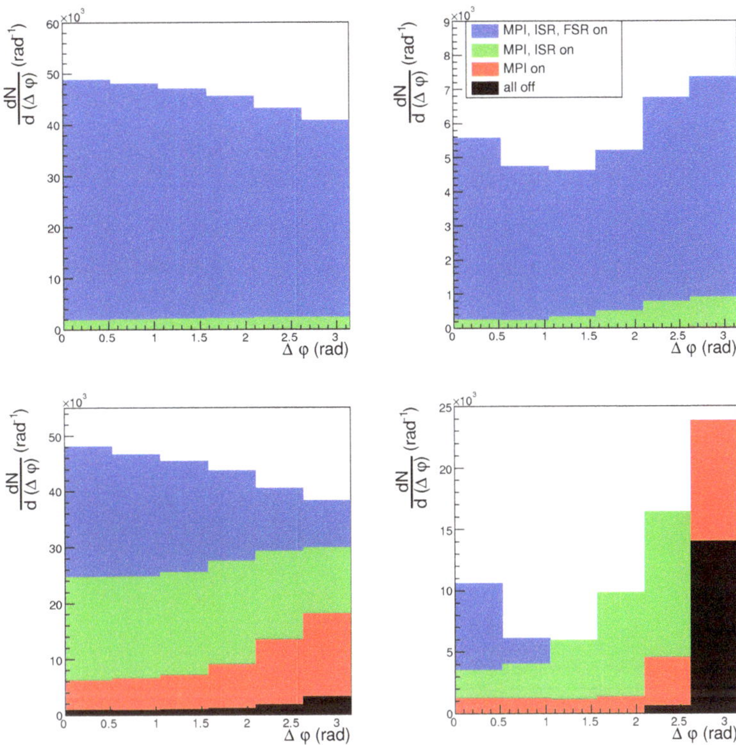

Figure 5. The azimuthal correlation of c–c̄ pairs where $p_T < 4\,\mathrm{GeV}/c$ and $p_T > 4\,\mathrm{GeV}/c$ (left and right columns, respectively), and the top row shows the low ρ range, and the bottom row shows the high ρ range. The different parton-level process settings are presented with different colours.

It is to be noted that although we use experimentally available rapidity and transverse momentum windows in the selection of the correlated pairs, we do not reconstruct the final state, and instead used Monte Carlo truth information to investigate the correlation at the parton level. This allows for the exploration of the connection between the partonic processes and the emerging final state, without having to deal with the effect of hadronization. In experiment, correlations of c–c̄ pairs may be accessible either through correlations of charmed hadrons (e.g., D^0–$\overline{D^0}$), or through the correlations of reconstructed pairs of jets containing a charmed hadrons. In the first case, the results will be influenced by jet fragmentation into hadrons, while in the latter case the the jet definition will affect the outcome (e.g., because small-angle parton pairs may be reconstructed as a single jet). Such future results may, therefore, be compared to simulations adapted to the specific experimental conditions, which may be the subject of a later study.

4. Conclusions

In this work, we explored the azimuthal correlation of charm quarks and antiquarks in PYTHIA8-simulated proton–proton collisions with respect to final-state charged hadron multiplicity, transverse sphericity, and flattenicity. We investigated the event-activity and event-shape dependent results in terms of different QCD heavy flavour creation processes, as well as parton-level processes. We observed that flattenicity is the most selective for the different QCD processes. By selecting events with low and high flattenicity, on a statistical basis we were able to differentiate between parton-level production processes, just by observing the event shape. Moreover, by selecting low-flattenicity events we can also differentiate c–c̄ pairs coming from events with final-state radiation.

Using the above-mentioned methods, it will be possible to select certain QCD processes in future heavy-quark (such as D^0–$\overline{D^0}$ or jet–jet) azimuthal correlation measurements in the LHC Run3 data. The results also outline a method for the detailed validation of heavy-flavour production models with data.

Author Contributions: Conceptualization, R.V.; methodology, R.V. and E.F.; software, A.H. and E.F.; validation, R.V.; formal analysis, A.H.; writing–original draft preparation, A.H.; writing—review and editing, R.V. and A.H.; visualization, A.H.; supervision, R.V.; project administration, R.V.; funding acquisition, R.V. All authors have read and agreed to the published version of the manuscript.

Funding: This work has been supported by the NKFIH grants OTKA FK131979 and K135515, as well as by the 2021-4.1.2-NEMZ_KI-2022-00007 project.

Data Availability Statement: Simulated data are available from the authors upon request.

Conflicts of Interest: The authors declare no conflict of interest.

References

1. Adcox, K.; Adler, S.S.; Afanasiev, S.; Aidala, C.; Ajitanand, N.N.; Akiba, Y.; Al-Jamel, A.; Alexander, J.; Amirikas, R.; Aoki, K.; et al. Formation of dense partonic matter in relativistic nucleus-nucleus collisions at RHIC: Experimental evaluation by the PHENIX collaboration. *Nucl. Phys. A* **2005**, *757*, 184–283. [CrossRef]
2. Adams, J.; Aggarwal, M.M.; Ahammed, Z.; Amonett, J.; Anderson, B.D.; Arkhipkin, D.; Averichev, G.S.; Badyal, S.K.; Bai, Y.; Balewski, J.; et al. Experimental and theoretical challenges in the search for the quark gluon plasma: The STAR Collaboration's critical assessment of the evidence from RHIC collisions. *Nucl. Phys. A* **2005**, *757*, 102–183. [CrossRef]
3. Khachatryan, V.; Sirunyan, A.M.; Tumasyan, A.; Adam, W.; Bergauer, T.; Dragicevic, M.; Erö, J.; Fabjan, C.; Friedl, M.; Frühwirth, R.; et al. Observation of Long-Range Near-Side Angular Correlations in Proton-Proton Collisions at the LHC. *J. High Energy Phys.* **2010**, *9*, 91. [CrossRef]
4. Acharya, S.; Adamová, D.; Adhya, S.P.; Adler, A.; Adolfsson, J.; Aggarwal, M.M.; Aglieri Rinella, G.; Agnello, M.; Agrawal, N.; Ahammed, Z.; et al. Investigations of Anisotropic Flow Using Multiparticle Azimuthal Correlations in pp, p-Pb, Xe-Xe, and Pb-Pb Collisions at the LHC. *Phys. Rev. Lett.* **2019**, *123*, 142301. [CrossRef] [PubMed]
5. Bierlich, C.; Gustafson, G.; Lönnblad, L. Collectivity without plasma in hadronic collisions. *Phys. Lett. B* **2018**, *779*, 58–63. [CrossRef]
6. Ortiz, A.; Bencedi, G.; Bello, H. Revealing the source of the radial flow patterns in proton–proton collisions using hard probes. *J. Phys. G* **2017**, *44*, 065001. [CrossRef]
7. Norrbin, E.; Sjostrand, T. Production and hadronization of heavy quarks. *Eur. Phys. J. C* **2000**, *17*, 137–161. [CrossRef]
8. Ilten, P.; Rodd, N.L.; Thaler, J.; Williams, M. Disentangling Heavy Flavor at Colliders. *Phys. Rev. D* **2017**, *96*, 054019. [CrossRef]
9. Vértesi, R.; Bencédi, G.; Misák, A.; Ortiz, A. Probing the interaction of semi-hard quarks and gluons with the underlying event in light- and heavy-flavor triggered proton-proton collisions. *Eur. Phys. J. A* **2021**, *57*, 301. [CrossRef]
10. Ortiz, A.; Paic, G. A look into the "hedgehog" events in pp collisions. *Rev. Mex. Fis. Suppl.* **2022**, *3*, 040911. [CrossRef]
11. Ortiz, A.; Paić, G.; Cuautle, E. Mid-rapidity charged hadron transverse spherocity in pp collisions simulated with Pythia. *Nucl. Phys. A* **2015**, *941*, 78–86. [CrossRef]
12. Acharya, S.; Adamová, D.; Adler, A.; Adolfsson, J.; Aglieri Rinella, G.; Agnello, M.; Agrawal, N.; Ahammed, Z.; Ahmad, S.; Ahn, S.U.; et al. Investigating charm production and fragmentation via azimuthal correlations of prompt D mesons with charged particles in pp collisions at $\sqrt{s} = 13$ TeV. *Eur. Phys. J. C* **2022**, *82*, 335. [CrossRef]
13. Sjostrand, T.; Mrenna, S.; Skands, P.Z. A Brief Introduction to PYTHIA 8.1. *Comput. Phys. Commun.* **2008**, *178*, 852–867. [CrossRef]

 *universe*

Article

Synthesis of Elements in Compact Stars in Pycnonuclear Reactions with Carbon Isotopes: Quasibound States vs. States of Zero-Points Vibrations

Sergei P. Maydanyuk [1,*], Gyorgy Wolf [1] and Kostiantyn A. Shaulsky [2]

[1] Wigner Research Centre for Physics, 1121 Budapest, Hungary; wolf.gyorgy@wigner.hu
[2] Institute for Nuclear Research, National Academy of Sciences of Ukraine, 03680 Kyiv, Ukraine; shaulskyi@kinr.kiev.ua
* Correspondence: sergei.maydanyuk@wigner.hu; Tel.: +36-20-554-5077

Abstract: (1) Purpose: Conditions of formation of compound nuclear systems needed for synthesis of heavy nuclei in pycnonuclear reactions in compact stars are studied on a quantum mechanical basis. (2) Methods: The method of multiple internal reflections is applied for pycnonuclear reactions in compact stars with new calculations of quasibound spectra and spectra of zero-point vibrations. (3) Results: Peculiarities of the method are analyzed for reaction with isotopes of Carbon. The developed method takes into account continuity and conservation of quantum flux (describing pycnonuclear reaction) inside the full spacial region of reaction, including the nuclear region. This gives the appearance of new states (called quasibound states) in which compound nuclear systems of Magnesium are formed with the largest probability. These states have not been studied yet in synthesis of elements in stars. Energy spectra of zero-point vibrations and spectra of quasibound states are estimated with high precision for reactions with isotopes of Carbon. For the first time, the influence of plasma screening on quasibound states and states of zero-point vibrations in pycnonuclear reactions has been studied. (4) Conclusions: The probability of formation of a compound nucleus in quasibound states in pycnonuclear reaction is essentially larger than the probability of formation of this system in states of zero-point vibrations studied by Zel'dovich and followers. Therefore, synthesis of Magnesium from isotopes of Carbon is more probable through the quasibound states than through the states of zero-point vibrations in compact stars. Energy spectra of zero-point vibrations are changed essentially after taking plasma screening into account. Analysis shows that from all studied isotopes of Magnesium, only ^{24}Mg is stable after synthesis at an energy of relative motion of 4.881 MeV of the incident nuclei ^{12}C.

Keywords: pycnonuclear reaction; compact star; neutron star; multiple internal reflections; coefficients of penetrability and reflection; fusion; quasibound state; energy of zero-point vibrations; compound nucleus; dense nuclear matter; tunneling

Citation: Maydanyuk, S.P.; Wolf, G.; Shaulsky, K.A. Synthesis of Elements in Compact Stars in Pycnonuclear Reactions with Carbon Isotopes: Quasibound States vs. States of Zero-Points Vibrations. *Universe* **2023**, *9*, 354. https://doi.org/10.3390/universe9080354

Academic Editors: Máté Csanád, Péter Kovács, Sándor Lökös and Dániel Kincses

Received: 4 June 2023
Revised: 22 July 2023
Accepted: 24 July 2023
Published: 29 July 2023

1. Introduction

The phenomenon of nuclear burning occurs in the cold and dense cores of white dwarfs [1] and crusts of neutron stars [2,3]. Such a phenomenon, known as a pycnonuclear reaction [4], is a reaction at sufficiently high densities in stars where zero-point vibrations of nuclei in the lattice sites lead to an essential increasing rate of formation of more heavy nuclei. Insight into this phenomenon was provided by Zel'dovich, who estimated zero-point energy as the energy of the ground state of the harmonic oscillator potential, which is formed near the middle point between two nuclei located in lattice sites [5]. Rates of reactions at such zero-point energies are calculated for some nuclei in compact stars [6].

Fusion is the key process in pycnonuclear reactions. In this process, a new nucleus with a larger mass is produced from the two closest nuclei in the lattice sites. This question was analyzed for reactions with nuclei of different charges and masses [7]. In that paper, the authors calculated the astrophysical *S*-factors for Carbon–Oxygen and Oxygen–Oxygen fusion reactions, wherein a microscopic basis was used. In Ref. [8], *S*-factors were calculated for 946 fusion reactions including stable and neutron-rich isotopes of C, O, Ne, and Mg at energies in the range of 2 to $\approx$18–30 MeV. Results in that paper can be converted to thermonuclear or pycnonuclear reaction rates to simulate stellar burning at high temperatures and nucleosynthesis in high-density environments. A large collection of astrophysical *S*-factors and their compact representation for isotopes of Be, B, C, N, O, F, Ne, Na, Mg, and Si were presented in Ref. [9]. Finally, a large database of *S*-factors was formed for about 5000 nonresonant fusion reactions. The structure of the multi-component matter (a regular lattice, a uniform mix, etc.) in these reactions, plasma screening [10], and rates of reactions in a wide range of temperatures and stellar densities [7,11] have been studied by many researchers.

It has been known that cross-sections of reactions are essentially changed after taking conservation of quantum fluxes into account in the internal region of the nuclear system [12–14]. This question has been studied for α decays of nuclei and captures of α-particles by nuclei. For example, nuclear processes during capture before fusion depend on the shape of the nuclear potential [13,14]. Such changes are controlled by additional independent parameters appearing from the fully quantum study. In the fully quantum study, different scenarios of capture (before fusion) can be modelled. Corresponding cross-sections are different by up to four times at the same beam energies of α-particles in experiments. Often, approaches used with the basis of WKB-approximation neglect these quantum phenomena. It is important to note that this dependence of cross-sections in the fully quantum study is not small. For example, it can be larger essentially than the inclusion of nuclear deformations to the calculation of cross-sections without such quantum parameters. Up to now, the method in Ref. [13] has been the most accurate for the description of experimental data for α-capture (this calculation is in Figure 3b in Ref. [14] for $\alpha + {}^{44}\text{Ca}$ in comparison with experimental data [15]).

In the fully quantum study, the accuracy of the determination of penetrability of the barrier and cross-section is about 10^{-14}, while such an accuracy in the WKB-approximation is about 10^{-1}–10^{-3} [13,14]. Pycnonuclear processes are at essentially low energies. In this situation, deep tunneling under the barrier exists only where the semiclassical approximation is not applicable [16]. This indicates the importance of developing fully quantum methods outside of semiclassical approximations. These quantum effects have not been studied yet by other researchers for pycnonuclear reactions in stars. In Ref. [17], investigation of these questions on the fully quantum basis was initiated, for example, for the reaction of ${}^{12}\text{C} + {}^{12}\text{C}$. The interest in that reaction is explained by its impact on nucleosynthesis, energy production, and other questions in stellar evolution [11,18]. In addition, this reaction has a significant impact on the evolution and structure of massive stars with $M \geq M_{\odot}$ ($M_{\odot}$ is the Solar mass). ${}^{12}\text{C} + {}^{12}\text{C}$ fusion is known as a pycnonulear reaction that reignites a Carbon–Oxygen white dwarf into a type Ia supernova explosion. However, it could be useful to obtain a more complete picture for the systematic analysis of nuclear processes and fusion for reactions with isotopes of Carbon. Therefore, in this paper, we perform such an investigation for pycnonuclear reactions with Carbon.

The paper is organized in the following way. In Section 2, a new generalized formalism of the multiple internal reflections is reviewed with focus on new elements for fusion and quasibound states in pycnonuclear reactions. In Section 3, reactions with isotopes of Carbon on the basis of the method are studied using calculations of penetrabilities of the potential barriers, probabilities of formation of the compound nucleus, estimation of energies for zero-point vibrations and quasibound states, etc. In Section 4, the influence of plasma screening on properties of the pycnonuclear reaction is studied in the example of ${}^{12}\text{C} + {}^{12}\text{C}$. Conclusions are summarized in Section 6.

2. Method of Quantum Mechanics for Nucleus–Nucleus Scattering with Fusion

We will study the capture of one nucleus with smaller mass by another nucleus with larger mass. This process can be studied on the basis of the solution of the Schrödinger equation with radial potential, which has a barrier approximated by a large number N of rectangular steps:

$$
V(r) = \begin{cases}
V_1 & \text{at } r_{\min} < r \le r_1 & \text{(region 1),} \\
\cdots & \cdots & \cdots \\
V_{N_{\text{cap}}} & \text{at } r_{N_{\text{cap}}-1} \le r \le r_{\text{cap}} & \text{(region } N_{\text{cap}}\text{),} \\
\cdots & \cdots & \cdots \\
V_N & \text{at } r_{N-1} \le r \le r_{\max} & \text{(region } N\text{),}
\end{cases}
\tag{1}
$$

where V_j are constants ($j = 1 \ldots N$). $r_1 \ldots r_N$ are parameters of the discretization scheme with constant step used in computer calculations. One can calculate these parameters as follows:

$$
\begin{aligned}
&\Delta r = \frac{r_{\max} - r_{\min}}{N}, \\
&r_1 = \Delta r \cdot 1 + r_{\min}, \quad r_{N-1} = \Delta r \cdot (N-1) + r_{\min}, \\
&r_2 = \Delta r \cdot 2 + r_{\min}, \quad r_N = \Delta r \cdot N + r_{\min} = r_{\max}. \\
&r_i = \Delta r \cdot i + r_{\min},
\end{aligned}
\tag{2}
$$

The solution of the radial wave function for the above barrier energies is:

$$
\chi(r) = \begin{cases}
\alpha_1 e^{ik_1 r} + \beta_1 e^{-ik_1 r}, & \text{at } r_{\min} < r \le r_1, \\
\alpha_2 e^{ik_2 r} + \beta_2 e^{-ik_2 r}, & \text{at } r_1 \le r \le r_2, \\
\cdots & \cdots \qquad\qquad \cdots \\
\alpha_{N-1} e^{ik_{N-1} r} + \beta_{N-1} e^{-ik_{N-1} r}, & \text{at } r_{N-2} \le r \le r_{N-1}, \\
e^{-ik_N r} + A_R e^{ik_N r}, & \text{at } r_{N-1} \le r \le r_{\max},
\end{cases}
\tag{3}
$$

where α_j, β_j, and A_R are unknown amplitudes and $k_j = \frac{1}{\hbar}\sqrt{2m(\tilde{E} - V_j)}$ are wave numbers. We will present the solution of this problem on the basis of the method of multiple internal reflections (see Refs. [19,20], references therein).

Note that, previously, the process of the capture of α-particles on nuclei was studied by us in Ref. [13], where we presented details of our formalism, demonstrated its accuracy in comparison with other existing methods, and used tests to check calculations. However, in Ref. [13], it was not taken into account that after tunneling through the barrier, further propagation of waves inside the internal region of potential exists. This aspect requires important modification of the formalism and estimations that were studied in Ref. [14]. In the current paper, we use results of the study in Ref. [14]. According to that research, we will indicate the region with the number N_{capture} as the place where the capture of the particle by the nucleus takes place with the largest probability.

In each region of potential, we calculate summed amplitudes as:

$$
\tilde{T}_{j-1}^- = \frac{\tilde{T}_j^- T_{j-1}^-}{1 - R_{j-1}^- \tilde{R}_j^+}, \quad
\tilde{R}_{j-1}^+ = R_{j-1}^+ + \frac{T_{j-1}^+ \tilde{R}_j^+ T_{j-1}^-}{1 - \tilde{R}_j^+ R_{j-1}^-}, \quad
\tilde{R}_{j+1}^- = R_{j+1}^- + \frac{T_{j+1}^- \tilde{R}_j^- T_{j+1}^+}{1 - R_{j+1}^+ \tilde{R}_j^-},
\tag{4}
$$

where:

$$
\begin{aligned}
T_j^+ &= \frac{2k_j}{k_j + k_{j+1}} e^{i(k_j - k_{j+1})r_j}, \quad
T_j^- = \frac{2k_{j+1}}{k_j + k_{j+1}} e^{i(k_j - k_{j+1})r_j}, \\[6pt]
R_j^+ &= \frac{k_j - k_{j+1}}{k_j + k_{j+1}} e^{2ik_j r_j}, \qquad\quad
R_j^- = \frac{k_{j+1} - k_j}{k_j + k_{j+1}} e^{-2ik_{j+1} r_j}.
\end{aligned}
\tag{5}
$$

All amplitudes $\tilde{R}^{+}_{N-2}\ldots\tilde{R}^{+}_{N_{\text{cap}}}$ and $\tilde{T}^{-}_{N-2}\ldots\tilde{T}^{-}_{N_{\text{cap}}}$ are calculated on the basis of these recurrent relations, above where at the start one can use:

$$\tilde{R}^{+}_{N-1} = R^{+}_{N-1}, \qquad \tilde{T}^{-}_{N-1} = T^{-}_{N-1}. \tag{6}$$

On the basis of such amplitudes, we calculate summed amplitudes α_j and β_j as:

$$\beta_j \equiv \sum_{i=1} \beta_j^{(i)} = \frac{\tilde{T}^{-}_j}{1 - \tilde{R}_{j-1}\tilde{R}^{+}_j}, \quad \alpha_j \equiv \sum_{i=1} \alpha_j^{(i)} = \frac{\tilde{R}_{j-1}\tilde{T}^{-}_j}{1 - \tilde{R}_{j-1}\tilde{R}^{+}_1}. \tag{7}$$

Summed amplitude $A_{T,\text{bar}}$ of transition through the barrier or summed amplitude $A_{R,\text{bar}}$ of reflection from the barrier are determined as all waves transmitted through the potential region with the barrier from r_{cap} to r_{N-1} or reflected from this potential region as:

$$A_{T,\text{bar}} = \tilde{T}^{-}_{N_{\text{cap}}}, \quad A_{R,\text{bar}} = \tilde{R}^{-}_{N-1}, \quad \text{at } \tilde{R}^{-}_{N_{\text{cap}}} = R^{-}_{N_{\text{cap}}}. \tag{8}$$

The method of multiple internal reflections also allows us to determine resonant and potential scatterings. Here, potential scattering can be defined on the basis of summed amplitude $A_{R,\text{ext}}$ of all waves reflected from the external barrier region, i.e., the region between the external turning point $r_{\text{tp,ext}}$ and r_{N-1}, and propagated outside as:

$$A_{R,\text{ext}} = \tilde{R}^{-}_{N-1}, \quad \text{at } \tilde{R}^{-}_{N_{\text{tp,ext}}} = R^{-}_{N_{\text{tp,ext}}}. \tag{9}$$

Resonant scattering can be defined on the basis of the summed amplitude $A_{R,\text{tun}}$ of all waves that are reflected from the potential region between point r_{cap} and the external turning point $r_{\text{tp,ext}}$ as:

$$A_{R,\text{tun}} = A_{R,\text{bar}} - A_{R,\text{ext}}. \tag{10}$$

The coefficient of penetrability T_{bar} and the coefficient of reflection R_{bar} concerning the potential barrier region, the coefficient R_{ext} of reflection from the external part of the barrier, and the coefficient R_{tun} of reflection from the pure barrier region are defined as:

$$T_{\text{bar}} = \frac{k_{\text{cap}}}{k_N}\left\|A_{T,\text{bar}}\right\|^2, \quad R_{\text{bar}} = \left\|A_{R,\text{bar}}\right\|^2, \quad R_{\text{ext}} = \left\|A_{R,\text{ext}}\right\|^2, \quad R_{\text{tun}} = \left\|A_{R,\text{tun}}\right\|^2. \tag{11}$$

A useful characteristic is amplitude of oscillations, defined concerning the point of capture with the number N_{cap} as:

$$A_{\text{osc}}(N_{\text{cap}}) = \frac{1}{1 - \tilde{R}^{-}_{N_{\text{cap}}-1}\tilde{R}^{+}_{N_{\text{cap}}}}. \tag{12}$$

In the standard test of quantum mechanics:

$$T_{\text{bar}} + R_{\text{bar}} = 1 \tag{13}$$

is naturally used in the formalism of multiple internal reflections.

According to the formalism of the method of multiple internal reflections [17], the probability of the existence of a compound nucleus is defined, as the integral over the region between two internal turning points, as:

$$P_{\text{cn}} \equiv \int_{r_{\text{int,1}}}^{r_{\text{int,2}}} \|\chi(r)\|^2\, dr = \sum_{j=1}^{n_{\text{int}}}\left\{\left(\|\alpha_j\|^2 + \|\beta_j\|^2\right)\Delta r + \frac{\alpha_j\beta_j^{*}}{2ik_j}e^{2ik_jr}\Big|_{r_{j-1}}^{r_j} - \frac{\alpha_j^{*}\beta_j}{2ik_j}e^{-2ik_jr}\Big|_{r_{j-1}}^{r_j}\right\}. \tag{14}$$

The solutions presented above are essentially simplified for the simplest barrier in Equation (1) in Ref. [14]. We write down $P_{\text{cn}}(E)$ as in Ref. [14] (see Equations (6) and (7)):

$$P_{\text{cn}}^{(\text{without fusion})} = P_{\text{osc}}\, T_{\text{bar}}\, P_{\text{loc}},$$

$$P_{\text{osc}} = \|A_{\text{osc}}\|^2 = \frac{(k+k_1)^2}{2k^2(1-\cos(2k_1 r_1)) + 2k_1^2\,(1+\cos(2k_1 r_1))},$$

$$T_{\text{bar}} \equiv \frac{k_1}{k_2}\,\|T_1^-\|^2,$$

$$P_{\text{loc}} = 2\frac{k_2}{k_1}\left(r_1 - \frac{\sin(2k_1 r_1)}{2k_1}\right). \tag{15}$$

For fast fusion for the simplest barrier, we obtain:

$$P_{\text{cn}}^{(\text{fast fusion})} = \left\|\sum_{i=1}\beta_1^{(i)}\right\|^2 \int_0^{r_1}\left\|R_0 e^{ik_1 r} + e^{-ik_1 r}\right\|^2 dr = \|T_1^-\|^2 r_1 = \frac{k_2\, r_1}{k_1}\,T_{\text{bar}}. \tag{16}$$

The fusion cross-section σ is defined as (see Ref. [13] for details):

$$\sigma_{\text{fus}}(E) = \sum_{l=0}^{+\infty}\sigma_l(E), \quad \sigma_l = \frac{\pi\hbar^2}{2mE}\,(2l+1)\,f_l(E)\,P_{\text{cn}}(E). \tag{17}$$

Here, E is the energy of the relative motion between two nuclei, σ_l is the partial cross-section at l, and P is the probability of formation of a compound nuclear system as defined in Equation (14) or (16). In this formula, an additional factor $f_l(E)$ is included, which is needed to connect the old factor of fusion P_l and the new probability $P_{\text{cn}}(E)$ and penetrability of the barrier region $T_{\text{bar},l}(E)$. This coefficient can be written down in explicit form for complete fusion:

$$f(E) = \frac{k_{\text{cap}}}{k_N\,\|r_{\text{cap}} - r_{\text{tp,in,1}}\|}. \tag{18}$$

The formalism developed above allows us to model different scenarios of fusion. For example, for the formation of the compound nucleus with slow fusion (i.e., without instantaneous fusion), we vary fusion coefficients in the region between points r_{cap} and $r_{\text{int,2}}$.

3. Analysis

We will study the reactions $^X\text{C} + {}^X\text{C} = {}^{2X}\text{Mg}$ [11] ($X = 10, 12, 14, 18, 20, 22, 24$) in this paper. The first indications of the possibility to synthesize more heavy elements from Carbon isotopes can be found in the research of Hamada and Salpeter [21], based on pycnonuclear reaction rates derived by Cameron [4]. Hamada and Salpeter estimated a density of $6 \times 10^9\ \text{g} \times \text{cm}^{-3}$ via a pycnonuclear process where nuclei of ^{12}C are transformed into ^{24}Mg at low energies. Then, estimates of densities of the stellar medium for those reactions were improved [1]. Note that there were uncertainties in the estimation of densities in those calculations. Moreover, estimations of rates can be changed to include temperatures and crystal imperfections in analysis. Summarizing, the critical density for Carbon was found to be $5 \times 10^{10}\ \text{g} \times \text{cm}^{-3}$. We will focus on the understanding of new quantum phenomena, which exist in pycnonclear reactions and have not been studied yet by other researchers. As the inclusion of such effects can significantly change the rates of reactions and even the picture of participating mechanisms, for brevity of calculations, we will use the density obtained by Hamada and Salpeter for the analysis of isotopes of Carbon.

3.1. Potential of Interaction for Nuclei in Lattice Sites

The potential of interactions between isotopes of Carbon ^XC is defined as:

$$V(r) = v_c(r) + v_N(r) + v_{l=0}(r), \tag{19}$$

where $v_c(r)$, $v_N(r)$, and $v_l(r)$ are Coulomb, nuclear, and centrifugal components that have the form:

$$v_N(r) = -\frac{V_R}{1 + \exp\left\{\dfrac{r - R_R}{a_R}\right\}}, \quad v_l(r) = \frac{l\,(l+1)}{2mr^2},$$

$$v_c(r) = \begin{cases} \dfrac{Z_1 Z_2 e^2}{r}, & \text{at } r \geq R_c, \\[2ex] \dfrac{Z_1 Z_2 e^2}{2R_c}\left\{3 - \dfrac{r^2}{R_c^2}\right\}, & \text{at } r < R_c. \end{cases} \tag{20}$$

Here, V_R is the strength of the nuclear term, defined as:

$$V_R = -75.0 \text{ MeV}. \tag{21}$$

R_c is the Coulomb radius of the nuclear system, R_R is the nuclear radius of the nuclear system, m is the reduced mass defined in Equation (30), and a_R is the diffusion parameter. We define these parameters as [17,22]:

$$R_R = r_R\left(A_1^{1/3} + A_2^{1/3}\right), \quad R_c = r_c\left(A_1^{1/3} + A_2^{1/3}\right), \quad a_R = 0.44 \text{ fm},$$
$$r_R = 1.30 \text{ fm}, \qquad\qquad r_c = 1.30 \text{ fm}. \tag{22}$$

These potentials for isotopes of Carbon are presented in Figure 1.

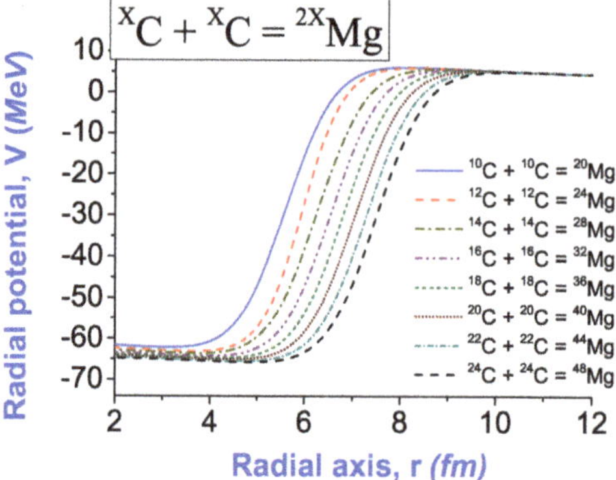

Figure 1. Potentials of interaction between two nuclei of Carbon ^{X}C (potentials and parameters are defined in Equations (19)–(21)).

A small difference between the shapes of the internal wells of the potentials is clearly visible in this figure (this internal well is absent in potentials used in Ref. [6], for example). For brevity, we include maximums of barriers and minimums of wells for potentials of interaction between nuclei in Table 1.

Table 1. Minimums of wells and maximums of barriers of potentials of interactions between two isotopes of Carbon, as well as distance R_0 between nuclei and their concentration n_A (isotopes of Carbon are chosen in accordance with Ref. [8] on the systematic study of astrophysical S-factors in fusion reactions for C, O, Ne, Mg; parameters are determined for density $\rho_0 = 6 \times 10^9\,\frac{\text{g}}{\text{cm}^3}$).

Reaction ^{X}C + ^{X}C	$r_{\min}$, fm	$V_{\min}$, MeV	$r_{\max}$, fm	$V_{\max}$, MeV	R_0, fm	n_A, 10^{-7} fm^{-3}
^{10}C + ^{10}C	3.36	−62.157	7.98	+6.249	87.06	3.61702731
^{12}C + ^{12}C	3.64	−63.018	8.33	+5.972	92.52	3.01418941
^{14}C + ^{14}C	3.92	−63.702	8.68	+5.743	97.40	2.58359092
^{16}C + ^{16}C	4.20	−64.258	8.96	+5.552	101.83	2.26064206
^{18}C + ^{18}C	4.48	−64.726	9.24	+5.386	105.91	2.00945961
^{20}C + ^{20}C	4.62	−65.133	9.52	+5.242	109.69	1.80851365
^{22}C + ^{22}C	4.90	−65.483	9.80	+5.115	113.23	1.64410331
^{24}C + ^{24}C	5.04	−65.792	10.08	+5.001	116.57	1.50709470

3.2. Space Location of Nuclei in Lattice Sites

Following the logic in Ref. [6] (see p. 90, Figure 3.5 in that book), the distance between the two closest nuclei located in lattice sites is $2\,R_0$. We place the "incident" nucleus between these nuclei. Such a distance can be derived as:

$$\rho_0 = \frac{m_A}{V_A} = \frac{A\,m_u}{4/3\,\pi\,R_0^3} \tag{23}$$

or:

$$R_0 = \left(\frac{A\,m_u}{4/3\,\pi\,\rho_0}\right)^{1/3}. \tag{24}$$

Here, ρ_0 is the density in the sphere surrounding one nucleus of the lattice site, V_A is the volume inside this sphere, A is the mass number of the nucleus, m_A is the mass of the nucleus, and m_u is the mass of the nucleon. One can calculate the concentration of nuclei n_A as:

$$n_A = \frac{\rho_0}{A\,m_u}. \tag{25}$$

For analysis of the pycnonuclear reactions ${}^X\mathrm{C} + {}^X\mathrm{C} = {}^{2X}\mathrm{Mg}$, we choose to use the density estimated in Ref. [6]:

$$\rho_0 = 6 \times 10^9\,\frac{\mathrm{g}}{\mathrm{cm}^3}. \tag{26}$$

The derived distance R_0 and concentration n_A for different isotopes of Carbon at such a density are given in Table 1.

3.3. Energy Spectra of Zero-Point Vibrations of Nuclei in Lattice Sites

A nucleus located in a lattice site and located between two nuclei with adjacent sites can oscillate and has a discrete spectrum of energy from such oscillations. The approach to determine the energy levels of such a spectrum was investigated by Zel'dovich and other researchers. In this approach, the energy of zero-point vibrations of the nucleus in the lattice site is calculated as [6] (see Equations (3.7.19) and (3.7.20)):

$$E_0^{(\mathrm{zero})} = \frac{\hbar w}{2} = \frac{\hbar\,Ze}{\sqrt{m\,R_0^3}}, \quad \Delta E = \frac{2\,Z^2 e^2}{R_0}, \quad E_{\mathrm{full}} = E_0^{(\mathrm{zero})} + \Delta E. \tag{27}$$

Here, $E_0^{(\mathrm{zero})}$ is the energy of the ground state of the harmonic oscillator relative to the potential minimum of this oscillator, ΔE is the shift of the oscillator relative to the zero value of the potential of interaction between nuclei (i.e., the distance between the minimum of the oscillator and the zero value of the potential of the interaction), and E_{full} is the energy value of the ground state in the system relative to the zero value of the potential. For example, for the reaction ${}^{12}\mathrm{C} + {}^{12}\mathrm{C} = {}^{24}\mathrm{Mg}$, we obtain:

$$E_0^{(\mathrm{zero})} = 0.02180806\ \mathrm{MeV}, \quad \Delta E = 0.56787237\ \mathrm{MeV}, \quad E_{\mathrm{full}}^{(\mathrm{zero\ mode})} = 0.58968043\ \mathrm{MeV}. \tag{28}$$

We call such a state the *state of zero-point vibrations of nuclei* (or the *state of zero mode*).

However, the harmonic oscillator has not only the ground state but the full discrete energy spectrum, which is calculated as:

$$E_n^{(\mathrm{zero})} = (2n+1) \cdot \frac{\hbar w}{2} = (2n+1) \cdot E_{n=0}^{(\mathrm{zero})} = (2n+1)\,\frac{\hbar\,Ze}{\sqrt{m\,R_0^3}}. \tag{29}$$

The energy spectrum can be written down via density of matter ρ_0 instead of distance R_0. Using Equation (24) and the formula for reduced mass:

$$R_0 = \left(\frac{A\,m_u}{4/3\,\pi\,\rho_0}\right)^{1/3}, \quad m = m_{\mathrm{p}}\,\frac{A_1\,A_2}{A_1 + A_2}, \tag{30}$$

from Equation (27), we obtain (let us consider the case of the same nuclei in the lattice: $A_1 = A_2$, and $A = A_1$):

$$E_0^{(\text{zero})} = c_1 \cdot \frac{Z}{A} \sqrt{\rho_0}, \quad c_1 = \hbar e \sqrt{\frac{8\pi}{3\, m_u m_p}},$$

$$\Delta E = c_2 \cdot Z^2 \left(\frac{\rho_0}{A}\right)^{1/3}, \quad c_2 = 2e^2 \left(\frac{4\pi}{3m_u}\right)^{1/3}. \tag{31}$$

We find new interesting property for nuclei of type $2Z = A$:

$$E_0^{(\text{zero})} = \frac{c_1}{2} \sqrt{\rho_0}, \quad \Delta E = c_2 \cdot Z^2 \left(\frac{\rho_0}{2Z}\right)^{1/3}. \tag{32}$$

Thus, according to this property, the spectra $E_n^{(\text{zero})}$ are the same for nuclei ^{8}Be, ^{10}B, ^{12}C, ^{14}N, ^{16}O, ^{18}F, ^{20}Ne, ^{22}Na, ^{24}Mg, ^{26}Si, etc. Those depend only on the chosen density in the stellar medium. In Table 2, energy values are presented for the first 10 states of zero-point vibrations calculated by Equation (27) for reactions ^{X}C $+$ ^{X}C.

Table 2. Energy levels for the first 10 states of zero-point vibtations calculated by Equations (27) for reactions ^{X}C $+$ ^{X}C.

No.	Energy, $E_n^{(\text{zero})}$, MeV	Energy, $E_{\text{full}}^{(\text{zero})}$, MeV
1	0.021808061833736	0.589680437522993
2	0.065424185501208	0.633296561190465
3	0.109040309168680	0.676912684857937
4	0.152656432836153	0.720528808525410
5	0.196272556503626	0.764144932192882
6	0.239888680171098	0.807761055860354
7	0.283504803838570	0.851377179527827
8	0.327120927506043	0.894993303195299
9	0.370737051173515	0.938609426862772
10	0.414353174840987	0.982225550530244

Energies of states of zero-point vibrations can be reestimated on the basis of the method of multiple internal reflections. For that, let us write down the radial wave function in the asymptotic region:

$$\chi(r) = e^{-ikr} + A_R\, e^{+ikr}. \tag{33}$$

Following quantum mechanics, the full wave function should be zero at point R_0 (for odd states) or be maximal in the module at that point (for even states):

$$\begin{aligned}
(1) \quad & \chi(R_0) = e^{-ikR_0} + A_R\, e^{+ikR_0} = e^{-ikR_0} + e^{+ikR_0}, \quad A_R = +1, \\
(2) \quad & \chi(R_0) = e^{-ikR_0} + A_R\, e^{+ikR_0} = e^{-ikR_0} - e^{+ikR_0}, \quad A_R = -1.
\end{aligned} \tag{34}$$

This requirement gives discreteness of the spectrum of energy for such states. Energy levels can be found if we impose a condition on the imaginary part of such an amplitude to equal zero:

$$\begin{aligned}
\text{even states:} \quad & A_R = +1, \quad \text{Re}(A_R) = +1, \quad \text{Im}(A_R) = 0, \\
\text{odd states:} \quad & A_R = -1, \quad \text{Re}(A_R) = -1, \quad \text{Im}(A_R) = 0.
\end{aligned} \tag{35}$$

The energies for states of zero-point vibrations for Carbon isotopes ^{X}C $+$ ^{X}C $\rightarrow$ ^{2X}Mg are given in Table 3.

Table 3. Energies of zero-point vibrations $E_{zero}^{(mir)}$ for reactions $^{X}C + {}^{X}C$ (presented data are in MeV, below 5 MeV) calculated by the method of multiple internal reflections (see Section 3.3 for details). Distances between each two adjacent energies are essentially different from the energy spectrum of the harmonic oscillator (see Equation (31), Table 2). There are few energies below the energy of the zero-point vibrations in the ground state $E_{full,0}^{(zero)}$ derived by the approach of Zel'dovich and his colleagues (see Equation (27)). These energies are derived with accuracy, which can be estimated from condition $|\operatorname{Re}(A_R)| \approx 1$. Additional estimation of accuracy of the calculated amplitude can be done by checking the condition of $[\operatorname{Re}(A_R)]^2 + [\operatorname{Im}(A_{rmR})]^2 = 1$. Summation of $[\operatorname{Re}(A_R)]^2 + [\operatorname{Im}(A_{rmR})]^2$ is an additional estimation of accuracy for the method of MIR in determination of obtained digits of the amplitude.

No.	$^{10}C + {}^{10}C$	$^{12}C + {}^{12}C$	$^{14}C + {}^{14}C$	$^{16}C + {}^{16}C$
1	0.517434869739479	0.517434869739479	0.517434869739479	0.527054108216433
2	0.536673346693387	0.536673346693387	0.536673346693387	0.536072144288577
3	0.546292585170341	0.546292585170341	0.546292585170341	0.545090180360721
4	0.565531062124249	0.555911823647295	0.555911823647295	0.554108216432866
5	0.584769539078156	0.575150300601202	0.565531062124249	0.572144288577154
6	0.613627254509018	0.594388777555110	0.575150300601202	0.581162324649299
7	0.642484969939880	0.613627254509018	0.594388777555110	0.599198396793587
8	0.680961923847695	0.642484969939880	0.623246492985972	0.626252505010020
9	0.738677354709419	0.680961923847695	0.652104208416834	0.653306613226453
10	0.815631262525050	0.729058116232465	0.690581162324649	0.689378757515030
11	0.950300601202405	0.806012024048096	0.729058116232465	0.734468937875752
12	1.27735470941884	0.911823647294589	0.796392785571142	0.797595190380762
13	2.23927855711423	1.11382765531062	0.892585170340681	0.878757515030060
14	3.69178356713427	2.76833667334669	1.04649298597194	1.02304609218437
15	—	4.08617234468938	1.64288577154309	1.39278557114228
16	—	—	3.04729458917836	1.99699398797595
17	—	—	4.28817635270541	3.20541082164329
18	—	—	—	4.37775551102204

No.	$^{18}C + {}^{18}C$	$^{20}C + {}^{20}C$	$^{22}C + {}^{22}C$	$^{24}C + {}^{24}C$
1	0.517434869739479	0.527054108216433	0.536673346693387	0.517434869739479
2	0.527054108216433	0.546292585170341	0.546292585170341	0.527054108216433
3	0.536673346693387	0.555911823647295	0.555911823647295	0.536673346693387
4	0.546292585170341	0.565531062124249	0.565531062124249	0.584769539078156
5	0.555911823647295	0.584769539078156	0.575150300601202	0.594388777555110
6	0.565531062124249	0.604008016032064	0.594388777555110	0.613627254509018
7	0.575150300601202	0.623246492985972	0.604008016032064	0.632865731462926
8	0.584769539078156	0.642484969939880	0.623246492985972	0.661723446893788
9	0.594388777555110	0.671342685370741	0.652104208416834	0.690581162324649
10	0.613627254509018	0.709819639278557	0.680961923847695	0.729058116232465
11	0.632865731462926	0.748296593186373	0.719438877755511	0.767535070140281
12	0.661723446893788	0.806012024048096	0.757915831663327	0.825250501002004
13	0.738677354709419	0.882965931863727	0.815631262525050	0.911823647294589
14	0.796392785571142	1.00801603206413	0.892585170340681	1.02725450901804
15	0.882965931863727	1.27735470941884	1.01763527054108	1.29659318637275
16	1.00801603206413	2.24889779559118	1.27735470941884	2.20080160320641
17	1.30621242484970	3.26853707414830	2.24889779559118	3.14348697394790
18	2.18156312625251	4.29779559118237	3.22044088176353	4.08617234468938
19	3.26853707414830	—	4.21122244488978	—

3.4. Probabilities of Formation of Compound Nuclei in Pycnonuclear Reactions

Our analysis has shown that the coefficients of penetrability and reflection increase monotonously with the energy of the incident nucleus [17]. This means that penetrability and reflection themselves cannot indicate the possible existence of some definite states of more heavy nuclei synthesized in pycnonuclear reactions in stars. Such behavior of these characteristics is in agreement with the analysis of the capture of α-particles by nuclei [13,14].

Another important quantum characteristic is the probability of formation of acompound nucleus, which can be created during the studied reactions with nuclei. In Figure 2, we present such probabilities for isotopes of Carbon calculated by our method.

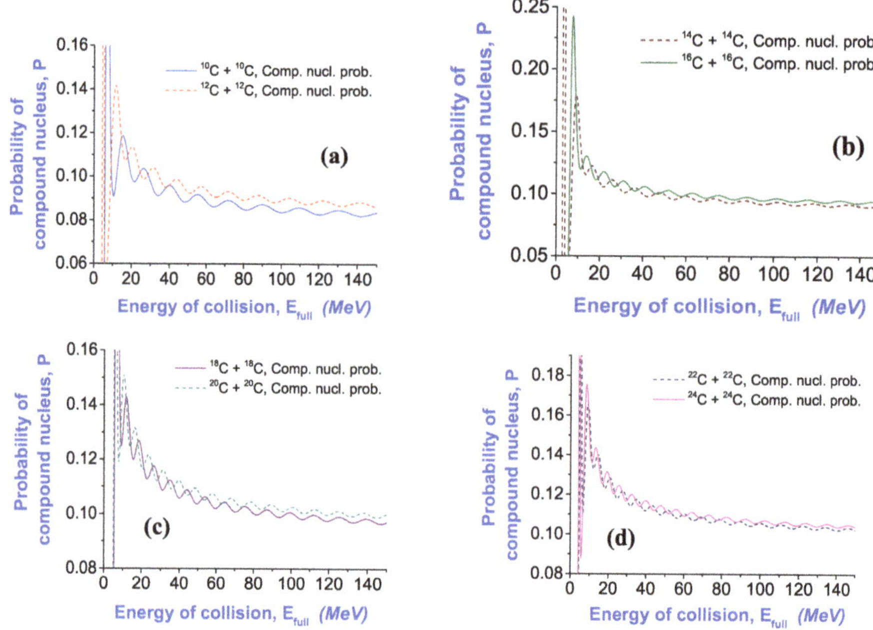

Figure 2. Probabilities of formation of a compound nucleus P_{cn} in dependence on energy for incident isotopes of Carbon in reactions $^{10}C + {}^{10}C$, $^{12}C + {}^{12}C$ (**a**), $^{14}C + {}^{14}C$, $^{16}C + {}^{16}C$ (**b**), $^{18}C + {}^{18}C$, $^{20}C + {}^{20}C$ (**c**), and $^{22}C + {}^{22}C$, $^{24}C + {}^{24}C$ (**d**) in lattice (potentials and parameters are defined in Equations (19)–(21)). One can clearly see the presence of maxima of such probabilities for all studied isotopes. Energies corresponding to maxima of such probabilities are given in Table 4.

In these figures, one can clearly see the presence of maxima in probabilities at certain definite energies for all studied isotopes of Carbon. This means that at such energies, compound nuclei are formed with maximum probability. Another conclusion from such calculations is that the formation of ^{X}Mg nuclei is much more probable at such energies than at energies of zero-mode vibrations. These maxima are explained by strict requirements of quantum mechanics [16], which take into account the further propagation of quantum fluxes in the potential region, in contrast to the existing modern description of pychonuclear reactions, where these fluxes are omitted in the nuclear region from the internal turning point.

In Table 4, we present the energies of the quasibound states for reactions with isotopes of Carbon up to 150 MeV.

Only first quasibound energies for $^{10}C + {}^{10}C$, $^{12}C + {}^{12}C$, and $^{24}C + {}^{24}C$ are smaller than the barrier maximums for these nuclear systems. This means that at such energies, compound nuclear systems are the most stable and are transformed to new synthesized isotopes of Magnesium ^{20}Mg, ^{24}Mg, and ^{48}Mg with large probability. There is a simple way to estimate half-lives of these obtained heavier nuclei using Gamow's approach (well developed in the problem of nuclear decays) or the method of multiple internal reflections for higher precision (we omit these calculations in this paper).

Table 4. Energies of the quasibound states of the compound nuclear systems in reactions with isotopes of Carbon ^{10}C, ^{12}C, ^{14}C, ^{16}C, ^{18}C, ^{20}C, ^{22}C, and ^{24}C, calculated by the method of multiple internal reflections up to 150 MeV (accuracy of about 10^{-14} in checking test $|T_{\text{bar}} + R_{\text{bar}}| = 1$ is obtained for each calculation). Comparing these energies with maximums of the potential barriers for all studied systems given in Table 1, we find that only first quasibound energies for ^{10}C + ^{10}C, ^{12}C + ^{12}C, and ^{24}C + ^{24}C are smaller than barrier maximums for these nuclear systems. That means that at such energies, the compound nuclear systems have barriers that prevent decays from going through the tunneling phenomenon.

No.	^{10}C + ^{10}C	^{12}C + ^{12}C	^{14}C + ^{14}C	^{16}C + ^{16}C	^{18}C + ^{18}C	^{20}C + ^{20}C	^{22}C + ^{22}C	^{24}C + ^{24}C
1	0.63471	4.88176	9.06212	7.27054	6.37475	5.47896	5.18036	4.58317
2	15.33267	11.45090	16.52705	13.83968	11.74950	10.55511	9.06212	8.46493
3	26.38076	20.40882	25.78357	21.90180	18.31864	16.52705	14.43687	13.24248
4	40.11623	31.45691	36.23447	30.85972	26.38076	23.69339	20.70741	19.21443
5	55.04609	43.69940	47.58116	40.71343	34.74148	31.15832	27.57515	25.48497
6	71.76754	57.13627	59.82365	51.46293	43.99800	39.51904	35.04008	32.35271
7	89.68337	71.76754	72.96192	62.80962	53.85170	48.47695	42.80361	39.81764
8	109.39078	87.29459	86.99599	74.75351	64.30261	57.73347	51.16433	47.58116
9	130.29259	104.01603	101.62725	87.59319	75.35070	67.88577	60.12224	55.64329
10	–	121.93186	117.45291	101.03006	86.99599	78.03808	69.37000	64.30261
11	–	–	134.17435	115.36273	98.93988	89.08617	79.23246	72.96192
12	–	–	–	130.29259	112.07816	100.43287	89.08617	82.51703
13	–	–	–	146.11824	125.21643	112.37675	99.83567	92.07214
14	–	–	–	–	139.25050	124.91784	110.88377	102.22445
15	–	–	–	–	–	137.75752	122.23046	112.67535
16	–	–	–	–	–	–	134.17435	123.42485
17	–	–	–	–	–	–	146.11824	134.77154
18	–	–	–	–	–	–	–	146.11824

4. Plasma Screening in Nuclear Reactions

It is well known that nuclear reactions at high densities of matter in compact stars are essentially modified due to plasma screening effects [1]. Therefore, a natural question appears as to how many of the results presented above are changed after taking effects of plasma screening into account. We will follow Ref. [10], where effects of plasma screening in thermonuclear fusion reactions in dense nuclear matter in stars were studied. Here, in addition to physical analysis, the authors provided a clear formalism for use and implementations into other research. Therefore, we will estimate the influence of plasma screening on pycnonuclear reactions on the basis of isotope ^{12}C, and we use that research as a basis for our analysis.

The methodology of the influence of electron clouds on the studied nuclear process is presented in Ref. [10], and we follow this approach. In frameworks of model [10], nuclear reactions are studied under the influence of strong plasma screening. At the first stage, authors introduce the Coulomb potential for colliding nuclei in the standard form as $U_C(r) = Z_1 Z_2 e^2 / r - H(r)$, where $H(r)$ is the mean-field plasma screening potential and Z_i is the electric charge of the nucleus with the number i ($i = 1, 2$). Potential $H(r)$ is determined by the ion-sphere model proposed by Salpeter [23].

$H(r)$ is produced by an electron cloud near nuclei (ions) (see Onsager molecules, e.g., Ref. [24], references therein). Following model [10], the electron cloud is considered as an incompressible uniformly charged liquid drop. This drop has a constant volume, but variable shape. The charge of this drop fully compensates for the charge of the interacting nuclei. The electron drop acts as a Wigner–Seitz cell with tunneling ions.

In the approach of [10], the authors also calculate the astrophysical S-factors and the reaction rates for thermonuclear reactions by including the screening potential in the total potential. The new modified S-factor is determined not only by nuclear interactions but also by parameters of dense matter. With the estimation of new rates, the authors found factors of the plasma screening enhancement.

Following Ref. [10] (see Equation (7) in that paper), we define the Coulomb potential $U_C(r)$ for colliding nuclei in the standard form:

$$U_{C,\,full}(r) = v_C(r) + H(r), \tag{36}$$

where $v_C(r)$ is the pure Coulomb potential without screening and $H(r)$ is the mean-field plasma screening potential. In contrast to Ref. [10], we calculate the pure Coulomb potential $v_C(r)$ on the basis of Equations (20)–(22) (here, the Coulomb potential in the nuclear region at $r < R_C$ is different from the corresponding potential in Equation (7) in Ref. [10]), and we use nuclear potential $v_N(r)$ in Equation (20) (we set $l = 0$). In definition of the screening part of potential, we follow Ref. [10] and use (see Equations (10) and (11) in that paper):

$$H(r) = E_{12}\, h(x), \quad x = \frac{r}{a_{12}}, \tag{37}$$

where:

$$h(x) = b_0 + b_2\, x^2 + b_4\, x^4 + \ldots \tag{38}$$

At $Z_1/Z_2 = 1$, parameters b_0, b_2, and b_4 are derived in Ref. [10] as:

$$b_0 = 1.0573, \quad b_2 = -0.25, \quad b_4 = 0.0394. \tag{39}$$

Other parameters are (see Equations (2) and (4) in Ref. [10]):

$$a_e = \left(\frac{3}{4\pi\, n_e}\right)^{1/3}, \quad a_j = Z_j^{1/3}\, a_e, \tag{40}$$

$$a_{12} = \frac{a_1 + a_2}{2}, \quad E_{12} = \frac{Z_1 Z_2\, e^2}{a_{12}}, \tag{41}$$

where n_e is concentration of electrons.

On the basis of Equation (25), we calculate the concentration of nuclei n_A at the studied density as:

$$^{12}\mathrm{C} + {}^{12}\mathrm{C}, \quad \rho_0 = 6 \times 10^9\, \frac{\mathrm{g}}{\mathrm{cm}^3} : \quad n_A = 3.014\,18 \times 10^{-7}\ \mathrm{fm}^{-3}. \tag{42}$$

This can be understood as each $^{12}\mathrm{C}$ nucleus having size of about 200 fm in volume. From here, we find the concentration of electrons:

$$n_e = Z \cdot n_A, \quad n_e = 1.808\,51 \times 10^{-6}\ \mathrm{fm}^{-3} \tag{43}$$

and from Equations (38) and (39), we obtain:

$$a_{12} = 92.522\,41\ \mathrm{fm}. \tag{44}$$

As it is indicated in Ref. [10], Equation (38) should be used at $x \ll 2$. We estimate that this condition is fulfilled in the full region of study of reaction $^{12}\mathrm{C} + {}^{12}\mathrm{C}$ at the chosen density, so we use Equation (38) for the description of the screening part of the potential.

The potential of interactions taking into account screening, calculated by such an approach, is shown in Figure 3.

From this figure, one can see that the screening does not change the potential much at the density of matter under consideration. Therefore, one can suppose that the screening does not influence essentially the results of quasibound states and energies above. However, energies of zero-point vibrations are essentially smaller than quasibound energies, and one can suppose that the energy spectrum of zero-point vibrations will be changed after the inclusion of plasma screening in calculations.

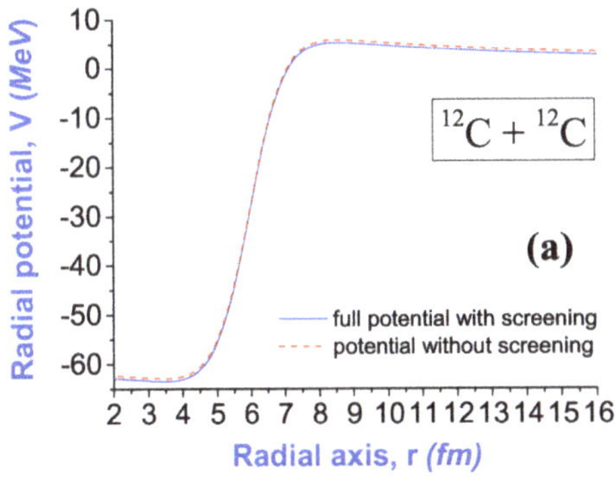
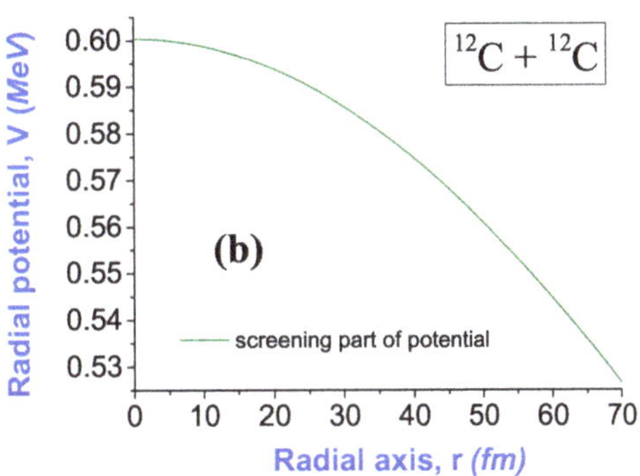

Figure 3. Potential of interaction between two ^{12}C nuclei with the inclusion of screening in comparison with the same potential without screening (**a**) and the part of the potential describing screening (**b**) (the potential and parameters are defined in Equations (19)–(21), and the screening part of the potential is defined in Equations (36)–(38)).

In Table 5, values of energies of zero-point vibrations calculated for the reaction ^{12}C + ^{12}C are presented with and without taking into account screening.

Table 5. Energies for zero-point vibrations $E_{\text{zero}}^{(\text{mir})}$ (values are presented in MeV, below 5 MeV) calculated for the reaction ^{12}C + ^{12}C.

No.	^{12}C + ^{12}C with Screening	^{12}C + ^{12}C without Screening
1	0.209619238476954	0.517434869739479
2	0.267334669338677	0.536673346693387
3	0.344288577154309	0.546292585170341
4	0.478957915831663	0.555911823647295
5	0.786773547094188	0.575150300601202
6	2.21042084168337	0.594388777555110
7	3.83607214428858	0.613627254509018
8		0.642484969939880
9		0.680961923847695
10		0.729058116232465
11		0.806012024048096
12		0.911823647294589
13		1.11382765531062
14		2.76833667334669
15	—	4.08617234468938

From this table, one can see that the energy spectrum of the zero-point vibrations for the reaction ^{12}C + ^{12}C is modified essentially after taking plasma screening into account.

5. Influence of Vibration of External Nuclei on Calculation of Quasibound States

Vibrations of the nucleus can be understood as oscillations of the particle inside the potential well in quantum mechanics (for example, see Ref. [16], p. 91). Here, the harmonic oscillator provides a clear example. The ground energy level of this particle inside the potential of the harmonic oscillator is not zero (exited energy levels are also not zero) due to the quantum nature of this phenomenon. Non-zero frequencies correspond to such energies. At the same time, quantum mechanics provides a formalism to calculate the most probable location of this particle, which is at the coordinates of the minimum of the potential well. This picture corresponds to the most probable position of the studied nucleus with non-zero frequencies (i.e., the position of nucleus is fixed in general logic). In other words, vibrations of the nucleus (inside some external field) can be studied in quantum mechanics as a particle oscillating inside the potential well, and quantum mechanics provides a formalism to calculate non-zero frequencies and the most probable location of this particle.

If we consider the nucleus to be between the two closest nuclei (we will call them *"external nuclei"*), one can find that this nucleus is located inside the Coulomb fields of the external nuclei. The summation of the Coulomb potentials of the external nuclei gives the harmonic oscillator enough of a small middle region between the external nuclei, which is the approximation for full potential (where we neglect interactions with external nuclei at closer distances). In such a way, we obtain a picture of the oscillation of the particle inside the harmonic potential well, and we report the vibrations of the corresponding nucleus.

However, we should be reminded that nuclear interactions of nuclei exist, which influence the phenomenon described above. In particular, this is crucial in the study of nuclear scattering. From our experience, the role of the nuclear part of the potentials is increased at low energies (this is the case of pycnonuclear reactions). Therefore, we obtain motivation to take into account the influence of nuclear forces from the external nuclei on the process of oscillation of the middle nucleus (the particle inside the more complex potential well, which continues to exist).

After the inclusion of the nuclear parts of the potentials from external nuclei, we obtain two additional (more deep) wells outside the well of harmonic oscillator type. Now, potential barriers appear, and there is a possibility to transfer a particle through any barrier with not-zero probability. This is estimated on the basis of penetrability, which we calculate with high precision, and we propose tests to check the calculations by other researchers. In addition, we find that after tunneling (or transferring the region of the barrier at above-barrier energies), the joint nuclear system (from the middle nucleus and one external nucleus) can exist with higher probability, which we describe and estimate via the formalism of quasibound states.

If we study the interaction of the middle nucleus with one external nucleus, quantum mechanics provides a strict formalism, where the particle with reduced mass moves in the external potential (for example, see Ref. [16], pp. 133÷136). The vibrating effects of both nuclei are included in such a model, which can be studied via estimation of energy levels, frequencies, properties of the wave function of this particle, etc. The influence of the second external nucleus can be included also as correction with the addition of a second potential with a second barrier.

However, outside such a correction above, more self-consistent study of joint vibrations of all external nuclei and the middle nucleus can be performed as the next step in this research line (this is a three-body problem in quantum mechanics). This study is omitted in the current manuscript.

6. Conclusions and Perspectives

The question of conditions needed for the most probable formation of compound nuclei (as the first stage needed for the synthesis of more heavy elements) in pycnonuclear reactions in compact stars is investigated in this paper. The method is based on the formalism of multiple internal reflections, constructed for the study of quantum phenomena with details, high precision, and tests in nuclear decays [12,19,20] and nuclear captures by nuclei [13,14]. In this paper, we continue investigations of pycnonuclear reactions with isotopes of Carbon, started in Ref. [17] for $^{12}C + ^{12}C$. Conclusions of our analysis are the following.

- In this research, pycnonuclear processes are studied, taking the nuclear part of the potential of interactions between nuclei into account. The requirement of continuity of quantum flux (describing pycnonuclear reactions on the basis of quantum mechanics) gives new states in which the compound nuclear system of ^{2X}Mg is formed with the highest probability (see Figure 2). Following the logic in Refs. [13,14,17], we call such states *quasibound states in pycnonuclear reactions*. Note that these states have not been studied yet by other researchers in the study of the synthesis of elements in stars.

- As shown in Figure 2, the probability of formation of a compound nuclear system in quasibound states is essentially higher than the probability of formation of this system in states of zero-point vibrations studied by Zel'dovich [5] and followers of that idea. The synthesis of more heavy nuclei of Magnesium from isotopes of Carbon is essentially more probable in quasibound states than in states of zero-point vibrations. This leads to the revision (reconsideration) of pictures of the formation of heavy elements in compact stars to use quasibound states as the basis for synthesis. Note the perspective to study in more detail the method in this paper on the basis of experimental measurements in Ref. [25].

- Only the first quasibound energies for $^{10}C + {}^{10}C$, $^{12}C + {}^{12}C$, and $^{24}C + {}^{24}C$ (see Table 4) are smaller than the barrier maximums for these nuclear systems (see Table 1). Therefore, at such energies, the compound nuclear systems have barriers that prevent their decays from going through the tunneling phenomenon. At such energies, the compound nuclear systems are the most probable and the most long lived. These systems are transformed into new synthesized isotopes of Magnesium ^{20}Mg, ^{24}Mg, and ^{48}Mg with large probabilities. There is a simple way to estimate the half-lives of these obtained more heavy nuclei using Gamow's approach or the method of multiple internal reflections for higher precision. Note that other approaches cannot estimate the quasibound energies needed for the prediction of the synthesis of more stable nuclear systems by such a way described above. At the same time, the method of multiple internal reflections calculates such energies with high precision, also providing tests to check calculations. However, the analysis of binding energies for the obtained isotopes of Magnesium shows that only ^{24}Mg will be stable after synthesis.

- For the first time, the influence of plasma screening on quasibound states and states of zero-point vibrations in pycnonuclear processes has been studied. It is found that the energy spectrum of zero-point vibrations is essentially modified after taking plasma screening into account (see Table 5 for the reaction $^{12}C + {}^{12}C$).

Author Contributions: Conceptualization, S.P.M.; Methodology, S.P.M.; Software, S.P.M. and K.A.S.; Validation, G.W.; Formal analysis, G.W. and K.A.S.; Investigation, S.P.M. and G.W.; Writing—original draft, S.P.M.; Writing—review & editing, G.W.; Funding acquisition, G.W. All authors have read and agreed to the published version of the manuscript.

Funding: G.W. and S.P.M. thank the support of OTKA grant K138277.

Data Availability Statement: All data in Tables and Figures in this paper are calculated on the basis of the method described in this paper. Formalism of method is presented in short form. But there are citations in reference list where one can find more details about method, its tests, applicability to different tasks, history of developments in such a research line.

Acknowledgments: S.P.M. thanks the Wigner Research Centre for Physics in Budapest for warm hospitality and support. Authors are highly appreciative to V. S. Vasilevsky for fruitful discussions concerning mechanisms of formation of the compound nuclear system and scattering of nuclei and peculiarities of the method of multiple internal reflections, A. G. Magner for fruitful discussions concerning the aspects of nuclear matter in conditions of compact stars and in Earth, and V. Ambrus for fruitful discussions concerning physical issues of astrophysical *S*-factors and applicability of quantum methods in stars.

Conflicts of Interest: The authors declare no conflict of interest.

References

1. Salpeter, E.E.; Horn, H.M.V. Nuclear reaction rates at high densities. *Astrophys. J.* **1969**, *155*, 183.
2. Schramm, S.; Koonin, S.E. Pycnonuclear Fusion Rates. *Astrophys. J.* **1990**, *365*, 296; Erratum in *Astrophys. J.* **1991**, *377*, 343. [CrossRef]
3. Haensel, P.; Zdunik, J.L. Equation of state and structure of the crust of an accreting neutron star. *Astron. Astrophys.* **1990**, *229*, 117; Erratum in *Astron. Astrophys.* **2003**, *404*, L33.
4. Cameron, A.G.W. Pycnonuclear reactions and nova explosions. *Astrophys. J.* **1959**, *130*, 916. [CrossRef]
5. Zel'dovich, Y.B.; Guseynov, O.H. Collapsed stars in binaries. *Astrophys. J.* **1965**, *144*, 840. [CrossRef]

6. Shapiro, S.L.; Teukolsky, S.A. *Black Holes, White Dwarfs, and Neutron Stars: The Physics of Compact Objects*; Wiley: Weinheim, Germany, 2004; 645p.
7. Yakovlev, D.G.; Gasques, L.R.; Beard, M.; Wiescher, M.; Afanasjev, A.V. Fusion reactions in multicomponent dense matter. *Phys. Rev. C* **2006**, *74*, 035803.
8. Beard, M.; Afanasjev, A.V.; Chamon, L.C.; Gasques, L.R.; Wiescher, M.; Yakovlev, D.G. Astrophysical S factors for fusion reactions involving C, O, Ne, and Mg isotopes. *At. Data Nucl. Data Tables* **2010**, *96*, 541–566.
9. Afanasjev, A.V.; Beard, M.; Chugunov, A.I.; Wiescher, M.; Yakovlev, D.G. Large collection of astrophysical S-factors and its compact representation. *Phys. Rev. C* **2012**, *85*, 054615.
10. Kravchuk, P.A.; Yakovlev, D.G. Strong plasma screening in thermonuclear reactions: Electron drop model. *Phys. Rev. C* **2014**, *89*, 015802.
11. Gasques, L.R.; Afanasjev, A.V.; Aguilera, E.F.; Beard, M.; Chamon, L.C.; Ring, P.; Wiescher, M.; Yakovlev, D.G. Nuclear fusion in dense matter: Reaction rate and carbon burning. *Phys. Rev. C* **2005**, *72*, 025806.
12. Maydanyuk, S.P.; Belchikov, S.V. Calculations of penetrability of barriers in the proton decay problem: The fully quantum approach and initial condition of decay. *J. Phys. Stud.* **2011**, *14*, 40.
13. Maydanyuk, S.P.; Zhang, P.-M.; Belchikov, S.V. Quantum design using a multiple internal reflections method in a study of fusion processes in the capture of alpha-particles by nuclei. *Nucl. Phys. A* **2015**, *940*, 89–118.
14. Maydanyuk, S.P.; Zhang, P.-M.; Zou, L.-P. New quasibound states of the compound nucleus in α-particle capture by the nucleus. *Phys. Rev. C* **2017**, *96*, 014602.
15. Eberhard, K.A.; Appel, C.; Bangert, R.; Cleemann, L.; Eberth, J.; Zobel, V. Fusion cross sections for $\alpha + {}^{40,44}$Ca and the problem of anomalous large-angle scattering. *Phys. Rev. Lett.* **1979**, *43*, 107–110.
16. Landau, L.D.; Lifshitz, E.M. *Kvantovaya Mehanika, Kurs Teoreticheskoi Fiziki* (*Quantum Mechanics, Course of Theoretical Physics*); Nauka Publishing: Moscow, Russia, 1989; Pergamon: Oxford, UK, 1982; Volume 3, p. 768. (In Russian)
17. Maydanyuk, S.P.; Shaulskyi, K.A. Quantum design in study of pycnonuclear reactions in compact stars: Nuclear fusion, new quasibound states and spectroscopy. *Eur. Phys. J.* **2022**, *A58*, 220.
18. Chien, L.H.; Khou, D.T.; Cuong, D.C.; Phuc, N.H. Consistent mean-field description of the ^{12}C $+ {}^{12}$C optical potential at low energies and the astrophysical S factor. *Phys. Rev. C* **2018**, *98*, 064604.
19. Olkhovsky, V.S.; Maydanyuk, S.P. Method of multiple internal reflections in description of tunneling evolution through barriers. *Ukr. Phys. J.* **2000**, *45*, 1262–1269.
20. Maydanyuk, S.P.; Olkhovsky, V.S.; Zaichenko, A.K. The method of multiple internal reflections in description of tunneling evolution of nonrelativistic particles and photons. *J. Phys. Stud.* **2002**, *6*, 1–16.
21. Hamada, T.; Salpeter, E.E. Models for zero-temperature stars. *Astrophys. J.* **1961**, *134*, 683.
22. Maydanyuk, S.P.; Olkhovsky, V.S.; Mandaglio, G.; Manganaro, M.; Fazio, G.; Giardina, G. Bremsstrahlung emission of high energy accompanying spontaneous of ^{252}Cf. *Phys. Rev. C* **2010**, *82*, 014602.
23. Salpeter, E.E. Electron screening and thermonuclear reactions. *Aust. J. Phys.* **1954**, *7*, 373. [CrossRef]
24. Rosenfeld, Y.; Chabrier, G. Onsager molecules for the Yukawa potentials: Screening potentials and the Jancovici coefficient in strongly coupled electron-screened plasmas. *J. Stat. Phys.* **1997**, *89*, 283.
25. Fang, X.; Tan, W.P.; Beard, M.; deBoer, R.J.; Gilardy, G.; Jung, H.; Liu, Q.; Lyons, S.; Robertson, D.; Setoodehnia, K.; et al. Experimental measurement of ^{12}C $+ {}^{16}$O fusion at stellar energies. *Phys. Rev. C* **2017**, *96*, 045804.

Article

Phase Transitions in the Interacting Relativistic Boson Systems

Dmitry Anchishkin [1,2,3,†], Volodymyr Gnatovskyy [2,*,†], Denys Zhuravel [1,†], Vladyslav Karpenko [2,†], Igor Mishustin [3,†] and Horst Stoecker [3,4,†]

[1] Bogolyubov Institute for Theoretical Physics, 03143 Kyiv, Ukraine; dmytro.kiev@gmail.com (D.A.); lpbest@ukr.net (D.Z.)

[2] Faculty of Physics, Taras Shevchenko National University of Kyiv, 03127 Kyiv, Ukraine; karpych1717@gmail.com

[3] Frankfurt Institute for Advanced Studies, 60438 Frankfurt am Main, Germany; mishustin@fias.uni-frankfurt.de (I.M.); stoecker@fias.uni-frankfurt.de (H.S.)

[4] Johann Wolfgang Goethe University, 60438 Frankfurt am Main, Germany

* Correspondence: vgnatovskyy@ukr.net

† These authors contributed equally to this work.

Abstract: The thermodynamic properties of the interacting particle–antiparticle boson system at high temperatures and densities were investigated within the framework of scalar and thermodynamic mean-field models. We assume isospin (charge) density conservation in the system. The equations of state and thermodynamic functions are determined after solving the self-consistent equations. We study the relationship between attractive and repulsive forces in the system and the influence of these interactions on the thermodynamic properties of the bosonic system, especially on the development of the Bose–Einstein condensate. It is shown that under "weak" attraction, the boson system has a phase transition of the second order, which occurs every time the dependence of the particle density crosses the critical curve or even touches it. It was found that with a "strong" attractive interaction, the system forms a Bose condensate during a phase transition of the first order, and, despite the finite value of the isospin density, these condensate states are characterized by a zero chemical potential. That is, such condensate states cannot be described by the grand canonical ensemble since the chemical potential is involved in the conditions of condensate formation, so it cannot be a free variable when the system is in the condensate phase.

Keywords: relativistic boson system of particles and antiparticles; Bose–Einstein condensation

Citation: Anchishkin, D.; Gnatovskyy, V.; Zhuravel, D.; Karpenko, V.; Mishustin, I.; Stoecker, H. Phase Transitions in the Interacting Relativistic Boson Systems. *Universe* **2023**, *9*, 411. https://doi.org/10.3390/universe9090411

Academic Editors: Máté Csanád, Péter Kovács, Sándor Lökös and Dániel Kincses

Received: 28 July 2023
Revised: 21 August 2023
Accepted: 1 September 2023
Published: 7 September 2023

1. Introduction

Knowledge of the phase structure of meson systems in the regime of finite temperatures and isospin densities is crucial for understanding a wide range of phenomena, from nucleus–nucleus collisions to neutron stars, as well as cosmology. This field is an important part of hot and dense hadronic matter research. Meanwhile, the study of meson systems has its own specifics due to the possibility of the Bose–Einstein condensation of bosonic particles. The aim of this paper is to investigate thermodynamic properties of a bosonic many-particle system, specifically the character of the phase transitions during the Bose–Einstein condensation at high densities. The latter condition means that the interaction in the bosonic system plays a sufficient role.

Historically, the problem of the Bose–Einstein condensation in the system of interacting bosons has been studied, starting from the pioneering works of N.N. Bogolyubov [1], where he investigated non-ideal gas of bosons and managed to describe the excitations of the system of interacting bosons in terms of non-interacting quasi-particles. Starting from this approach, the investigation of interacting bosons at the temperatures close to zero yielded the powerful impulse from the mean-field approach. Indeed, if the interactions in the diluted atomic gases are sufficiently weak, it can be argued that the mean field is the condensate wave function itself, as it was argued in Refs. [1–3]. Bogolyubov developed this idea systematically to study Bose condensation and superfluidity. Then, neglecting the fluctuations altogether, it is possible to derive the equation of motion for the wave function of the mean field, i.e., for the condensate wave function. This is the nonlinear Schrodinger equation or Gross–Pitaevskii equation [4,5]. Afterwards, these approaches were supplemented by the number of fruitful generalizations.

However, these methods and approaches are not suitable for the study of the Bose–Einstein condensation at high densities. That has two main reasons. First, high densities imply that the possible condensate states occupy the region of high temperatures where the density of thermal particles can no longer be treated as a small fluctuation comparing with the density of condensate. Second, as was pointed out by Kerson Huang in his textbook [6], the real conservation law deals with the conserved quantity that is the number of particles minus the number of antiparticles. That is why any study of the Bose–Einstein condensation in the relativistic Bose gas must take antiparticles into account. Firstly, this was discussed in Ref. [7]. Moreover, as was shown in [8] in the case of the "weak" attraction in the system and a conserved charge, the particles only develop the condensate states, but the antiparticles are in the thermal phase for all temperature ranges beginning from zero temperature.

In the present study, we are focused first of all on meson systems. This field is an essential part of investigations of hot and dense hadronic matter, which is a subject of active research [9]. In our study, we name the bosonic particles "pions" conventionally. The preference is made because the charged π mesons are the lightest hadrons that couple to the isospin chemical potential. At the same time, the pions are the lightest nuclear boson particles and thus, an account for "temperature creation" of particle–antiparticle pairs is a task for quantum statistics widely exploited in the paper. The problem of the Bose–Einstein condensation of π mesons has been studied previously, starting from the pioneering works of A.B. Migdal and coworkers (see [10] for references). Formation of classical pion fields in heavy-ion collisions was discussed in Refs. [11–14], and the systems of pions and K mesons with a finite isospin chemical potential have been considered in more recent studies [15–19]. A scalar model of a bosonic system that develops a Bose–Einstein condensate with conservation of isospin (charge) was first studied in [7,20,21]. Various aspects of free and interacting systems of relativistic bosons are discussed further in Refs. [22–26]. First-principles lattice calculations provide interesting new results concerning dense pion systems [27,28].

The presented study is associated with the approach proposed in Ref. [8], where the boson system was considered when the attraction between particles is "weak". Here, we proceed to investigate the thermodynamic properties of interacting particle–antiparticle meson systems at the conserved isospin density in the framework of the canonical ensemble using the mean-field model (see Appendix A). In this paper, we study also the boson systems where the attractive interaction between particles is "strong". (The rigorous definitions of the "weak" and "strong" attractive interactions will be given further.) We regard a studied self-interacting many-particle system as a toy model that can help us understand the Bose–Einstein condensation and phase transitions over a wide range of temperatures and densities.

The paper is organized as follows. Section 2 shortly describes the thermodynamic properties and condensation in an ideal Boson gas at the particle-number conservation. In Section 3, we introduce a self-interacting scalar mean-field model, which is then used to investigate condensate creation in the bosonic system of particles and antiparticles. An analogous description of the bosonic system of particles and antiparticles, but in the framework of the thermodynamic mean-field model, is given in Section 4. Section 5 compares the results obtained in the former two approaches for describing the bosonic system and the condensate formation at zero total charge. The phase transitions in the particle–antiparticle system with conserved isospin (charge) density are studied in Section 6. Section 7 is a final one, where we compare the description of the Boson systems in the presence of condensate in the framework of the canonical ensemble and the grand canonical ensemble. Conclusions of the present study are given in Section 8.

2. Canonical Ensemble: Condensation in Ideal Boson Gas

As a referring point, let us give a reminder about the main properties of the Bose condensation in a single-component ideal gas at conserved particle-number density n. This is shown in Figure 1, where two samples of the particle-number density are presented, $n = 0.1, 0.2$ fm^{-3}. The red dashed line is the critical curve $n_{\text{lim}}^{(\text{id})}$ that determines the critical temperature T_c. The critical curve is the dependence of the particle-number density on the temperature at the maximum value of the chemical potential, which is equal to the particle mass, $\mu = m$ [1]. Thus, the formula that determines the critical curve reads

$$n_{\text{lim}}^{(\text{id})} = g \int \frac{d^3k}{(2\pi)^3} f_{\text{BE}}(E, \mu)\big|_{\mu=m} , \tag{1}$$

where $E = \omega_k = \sqrt{m^2 + k^2}$ and

$$f_{\text{BE}}(E, \mu) = \frac{1}{e^{(E-\mu)/T} - 1} . \tag{2}$$

In Figure 1 and further in the text, the dependence $n(T)$ given in Equation (3) is noted as $n = n_{\text{lim}}^{(\text{id})}(T)$. The solution of Equation (3) with respect to temperature for the given particle density n determines the critical temperature $T_c(n)$.

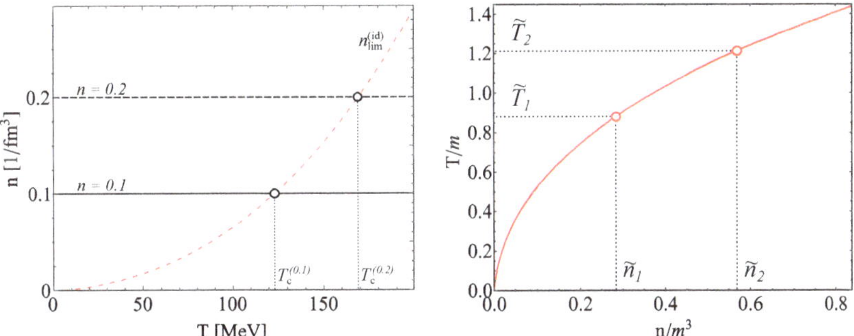

Figure 1. Left panel: particle-number density versus temperature in ideal single-component gas. The horizontal lines represent two constant particle density samples, $n = 0.1, 0.2$ fm^{-3}, which correspond to critical temperatures $T_c^{(0.1)}$ and $T_c^{(0.2)}$, respectively. Here, the critical curve $n_{\text{lim}}^{(\text{id})}(T)$ is defined in (3). **Right panel:** normalized critical temperature $\widetilde{T} = T/m$ vs. normalized particle density $\widetilde{n} = n/m^3$ in ideal single-component gas.

In the condensate phase, the generalization of Equation (3) is

$$n = n_{\text{cond}}(T) + g \int \frac{d^3k}{(2\pi)^3} f_{\text{BE}}(E, \mu)\big|_{\mu=m} . \tag{3}$$

The results of calculation of the energy density and heat capacity represented in Figure 2 evidently show that at the crossing point of the particle-number density and the critical curve, the phase transition of the second order occurs. Indeed, there is a finite discontinuity of the derivative of the heat capacity in the T_c points and a smooth behavior of the energy–density dependence in these points, i.e, there is no release of the latent heat.

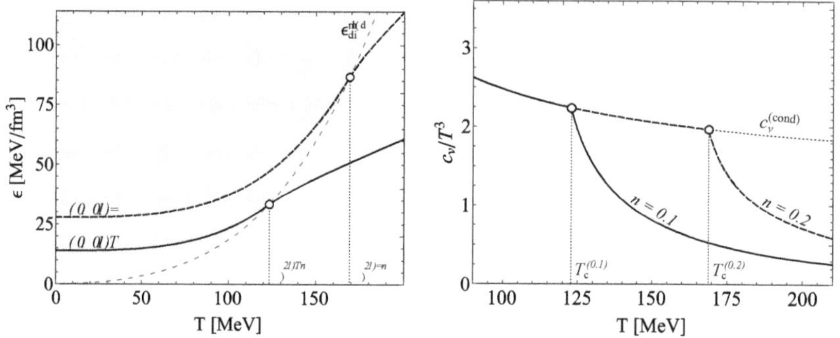

Figure 2. Left panel: energy density versus temperature for the same system and conditions as in the left panel. The red dashed line marked as $\varepsilon_{\text{lim}}^{(\text{id})}$ represents the energy density of the states that belong to the critical curve $n_{\text{lim}}^{(\text{id})}$ depicted in the upper panel. **Right panel:** heat capacity normalized to T^3 as a function of temperature in the ideal single-component gas where the particle-number density is kept constant.

Let us briefly discuss the results obtained for a single-component ideal gas, where the particle-number density n remains constant. First of all, we fix that when the line $n(T) = \text{const}$ intersects the critical curve $n_{\text{lim}}^{(\text{id})}(T)$, the system undergoes a phase transition of the second order or follows the Ehrenfest classification of the third order. It has long been known, see Ref. [29], that the Bose–Einstein condensation is indeed a third-order phase transition according to the first classification of general types of transitions between phases of matter, introduced by Paul Ehrenfest in 1933 [30,31]. Therefore, the obtained temperature T_c is really the temperature of the phase transition of the second order (according to modern terminology), and the density of condensate n_{cond} is the order parameter. In what follows, we will show that the same behavior is typical also in the case of interacting two-component systems at conserved charge density.

3. Self-Interacting Scalar Field

We start our consideration from the Lagrangian density of the self-interacting real scalar field

$$\mathcal{L}(x) = \frac{1}{2}\left[\partial_\mu \hat{\phi}(x)\partial^\mu \hat{\phi}(x) - m^2\,\hat{\phi}^2(x)\right] + \mathcal{L}_{\text{int}}[\hat{\phi}^2(x)], \tag{4}$$

where $x = (t, r)$. We adopt that

$$\hat{\phi}(r) = \phi_{\text{cond}} + \hat{\psi}(r), \quad \text{where} \quad \langle\,\hat{\psi}(r)\,\rangle = 0. \tag{5}$$

Here, we use the famous Bogolyubov's decomposition of the field operator into two contributions [1–3]

$$\hat{\phi}(r) = \frac{1}{\sqrt{V}}a_0 + \frac{1}{\sqrt{V}}\sum_{k\neq 0} a_k e^{ik\cdot r/\hbar}. \tag{6}$$

Due to the argument that at $T \to 0$ in a non-perfect Bose gas, the number of particles on the ground state N_0 approximately equals to the total number of particles N,

$$N_0 = \langle a_0^+ a_0 \rangle \approx N, \tag{7}$$

one can treat a_0 and a_0^+ as classical values.

Heisenberg representation:

$$\hat{\phi}(x) = e^{iHt}\hat{\phi}(r)e^{-iHt} = \phi_{\text{cond}} + \hat{\psi}(x) \quad \text{with} \quad \langle\hat{\psi}(x)\rangle = 0, \tag{8}$$

$$\left[\hat{\psi}(t,\mathbf{r}),\,\frac{\partial\hat{\psi}(t,\mathbf{r}')}{\partial t}\right] = \left[\hat{\phi}(t,\mathbf{r}),\,\frac{\partial\hat{\phi}(t,\mathbf{r}')}{\partial t}\right] = i\delta^3(\mathbf{r}-\mathbf{r}')\,. \tag{9}$$

Hence, the quantum fluctuations of the field $\hat{\psi}(x)$ have the same commutation relation as the complete field $\hat{\phi}(x)$. Expansion over solutions of the Klein–Gordon equation is

$$\hat{\psi}(x) = \int_{|\mathbf{p}|\neq 0}\frac{d^3p}{(2\pi)^3 2\omega_p}\left(a_{\mathbf{p}}\,e^{-ip\cdot x} + a_{\mathbf{p}}^{+}\,e^{ip\cdot x}\right)\Big|_{p^0=\omega_p}\,, \tag{10}$$

where

$$[a_k,\,a_p^{+}] = (2\pi)^3 2\omega_p\,\delta^3(\mathbf{k}-\mathbf{p})\,, \qquad [a_k,\,a_p] = 0\,. \tag{11}$$

For the field variance, we obtain the following decomposition:

$$\left\langle\hat{\phi}^2(x)\right\rangle = \left\langle\phi_{\text{cond}}^2 + 2\phi_{\text{cond}}\hat{\psi} + \hat{\psi}^2\right\rangle = \phi_{\text{cond}}^2 + \left\langle\hat{\psi}^2\right\rangle\,. \tag{12}$$

We see that the field variance is decomposed also on classical and quantum pieces.

3.1. The Effective Lagrangian in the Mean-Field Approximation

We are going to consider the Bose–Einstein condensation of the scalar field (for details see Ref. [32]),

$$\mathcal{L}(x) = \frac{1}{2}\left[\partial_\mu\hat{\phi}(x)\,\partial^\mu\hat{\phi}(x) - m^2\,\hat{\sigma}(x)\right] + \mathcal{L}_{\text{int}}(\hat{\sigma})\,, \tag{13}$$

where we introduced notation

$$\hat{\sigma}(x) = \hat{\phi}^2(x)\,. \tag{14}$$

We use the quantum statistical averaging of the operator $\hat{A}$:

$$\langle\hat{A}\rangle = \frac{1}{Z}\text{Tr}\left[e^{-\beta(\hat{H}-\mu\hat{N})}\,\hat{A}\right]\,, \qquad Z = \text{Tr}\left[e^{-\beta(\hat{H}-\mu\hat{N})}\right]\,. \tag{15}$$

Next, we introduce the mean value σ of the operator $\hat{\sigma}$

$$\sigma = \langle\hat{\sigma}\rangle\,, \qquad \delta\hat{\sigma} = \hat{\sigma} - \sigma\,. \tag{16}$$

Here, $\delta\hat{\sigma}$ is the deviation of the operator $\hat{\sigma}$ from its mean value. One can expand the Lagrangian (13) as the function on the variable $\hat{\sigma}$ around the point σ:

$$\mathcal{L}_{\text{int}}(\hat{\sigma}) \simeq \mathcal{L}_{\text{int}}(\sigma) + \delta\hat{\sigma}\,\mathcal{L}'_{\text{int}}(\sigma) = \mathcal{L}_{\text{int}}(\sigma) + \hat{\sigma}\,\mathcal{L}'_{\text{int}}(\sigma) - \sigma\,\mathcal{L}'_{\text{int}}(\sigma)\,, \tag{17}$$

where prime means the derivative with respect to σ. We come to the effective Lagrangian in the mean-field approximation

$$\mathcal{L}(x) \simeq \frac{1}{2}\left[\partial_\mu\hat{\phi}(x)\,\partial^\mu\hat{\phi}(x) - M^2(\sigma)\,\hat{\phi}^2(x)\right] + P_{\text{ex}}(\sigma)\,, \tag{18}$$

where we introduced the following notations

$$P_{\text{ex}}(\sigma) \equiv \mathcal{L}_{\text{int}}(\sigma) - \sigma\,\frac{\partial\mathcal{L}_{\text{int}}(\sigma)}{\partial\sigma}\,, \qquad \hat{M}^2(\sigma) = m^2 + 2\,U(\sigma)\,, \tag{19}$$

with

$$U(\sigma) \equiv -\frac{\partial\mathcal{L}_{\text{int}}(\sigma)}{\partial\sigma}\,. \tag{20}$$

The differential relation between the excess pressure $P_{\text{ex}}(\sigma)$ and the mean field $U(\sigma)$ follows from this definition

$$\sigma\,\frac{\partial U(\sigma)}{\partial\sigma} = \frac{\partial P_{\text{ex}}(\sigma)}{\partial\sigma}\,. \tag{21}$$

3.2. Hamiltonian Density in the Mean-Field Approximation

Momentum operator $\hat{\pi}$ satisfies the equal-time commutation relations

$$\hat{\pi}(x) = \partial_t\,\hat{\phi}(x)\,, \qquad \left[\hat{\phi}(t,\mathbf{r}),\,\hat{\pi}(t,\mathbf{r}')\right] = i\delta^3(\mathbf{r}-\mathbf{r}')\,. \tag{22}$$

The Hamiltonian density $\hat{\mathcal{H}} = \hat{\pi}\,\partial_t\hat{\phi} - \mathcal{L}$ reads

$$\hat{\mathcal{H}} \simeq \frac{1}{2}\left[\hat{\pi}^2(x) + \boldsymbol{\nabla}\hat{\phi}(x)\cdot\boldsymbol{\nabla}\hat{\phi}(x) + M^2(\sigma)\hat{\phi}^2(x)\right] - P_{\text{ex}}(\sigma)\,. \tag{23}$$

Using solutions of the Klein–Gordon equation,

$$\partial^\mu\partial_\mu\,\hat{\phi} + M^2(\sigma)\,\hat{\phi} = 0, \tag{24}$$

one can represent the scalar field $\hat{\phi}(x)$ as

$$\hat{\phi}(x) = g\int \frac{d^3k}{(2\pi)^3\sqrt{2\omega_k}}\left[a_k e^{-ik\cdot x} + a_k^+ e^{ik\cdot x}\right], \tag{25}$$

where $k^0 = \omega_k = \sqrt{k^2 + M^2(\sigma)}$ and the operators of creation and annihilation satisfy the standard commutation relations

$$\left[a_k, a_{k'}^+\right] = (2\pi)^3\delta(k-k'), \qquad \left[a_k, a_{k'}\right] = \left[a_k^+, a_{k'}^+\right] = 0\,. \tag{26}$$

As a first step, we consider a boson system at zero isospin (charge) density $n_I = 0$, i.e., the numbers of particles and antiparticles are equal. In this case, the Hamiltonian in the mean-field (MF) approximation reads

$$\hat{H} = \int d^3x\,\hat{\mathcal{H}} = V\left[g\int \frac{d^3k}{(2\pi)^3}\,\omega_k\,a_k^+ a_k - P_{\text{ex}}(\sigma)\right]. \tag{27}$$

In the MF approximation, the equilibrium momentum distribution coincides with that of an ideal gas of bosons with the effective mass $M(\sigma)$

$$n_k(\sigma) \equiv \langle a_k^+\,a_k\rangle = (e^{\beta\omega_k}-1)^{-1}\,, \qquad \beta = 1/T,\quad k_{\text{B}}=1,\quad \mu_I=0, \tag{28}$$

where $\omega_k = \sqrt{M^2(\sigma) + k^2}$ with $M^2(\sigma) = m^2 + 2U(\sigma)$.

The thermodynamical description of the system is obtained by means of solution of self-consistent equations for the thermal phase and condensate phase with respect to the scalar density $\sigma = \langle\hat{\phi}^2\rangle^{\,2}$. In the thermal phase, this equation reads

$$\sigma = g\int \frac{d^3k}{(2\pi)^3}\,\frac{n_k(\sigma)}{\omega_k}\,. \tag{29}$$

In the condensate phase, one should take into account the necessary condition for condensate creation $M^2(\sigma) = 0$ and include into the equation the density of the scalar condensate, then the equation becomes

$$\sigma = \sigma_{\text{cond}} + g\int \frac{d^3k}{(2\pi)^3}\,\frac{n_k(\sigma)}{\omega_k}\bigg|_{M^2(\sigma)=0}, \tag{30}$$

where, in the case of $\mu_I = 0$ (or $n_I = 0$), we are left with one canonical variable T. The last equation corresponds to the relation

$$\left\langle\hat{\phi}^2\right\rangle = \phi_{\text{cond}}^2 + \left\langle\hat{\psi}^2\right\rangle, \tag{31}$$

which we obtained as a result of the decomposition of the field operator (5) and specific features of the quantum fluctuations, see Equation (12).

Other thermodynamic quantities that characterize the quasi-particle boson system can be obtained in a regular way in the framework of the quantum statistics. The pressure reads

$$p = p_{\text{kin}}(T, \sigma) + P_{\text{ex}}(\sigma),$$ (32)

where the kinetic pressure in the thermal phase is

$$p_{\text{kin}}(T, \sigma) = \frac{g}{3} \int \frac{d^3k}{(2\pi)^3} \frac{k^2}{\omega_k} n_k(\sigma),$$ (33)

whereas the kinetic pressure in the condensate phase reads

$$p_{\text{kin}}(T, \sigma) = \frac{g}{3} \int \frac{d^3k}{(2\pi)^3} \frac{k^2}{\omega_k} n_k(\sigma)\Big|_{M^2(\sigma)=0}.$$ (34)

The energy density and entropy density $s = (\varepsilon + p)/T$ in the thermal phase read

$$\varepsilon = g \int \frac{d^3k}{(2\pi)^3} \omega_k n_k(\sigma) - P_{\text{ex}}(\sigma),$$ (35)

$$s = \frac{g}{T} \int \frac{d^3k}{(2\pi)^3} \left(\omega_k + \frac{k^2}{3\omega_k} \right) n_k(\sigma).$$ (36)

The energy density and entropy density in the condensate phase read

$$\varepsilon = \varepsilon_{\text{cond}} + g \int \frac{d^3k}{(2\pi)^3} \omega_k n_k(\sigma)\Big|_{M^2(\sigma)=0} - P_{\text{ex}}(\sigma),$$ (37)

$$s = s_{\text{cond}} + \frac{g}{T} \int \frac{d^3k}{(2\pi)^3} \left(\omega_k + \frac{k^2}{3\omega_k} \right) n_k(\sigma)\Big|_{M^2(\sigma)=0}.$$ (38)

3.3. Bosonic System with $\varphi^4 + \varphi^6$ Self-Interaction

For specific numerical calculations, we adopt the following parametrization of the interaction part of the Lagrangian

$$\mathcal{L}_{\text{int}}\left(\hat{\phi}^2(x) \right) = \frac{a}{4} \hat{\phi}^4(x) - \frac{b}{6} \hat{\phi}^6(x).$$ (39)

Then, the mean field and the excess pressure are

$$U(\sigma) = -\frac{1}{2}a\sigma + \frac{1}{2}b\sigma^2, \qquad P_{\text{ex}}(\sigma) = -\frac{a}{4}\sigma^2 + \frac{b}{3}\sigma^3,$$ (40)

where $\sigma = \langle \hat{\phi}^2 \rangle$. This means that attraction and repulsion between particles in the form of a mean field are simultaneously present in the system of bosons. The distribution function $n_k = \left[\exp\left(\sqrt{k^2 + M^2}/T \right) - 1 \right]^{-1}$ makes sense when the argument is positive, i.e.,

$$M^2(\sigma) = m^2 + 2U(\sigma) = m^2 - a\sigma + b\sigma^2 \geqslant 0.$$ (41)

The limiting case in this relation is the condition for the occurrence of scalar condensate:

$$M^2(\sigma) = m^2 - a\sigma + b\sigma^2 = 0.$$ (42)

Solutions of this equation are

$$\sigma_{1,2} = \frac{m}{\sqrt{b}}\left(\kappa \mp \sqrt{\kappa^2 - 1}\right), \tag{43}$$

where we introduce the dimensionless parameter κ:

$$\kappa = \frac{a}{2m\sqrt{b}}, \qquad \rightarrow \quad a = \kappa\, a_c, \quad a_c = 2m\sqrt{b}. \tag{44}$$

Thus, we conclude that when $\kappa \leq 1$, the quasi-particle effective mass becomes imaginary ($M^2 < 0$), and the system becomes unstable. The stability is restored by the formation of the Bose condensate.

4. An Interacting Boson System within the Thermodynamic Mean-Field Model

We are going to compare a description of the boson system at high densities in the field-theoretical and in the quantum-statistical approaches. The consideration of the latter one starts from separation of the free energy $F(T, N, V)$ into free and interaction parts as

$$F(T, N, V) = F_0 + F_{\text{int}}. \tag{45}$$

Or, for the free energy density as $\Phi(T, n) = \Phi_0 + \Phi_{\text{int}}$, where $\Phi(T, n) = F(T, N, V)/V$. Then, we introduce the following important notations (for details, see [33])

$$U(n, T) = \left[\frac{\partial \Phi_{\text{int}}(n, T)}{\partial n}\right]_T, \tag{46}$$

$$P_{\text{ex}}(n, T) = n\left[\frac{\partial \Phi_{\text{int}}(n, T)}{\partial n}\right]_T - \Phi_{\text{int}}(n, T). \tag{47}$$

These quantities are related to one another by the differential equality

$$n\frac{\partial U(n, T)}{\partial n} = \frac{\partial P_{\text{ex}}(n, T)}{\partial n}. \tag{48}$$

In these notations, the pressure in the system can be written as

$$p(T, \mu_I) = \frac{g}{3}\int \frac{d^3k}{(2\pi)^3}\frac{k^2}{\sqrt{m^2 + k^2}} f_{\text{BE}}\left(E_k(n), \mu_I\right) + P_{\text{ex}}(n), \tag{49}$$

where g is the degeneracy factor, $E_k(n) = \sqrt{m^2 + k^2} + U(n)$ is the effective single-particle energy, μ_I is the isospin chemical potential and f_{BE} is the Bose–Einstein distribution function

$$f_{\text{BE}}(E, \mu) = \left\{\exp\left[\frac{E - \mu}{T}\right] - 1\right\}^{-1}. \tag{50}$$

In the particle–antiparticle system, the Euler relation is $\varepsilon + p = Ts + \mu_I n_I$, where n_I is the isospin (charge) density. Let us first consider the case of zero charge density, i.e., $n_I = 0$, that corresponds to $\mu_I = 0$ in the grand canonical ensemble.

The mean-field model implies that the thermodynamic description of the system is obtained via a self-consistent approach. In our case, this is achieved by a self-consistent equation for the total particle density n, which should be solved separately in the thermal and condensate phases. In the thermal phase, this equation has a structure $n = n_{\text{th}}(T, n)$, and it should be solved with respect to the total particle density n for every fixed value of T,

$$n = g\int \frac{d^3k}{(2\pi)^3} f_{\text{BE}}\left(E_k(n)\right), \tag{51}$$

where $f_{\rm BE}(E) = [\exp(E/T) - 1]^{-1}$. The solution of Equation (51) in the thermal phase results in the explicit dependence $n = n(T)$, which in general differs from the ideal gas dependence, $n_0(T)$. Knowledge of the dependence $n(T)$ gives a possibility to obtain equation of state through a direct calculation of other thermodynamic quantities such as pressure, energy density, entropy density, etc.

In the condensate phase, one should take into account the condensation condition at $\mu_I = 0$, $U(n) + m = 0$ that leads to specific "critical" density n_c, which is a real root of this equation. The solution of this equation has the following structure: $n_c = n_{\rm cond}(T) + n_{\rm th}(T)$, where the density of the condensate component $n_{\rm cond}$ appears as a new degree of freedom. Thus, in the condensate phase, the self-consistent equation for $n_{\rm cond}$ reads

$$n_c = n_{\rm cond} + g \int \frac{d^3 k}{(2\pi)^3} f_{\rm BE}(E_{\rm kin}), \qquad (52)$$

where $E_{\rm kin} = \sqrt{m^2 + k^2} - m$.

4.1. Parametrization of the Interaction

To be closer to the field-theoretical approach, we use the following correspondence between the scalar density $\langle \phi^2 \rangle$ and particle number density n, which simply coincide with one another in the non-relativistic limit. Then, using the correspondence $\varphi^4 \rightarrow n^2$ and $\varphi^6 \rightarrow n^3$, we write the excess pressure and the corresponding mean field (see the differential relation (48)) as

$$P_{\rm ex}(n) = -\frac{1}{2} A n^2 + \frac{2}{3} B n^3, \qquad \rightarrow \qquad U(n) = -A n + B n^2, \qquad (53)$$

where the positive parameter A is responsible for attraction between particles and the positive parameter B for repulsion between particles in a boson system (for details see [10]). The parameter A will be varied, whereas the parameter B, associated with a hard-core repulsion, will be kept constant. It is advisable to parameterize A in the following way: let us use solutions of equation $U(n) + m = 0$, which determine the condition for a condensate creation (a similar algorithm was adopted in Refs. [10,34]). For the given mean field (53), there are two roots of this equation

$$n_1 = \sqrt{\frac{m}{B}} \left(\kappa - \sqrt{\kappa^2 - 1} \right), \qquad n_2 = \sqrt{\frac{m}{B}} \left(\kappa + \sqrt{\kappa^2 - 1} \right), \qquad (54)$$

where we introduce the dimensionless parameter κ:

$$\kappa \equiv \frac{A}{2\sqrt{m B}}. \qquad (55)$$

Then, one can parameterize the attraction coefficient as $A = \kappa A_c$ with $A_c = 2\sqrt{mB}$. As it is seen below, the parameter κ is the scale parameter that determines the phase structure of the system. We consider two intervals of the parameter κ: (1) a "weak" attraction that corresponds to $\kappa < 1$, i.e., $n_{1,2}$ are not the real roots, and (2) a "strong" attraction that corresponds to $\kappa > 1$, i.e., $n_{1,2}$ are the real roots. The critical value A_c is obtained when both roots coincide, i.e., when $\kappa = 1$, then $A = A_c = 2\sqrt{mB}$.

5. Condensation of Interacting Bosons at Finite Temperatures

In this section, we compare the numerical results obtained within two approaches, the field-theoretical approach, which is based on the scalar mean-field (SMF) model, and the quantum-statistical approach, which is based on the thermodynamic mean-field (TMF) model. Our purpose is to study an influence of the attraction and repulsion between particles on the thermodynamic properties of a Boson system, especially in the presence of the condensate. In both cases, we will keep constant the repulsive term while varying the attractive interaction by means of the parameter κ. We present the solutions of self-consistent equations for different values of the attraction coefficient a in the SMF model while fixing the repulsion coefficient as $b = 25\,m_\pi^{-2}$. The same is done for the TMF model, where we vary the attraction coefficient A while fixing the repulsion coefficient as $B/m_\pi = 10v_0^2$. It is necessary to note that these variations of the attraction coefficients are done in the same way in both approaches by means of the dimensionless parameter κ: in the SMF model as $a = \kappa a_c$, where $a_c = 2m\sqrt{b}$, and in the TMF model as $A = \kappa A_c$, where $A_c = 2\sqrt{mB}$. We name the boson particles "pions" and take their mass as $m = m_\pi = 139$ MeV for the degeneracy factor $g = 3$. In the SMF model, the critical curve is obtained when $M^2 = m^2 + 2U(n) = 0$, and it reads

$$\sigma_{\lim} = g \int \frac{d^3k}{(2\pi)^3} \left(e^{k/T} - 1 \right)^{-1} = \frac{g}{12}\, T^2 . \tag{56}$$

In the case of the TMF model, there is a similar condition for determination of the critical curve, $m + U(n) = 0$, that looks like a presence of the effective chemical potential $\mu = m$. Therefore, the critical curve in the case of the TMF model reads

$$n_{\lim} = g \int \frac{d^3k}{(2\pi)^3} \left\{ \exp\left[\frac{\sqrt{m^2 + k^2} - m}{T} \right] - 1 \right\}^{-1} . \tag{57}$$

The numerical calculations of the particle density vs, temperature for the SMF model and TMF model are presented in Figure 3 in the left and right panels, respectively. The calculations are done for different values of the attraction coefficients a and A, which are parameterized by parameter κ in both models. We name $\kappa < 1$ as the "weak" attraction and $\kappa > 1$ as the "strong" attraction. It is seen that for "weak" attraction, the scalar densities and the particle number densities are in the thermal phase. At $\kappa = \kappa_c = 1$, the density curves have one common point with the critical (red dashed line). The critical curve $\sigma_{\lim}(T)$ is depicted in Figure 3 in the left panel as a red dotted-dashed line, and the critical curve $n_{\lim}(T)$ is depicted in Figure 3 in the right panel as a red dashed line. In both approaches, at "strong" attraction, $\kappa > 1$, there is a first-order phase transition at $T = T_c$ with creation of the condensate. This is a result of competition of pressure corresponding to two different solutions of the self-consistent equation in the thermal and in the condensate phases.

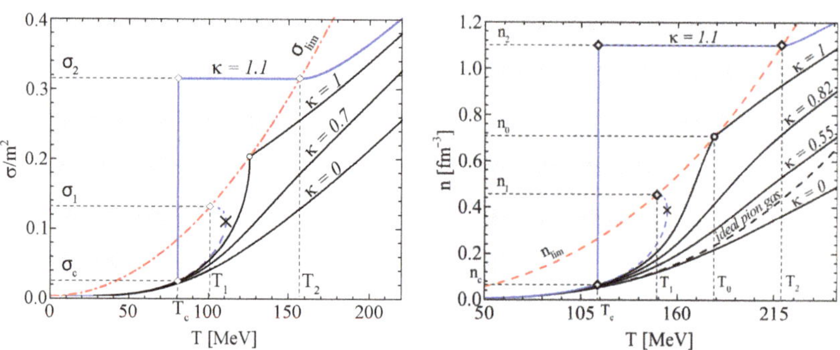

Figure 3. Left panel: scalar density vs. temperature, $b = 25\,m_\pi^{-2}$, $a = \kappa a_c$, $a_c = 2m\sqrt{b}$. **Right panel:** particle-number density vs. temperature, $B = 10m_\pi v_0^2$, $A = \kappa A_c$, $A_c = 2\sqrt{mB}$. In both panels, the shaded area indicates the states of the Bose–Einstein condensate. Crosses on both panels separate metastable and non-physical states.

In the SMF model, we solve Equation (30) to obtain the scalar density $\sigma = \sigma_{\text{therm}}(T)$ in the thermal (liquid–gas) phase and a corresponding pressure $p_{\text{lg}}(T)$. On the other hand, Equation (30) for the condensate (mix) phase [3] is characterized by two constant solutions $\sigma = \sigma_1$ and $\sigma = \sigma_2$, see Figure 3, the left panel. Then, we compare the pressure dependencies $p_{\text{mix}}^{(1)}(T)$ and $p_{\text{mix}}^{(2)}(T)$ corresponding to σ_1 and σ_2, respectively, with one another and with $p_{\mathrm{lg}}(T)$. The result of this comparison is depicted in Figure 4 in left panel as the solid blue line. It is seen that at $T = T_{\mathrm{c}}$, the pressure $p_{\mathrm{mix}}^{(2)}(T)$ becomes the largest, which determines the phase transition of the first order with creation of the scalar condensate (for details, see [32]).

The same analysis is made also for the TMF model. We solve Equation (51) to obtain the dependence of the particle density $n = n_{\mathrm{therm}}(T)$ in the thermal (liquid–gas) phase and a corresponding pressure $p_{\mathrm{lg}}(T)$. Equation (52) is characterized by two constant real roots n_{c1} and n_{c2} and two corresponding pressures. The result of this comparison is depicted in Figure 4 in the right panel as the solid blue line.

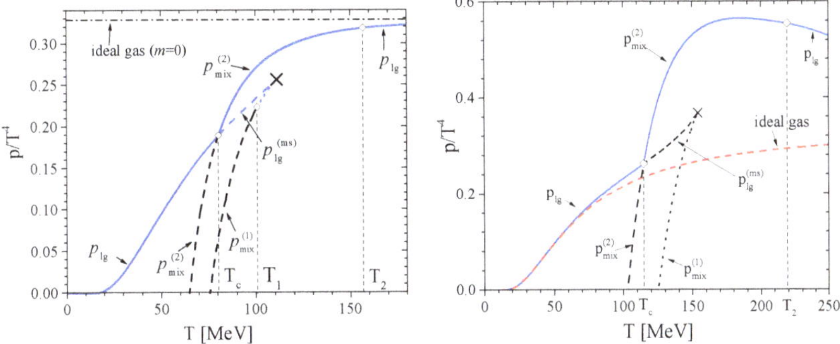

Figure 4. In both panels: Pressure vs. temperature for the supercritical attraction, $\kappa = 1.1$. The solid blue line that consists of two segments, p_{lg} and $p_{\mathrm{mix}}^{(2)}$, is the final equation of state, T_{c} is the critical temperature that indicates the phase transition of the first order. Crosses on both panels separate metastable and non-physical states. **Left panel:** The scalar mean-field model. The pressure $p_{\mathrm{mix}}^{(1)}$ corresponds to the scalar density σ_1. **Right panel:** The thermodynamic mean-field model. The pressure $p_{\mathrm{mix}}^{(1)}$ corresponds to the particle-number density n_1.

It is seen from the comparison of results depicted in the two panels in Figure 3, and in the two panels in Figure 4, that the two models show a very similar behavior. That is why in what follows, we consider only the TMF model, assuming that it gives a true thermodynamic description of the bosonic system at high densities.

6. Particle–Antiparticle System with Conserved Isospin (Charge) Density

6.1. Derivation of Basic Equations

Let us consider a homogeneous system with conserved charge (isospin). The description of such a system can be done within the canonical ensemble with the canonical variables (T, n_I). Here, $n_I = n^{(-)} - n^{(+)}$ is the difference between the densities of π^- and π^+ mesons. Note, now we use the thermodynamic mean-field (TMF) model for many-component boson systems, see Appendix A. As a first step, we consider the "weak" attraction between particles, i.e., $\kappa \leq 1$. In this case, there are two pairs of self-consistent equations. The first set of equations describes the system when both components, i.e., both the π^- and π^+ mesons, are in the thermal phase (high temperatures). The second set of equations describes the system at low temperatures, when the π^- mesons are in the condensate phase but the π^+ mesons are in the thermal (kinetic) phase (see details in Ref. [8]). At high temperatures, the set of equations reads

$$n = \int \frac{d^3k}{(2\pi)^3} \left[f_{\mathrm{BE}}\left(E(k,n),\mu_I\right) + f_{\mathrm{BE}}\left(E(k,n),-\mu_I\right) \right], \tag{58}$$

$$n_I = \int \frac{d^3k}{(2\pi)^3} \left[f_{\mathrm{BE}}\left(E(k,n),\mu_I\right) - f_{\mathrm{BE}}\left(E(k,n),-\mu_I\right) \right], \tag{59}$$

where the Bose–Einstein distribution function $f_{\mathrm{BE}}(E,\mu)$ is defined in (2) and $E(k,n) = \sqrt{m^2 + k^2} + U(n)$. These equations should be solved with respect to the particle density n and chemical potential μ_I. We use the same parameterization as in the case of zero charge density (see Section 4.1), it depends on the total particle-number density n: $P_{\mathrm{ex}}(n) = -(1/2)An^2 + (2/3)Bn^3 \;\to\; U(n) = -An + Bn^2$. Actually, this parameterization of the interaction is in analogy to the field-theoretical approach with a correspondence $\langle \varphi^+\varphi \rangle \leftrightarrow n$, then, in the same manner, one can write $\varphi^4 \to n^2$ and $\varphi^6 \to n^3$.

One of the main goals of our research is to investigate the influence of attraction and repulsion between particles on the thermodynamic properties of the bosonic system, especially in the presence of a condensate. In this study, we fix the repulsive interaction in the system while changing the attraction between particles. As in the case of zero isospin density [10], we use the same parameterization of the attraction coefficient A using the solutions (54) of equation $U(n) + m = 0$. Then, in the same manner, we introduce dimensionless coefficient $\kappa \equiv A/(2\sqrt{mB})$ that parameterizes the parameter A as $A = \kappa A_{\mathrm{c}}$ with $A_{\mathrm{c}} = 2\sqrt{mB}$. Below, we use parameter κ to vary attraction between particles.

If one of the components of the particle–antiparticle system is in the condensate phase (low temperatures) [4] , then self-consistent equations that determine the thermodynamic structure of the system read

$$
n \;=\; n_{\mathrm{cond}}^{(-)}(T) + n_{\lim}(T) + \int \frac{d^3k}{(2\pi)^3}\, f_{\mathrm{BE}}\big(E(k,n), -\mu_I\big)\Big|_{\mu_I = m + U(n)}, \tag{60}
$$

$$
n_I \;=\; n_{\mathrm{cond}}^{(-)}(T) + n_{\lim}(T) - \int \frac{d^3k}{(2\pi)^3}\, f_{\mathrm{BE}}\big(E(k,n), -\mu_I\big)\Big|_{\mu_I = m + U(n)}, \tag{61}
$$

where we assume that the condensed state of π^- mesons develops under the (necessary) condition

$$
m + U(n) - \mu_I = 0. \tag{62}
$$

We use notation

$$
n_{\lim}(T) \;=\; \int \frac{d^3k}{(2\pi)^3}\, f_{\mathrm{BE}}(\omega_k, \mu_I)\Big|_{\mu_I = m} \tag{63}
$$

for a density of the thermal particles at the onset of condensation (the critical curve).

6.2. Numerical Results: Second-Order Phase Transitions Generated by the Particles That Carry Dominant Charge

The solutions of Equations (58)–(61) are depicted in Figure 5 as the dependence of particle-number densities of π^- mesons (left panel) and π^+ mesons (right panel) at fixed isospin density $n_I = 0.1$ fm^{-3} and a set of attraction parameters $\kappa = 0, 0.6, 0.85, 0.96, 1$. The red dashed lines in both panels are the critical curves $n_{\lim}$, which reflect the maximal density of thermal π^- pions (left panel) or π^+ pions (right panel). The dashed area indicates the phase with the condensed particles. The open stars in the left panel indicate the Bose condensation as a phase transition of the second order in the π^- component, where $T_{\mathrm{c}}^{(-)}$ is the temperature of the Bose condensation of π^- mesons. The "dark" star in the right panel indicates a virtual-like second-order phase transition created by the π^+ meson subsystem at the attraction parameter $\kappa = 1$. Each "star" on the graphs corresponds to a second-order phase transition. Roughly speaking, each intersection of the particle density curve with the critical curve corresponds to a phase transition of the second order.

It turns out that the phase structure of π^- mesons (the particles with dominant charge) can be grouped into two types: (a) the curve $n = n^{(-)}(T)$ has one cross with the critical curve $n_{\lim}(T)$, and (b) the curve $n = n^{(-)}(T)$ has three crosses with the critical curve. The regular behavior or the type (a) occurs when the parameter κ belongs to the low interval $0 \le \kappa < \kappa_s$, where $\kappa_s \approx 0.93$. In this case, π^- mesons for $T < T_{\mathrm{c1}}^{(-)}$ are in the condensate phase, and in the temperature interval $T > T_{\mathrm{c1}}^{(-)}$ they are in the thermal phase, see Figure 5, the left panel.

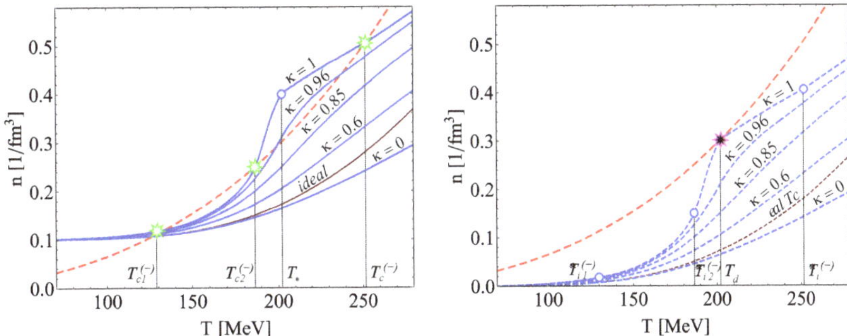

Figure 5. Left panel: The particle-number densities $n^{(-)}$ of π^- mesons versus temperature for the system of interacting π^+ - π^- mesons at fixed isospin density $n_I = 0.1$ fm^{-3} and the set of "weak" attraction parameters $\kappa = 0, 0.6, 0.85, 0.96, 1$. The red dashed curve $n_{\lim}$ reflects the maximal density of thermal π^- mesons (or π^+ mesons) in the ideal π^+ - π^- gas. The dashed area indicates the phase with the condensed particles. The open stars show the onset of phase transition of the second order of the π^- mesons. **Right panel:** The particle-number densities $n^{(+)}$ of π^+ mesons versus temperature at the same set of parameters as in the left panel. The "dark" star corresponding to the T_* temperature indicates a virtual second-order phase transition of the π^+ component without condensate formation.

Therefore, for the κ of type (a), the temperature of the phase transition T_c in the whole system is determined as $T_c = T_{c1}^{(-)}$, or it is a regular phase transition of the second order. Indeed, in Figure 6, in the left panel, one can clearly see a finite discontinuity of the derivative of heat capacity (left panel) and the absence of the latent heat (right panel) at the temperature T_c.

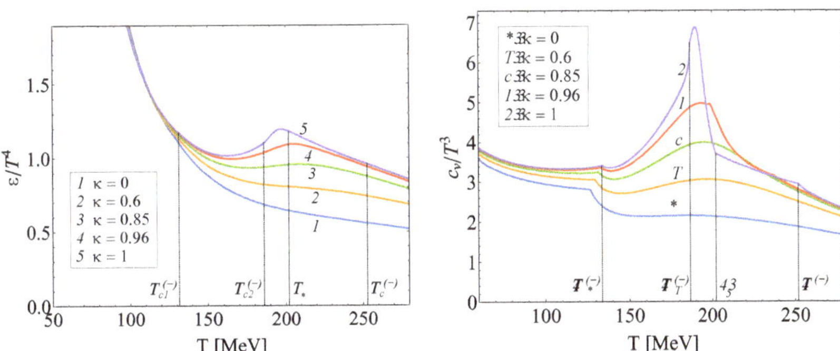

Figure 6. Left panel: Energy density versus temperature in the interacting particle–antiparticle system of pions at $\kappa = 0, 0.6, 0.85, 0.96, 1$. The isospin (charge) density is kept constant, $n_I = 0.1$ fm^{-3}. The points of the phase transition of the second order are indicated by the corresponding temperatures $T_{c1}^{(-)}$, $T_{c2}^{(-)}$, T_*, $T_c^{(-)}$. **Right panel:** heat capacity as a function of temperature for the same system and conditions as in the left panel.

In fact, in case (a), the dependence $n^{(-)}(T)$, which reflects the behavior of the density of π^- mesons (Figure 5, left panel), looks very similar to the behavior at a constant density of particles in a single-component system, at least in the condensate phase, that is, for temperatures $0 \leq T \leq T_c$, see Figure 1 in Section 2. On the other hand, the dependence $n^{(+)}(T)$ (Figure 5, right panel) that reflects behavior of the π^+ particle density looks very similar to the particle-density dependence at $n_I = 0$ and $\kappa < 1$, shown in Figure 3 in the right panel. Both of these features can be explained by the similar initial conditions at $T = 0$ and a slow creation of the thermal pion pairs at low temperatures.

When the attraction parameter κ increases and becomes type (b), i.e., $\kappa_s < \kappa \leq 1$, the phase structure of the charge-dominant component (π^- mesons) is more complex. In this case, the curve $n^{(-)}(T)$ consistently crosses the critical curve $n_{\lim}(T)$ three times at temperatures $T_{c1}^{(-)} < T_{c2}^{(-)} < T_c$, see Figure 5, the left panel. The obvious explanation of this phenomenon is due to the charge conservation. Indeed, for sufficiently high values of κ, say $\kappa > \kappa_s$, the π^+ density approaches the critical curve (see Figure 5, right panel) and simply "squeezes out" to the other side of the critical curve the π^- density since its values must be higher by n_I than π^+ density. That is, the states of the π^- mesons again "pass" into the condensate phase. As can be seen in Figure 6, each intersection of the curve $n^{(-)}(T)$ with the critical curve $n_{\lim}(T)$ corresponds to a phase transition of the second order. Indeed, in Figure 6, in the right panel, one can see a finite discontinuity of the derivative of heat capacity at temperatures $T_{c1}^{(-)}$, $T_{c2}^{(-)}$, T_* and T_c. At the same time, in the left panel in Figure 6, we see no jumps corresponding to the latent heat at these temperatures. Therefore, we can conclude that due to the conservation of charge, along with the regular phase transition of the second order, multiple "weak" phase transitions can also occur in a particle–antiparticle system.

At the same time, the antiparticle component of the system or π^+ mesons are in the thermal phase for the whole temperature range, see Figure 5, the right panel. Only at the critical value $\kappa = \kappa_c = 1$, the density $n^{(+)}(T)$ touches the critical curve $n_{\lim}(T)$ at the temperature $T = T_*$. For this special case where $\kappa = 1$, we have calculated the heat capacity and its derivative, see Figure 7. One can see that heat capacity (left panel) has a pronounced peak at a relatively high temperature of ~ 190 MeV.

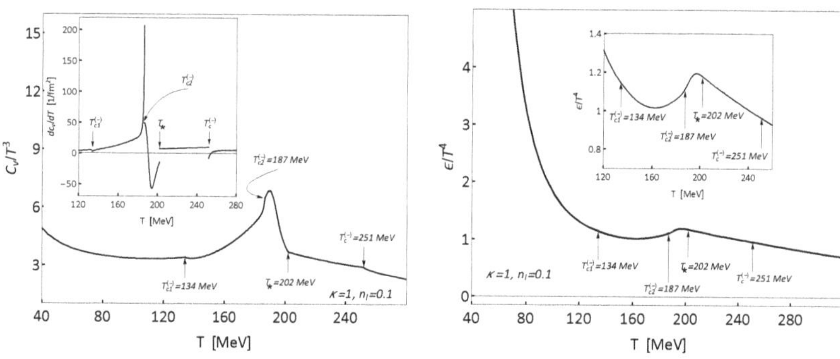

Figure 7. Left panel: Heat capacity normalized by T^3 as a function of temperature in the interacting particle–antiparticle system at $\kappa = 1$ (black solid curve). The isospin (charge) density is kept constant, $n_I = 0.1$ fm^{-3}. The derivative of heat capacity is shown in a small window. **Right panel:** Energy density normalized by T^4 versus temperature for the same system and conditions as in the left panel (black solid curve). The enlarged central area of the graphic is shown in a small window.

It is necessary to point out that the heat capacity and energy density are the physical quantities, which reflect the integrated behavior of the total particle–antiparticle system. That is why the curves $c_v(T)$ and $\varepsilon(T)$ "carry" specific peculiarities that are due to the joined behavior of the particles and antiparticles. This can be seen clearly in Figure 7 for $\kappa = 1$. Indeed, we see three phase transitions of second order at $T = T_{c1}$, T_{c2}, T_c that are due to behavior of π^- mesons at $\kappa = 1$. Meanwhile, for the π^+ meson subsystem at $\kappa = 1$, one can see the virtual second-order phase transition at $T = T_*$, marked as the filled star on the critical curve in Figure 5, right panel. It is a specific phase transition of the second order because there is no creation of the condensate in both directions from the temperature T_* [5]. The character of this phase transition is clearly seen in Figure 7, in the small window in left panel as a discontinuity of the heat-capacity derivative at $T = T_*$. At the same time, we see a smooth behavior of the energy density at this temperature, see the small window in Figure 7 in the right panel.

We notice that all crosses of the dependencies $n^{(-)}(T)$ and $n^{(+)}(T)$ with the critical curve $n_{\lim}(T)$ are exhibited as the finite discontinuity of the derivatives of heat capacity $c_v(T)$ at the temperatures $T = T_{c1}, T_{c2}, T_*, T_c$, see the left panel in Figure 7. In the right panel in this figure, we plot the energy density. One can recognize that it is really the second-order phase transitions at these four temperature points because the dependence of the energy density, $\varepsilon(T)$, is indeed continuous and without release of the latent heat.

Therefore, regarding thermodynamic behavior of the particle–antiparticle bosonic system at "weak" attraction ($\kappa \leq 1$), we identified the phase transitions of the second order at every cross point of the density $n^{(-)}(T)$ with the critical curve $n_{\lim}(T)$ defined in Equation (63). For parameter κ in the interval $0 \leq \kappa < \kappa_s$, we fix the onset of condensation at one temperature $T = T_c^{(-)}$, corresponding to a phase transition of the second order. However, for the values of parameter κ in the interval $\kappa_s < \kappa \leq 1$, we find the onset of condensation at three temperatures $T_{c1}^{(-)}$, $T_{c2}^{(-)}$ and T_c due to an oscillating behavior of the curve $n^{(-)}(T)$ around the line $n_{\lim}(T)$.

The density dependence $n^{(+)}(T)$ of π^+ mesons at $\kappa = 1$ provides a remarkable feature that we once noted above. As one can see in Figure 5, in the right panel, at the temperature $T_* = 202$ MeV, the curve $n^{(+)}(T)$, calculated at $\kappa = 1$, touches the critical curve $n_{\lim}(T)$, but it does not cross it. Let us look at this in some detail. For the value $\kappa = 1$, the roots (54) of equation $U(n) + m = 0$ coincide with one another: $n_1 = n_2 \equiv n_*$, where $n_* = \sqrt{m/B}$. At this density, because $U(n_*) + m = 0$, the condition (62) of the condensate creation leads to zero value of the chemical potential, i.e., $\mu_I = 0$, but $U(n_*) = -m$. Therefore, for the particle-density point $n = n_*$, the arguments in the Bose–Einstein distribution functions of the densities $n^{(+)}$ and $n_{\lim}$ coincide and equal to $(\omega_k - m)/T$. Hence, it is possible to calculate the temperature that corresponds to the total particle density n_* by solving the following equation: $n_{\lim}(T_*) = (n_* - n_I)/2$. One can see the behavior of the chemical potential vs. temperature at $\kappa = 1$ in Figure 8 in the left panel as the blue solid curve (the axis indicating the value of the chemical potential is on the right of the graph). The chemical potential drops down to zero at one point $T = T_*$, where the density of π^+ mesons touches the critical curve, see Figure 5, the right panel. As can be seen in Figure 7, in the left panel, the common point of the line $n^{(+)}(T)$ with the line $n_{\lim}(T)$ is sufficient to create a finite discontinuity of the derivative of the heat capacity with a continuous behavior of the energy density, that is, to cause a phase transition of the second order at the temperature T_*. We name this phenomenon the virtual phase transition of the second order because it does not lead to the creation of the condensate that plays a role of the order parameter.

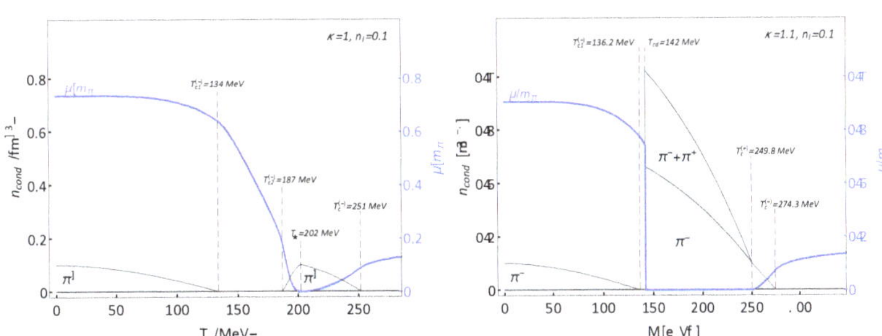

Figure 8. Left panel: Density of the condensate of π^- mesons as a function of temperature in the interacting particle–antiparticle gas at $\kappa = 1$. The isospin (charge) density is kept constant, $n_I = 0.1$ fm^{-3}. Shaded blue areas show the condensate states of π^- mesons. The blue solid line shows the behavior of the chemical potential. **Right panel:** The same as in the left panel but for $\kappa = 1.1$. The sail-like shaded area indicates the condensate states created by π^- mesons and by π^+ mesons at the same time. The gap of the chemical potential at $T = T_{cd}$ reflects phase transition of the first order, which creates the condensate of π^- and π^+ mesons.

At the end of this section, we can formulate the following theorem: *Each intersection of the particle density curve $n^{(\pm)}(T)$ with the critical curve $n_{\mathrm{lim}}(T)$ (or even touching the critical curve) leads to a phase transition of the second order at the temperature that characterizes this intersection point. At the temperature T_* corresponding to the point of touching, we encounter a virtual phase transition of the second order without the formation of a condensate, that is, without the formation of an order parameter.*

7. Canonical Ensemble vs. Grand Canonical Ensemble: Description of the Boson Systems in the Presence of a Condensate

7.1. Particle-Number Conservation in an Ideal Single-Component Bosonic System

Let us assume that in the case of the conserved charge, we want to describe the boson system in the framework of the grand canonical ensemble, where the canonical variables are (T, μ). As a starting point, let us consider an isolated ideal single-component boson gas with a conserved number of particles (next, in the framework of the grand canonical ensemble, we will consider a particle–antiparticle boson system at a conserved charge density).

It turns out that even in this case, the general procedure is not so unambiguous. First of all, one should adjust the chemical potential at high temperatures T, where no condensate is present in the system, at a given particle density n, which should be treated as a mean value. In the canonical ensemble, where the free variable is the particle density n, the chemical potential is found from equation

$$n = g \int \frac{d^3k}{(2\pi)^3} f_{\mathrm{BE}}(\omega_k, \mu) \,. \tag{64}$$

On the other hand, it can be represented vice versa: at some given temperature T' and chemical potential μ', by using Equation (64), one can calculate the mean value $\bar{n}$, which will be adopted as a conserved particle-number density in the canonical ensemble. However, further, for other temperatures than T', one has to know the chemical potential that provides the same particle density n. Again, it is necessary to solve Equation (64) with respect to the chemical potential to obtain a dependence $\mu(T, n)$. The solution of Equation (64) is represented in Figure 9 in the left panel for two densities $n = 0.1 \ \mathrm{fm}^{-3}$ and $n = 0.2 \ \mathrm{fm}^{-3}$, where the critical curve is depicted as $n_{\mathrm{lim}}^{(\mathrm{id})}$. It should be noted that in the condensed phase $T < T_\mathrm{c}$, the chemical potential is equal to the maximum value, which is the mass of particles $\mu = m$. Then, in the condensate phase, the variables (T, μ) determine only the density of thermal particles in this temperature interval, see two examples of curves in Figure 9 in the right panel. In addition to this, it should be noted that if the chemical potential participates in the condition of condensate formation, i.e., $\mu = m$, then from a formal point of view, it cannot be a free variable in the condensate phase.

Therefore, if we continue to investigate the conservation of the number of particles in a single-component ideal gas over a wide temperature interval, we must use the chemical potential profile depicted in Figure 9 in the left panel. Then, indeed, if we use this function $\mu(T, n)$ in Equation (64) to calculate the particle density, the resulting dependence $n(T)$ actually remains constant, $n(T) = \bar{n} = \mathrm{const}$. However, in fact, this is not the use of the grand canonical ensemble, where the two free variables (T, μ) should determine the thermodynamic state of the system, we see that the chemical potential profile is calculated with the help of some value of n. This especially applies to the condensate phase, where the chemical potential is limited by the condition of condensate formation, i.e., $\mu = m$.

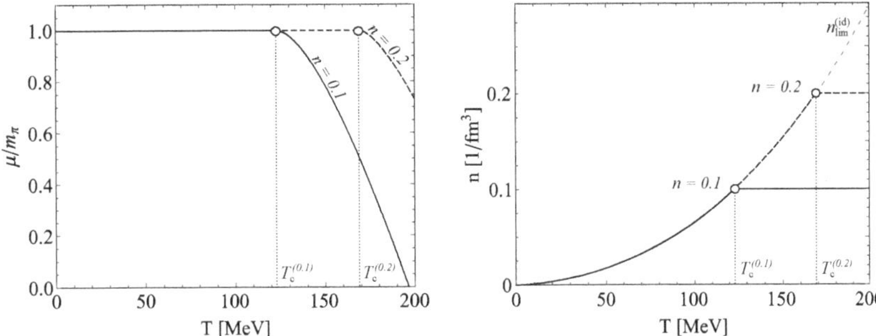

Figure 9. Left panel: Chemical potential vs. temperature in an ideal single-component boson gas at conserved mean value n of the particle-number density for two samples: $n = 0.1$ fm^{-3} with $T_c^{(0.1)}$ (the black solid line) and $n = 0.2$ fm^{-3} with $T_c^{(0.2)}$ (the black dashed line). The segment $\mu = m$ belongs to the condensate phase. **Right panel:** Density of thermal particles vs. temperature in an ideal single-component boson gas. The critical curve $n_{\text{lim}}^{(\text{id})}$ is defined in (3). (The same notations as in the left panel).

The picture obtained becomes even more striking when we study the conservation of charge in a relativistic ideal boson gas of particles and antiparticles at $n_I \neq 0$. Indeed, if we assume that particles and antiparticles are simultaneously in the condensate phase, then two conditions must be satisfied simultaneously: $m - \mu_I = 0$ and $m + \mu_I = 0$, where μ_I is the isospin chemical potential, which corresponds to n_I. This leads to two equations: $m = 0$ and $\mu_I = 0$. As we can see, the first equation is impossible or unphysical. That is, only one condition can be fulfilled, for example $m - \mu_I = 0$. Therefore, we can formulate the following theorem: *in a relativistic bosonic ideal gas of particles and antiparticles with a conserved charge $n_I \neq 0$, only one component of the system can form a condensate phase.* The sign of the excess charge, the modulus of which is equal to n_I, determines the answer, which component of the system, particles or antiparticles, can be in the condensate.

7.2. Charge Conservation in an Interacting Particle–Antiparticle Boson System

A similar paradoxical picture arises when describing an interacting particle–antiparticle bosonic system at a finite isospin (charge) density $n_I \neq 0$ within the grand canonical ensemble. With "strong" attraction, when the temperature rises from zero, the system has a different phase structure in different temperature intervals, as was the case with "weak" attraction.

As we saw in the previous Section 6.1, with "weak" attraction, the boson system has a different phase structure in different temperature intervals. With a "strong" interaction, an additional thermodynamic state arises, when both components, that is, particles and antiparticles, can simultaneously be in the condensate phase. Therefore, if $\kappa > 1$, it is necessary to sequentially solve three sets of equations, each of which corresponds to a certain thermodynamic phase:

(a) at low temperatures, when the charge-dominant component of the particle–antiparticle system is in the condensate phase [6] and the low-density component is only in the thermal phase, this is a set of Equations (60) and (61);

(b) when both components, i.e., mesons π^- and π^+, are in the condensate phase, it is necessary to modify set (a), see hereinafter;

(c) at high temperatures, it is a set of Equations (58) and (59), which defines the state when both components of the system, that is, particles and antiparticles, are only in the thermal phase.

There is a delicate issue when both particles and antiparticles undergo the Bose–Einstein condensation at the same time. In this case, in addition to the condensate condition (62) for π^- mesons, the argument of the distribution function for π^+ mesons must satisfy a similar condition to ensure that this component of the system is also present in the condensate at the same temperature T and chemical potential μ. Therefore, when both particles and antiparticles are in the condensate, we obtain two conditions simultaneously:

$$U(n) - \mu_I + m = 0, \tag{65}$$

$$U(n) + \mu_I + m = 0. \tag{66}$$

Then, Equations (60) and (61) should be modified to take these conditions into account. We must include a condensate component $n^{(+)}_{\text{cond}}$ of π^+ mesons, accounting for the fact that the density of thermal π^+ mesons is now $n_{\lim}(T)$, as well as the density of thermal π^- mesons. Hence, when both components are in the condensate, the set of self-consistent equations reads (case (b))

$$n = n^{(-)}_{\text{cond}}(T) + n^{(+)}_{\text{cond}}(T) + 2\,n_{\lim}(T), \tag{67}$$

$$n_I = n^{(-)}_{\text{cond}}(T) - n^{(+)}_{\text{cond}}(T). \tag{68}$$

It turns out that the solutions of sets (a) and (b) exist in the same temperature interval. Indeed, in addition to self-consistent solutions of equation (a), there are two other branches of solutions: $(n^{(-)}_1 = \text{const}, n^{(+)}_1 = \text{const})$ and $(n^{(-)}_2 = \text{const}, n^{(+)}_2 = \text{const})$, which satisfy Equations (67) and (68). It can be shown that the branch $(n^{(-)}_2 = (n_2 + n_I)/2$, $n^{(+)}_2 = (n_2 - n_I)/2$, where n_2 is the root (54) of equation $U(n) + m = 0$, is preferable because of the higher pressure corresponding to these states.

The competition between branches (a) and (b) is resolved in the standard way according to the Gibbs criterion: the state corresponding to the highest pressure is preferred in the thermodynamic realization. Using this rule we find the temperature T_{cd} from equation $p_{(a)}(T, n_I) = p_{(b)}(T, n_I)$, where the pressure $p_{(a)}(T, n_I)$ corresponds to solutions of the set of equation (a) and $p_{(b)}(T, n_I)$ to the set of equation (b). For temperatures above T_{cd}, the pressure that corresponds to the states determined by set (b) dominates, i.e., $p_{(b)}(T, n_I) > p_{(a)}(T, n_I)$. This leads to the transition from branch (a) to branch (b) of self-consistent solutions, which leads to a phase transition of the first order at the temperature $T = T_{\text{cd}}$.

The set of Equations (65) and (66) can be rewritten as

$$\mu_I = 0, \tag{69}$$

$$U(n) + m = 0. \tag{70}$$

Note, in Ref. [10], the system of pions was studied in the grand canonical ensemble at $\mu_I = 0$ in the mean-field approach, and the condition for the onset of the condensate phase leads to the same equation $U(n) + m = 0$.

Results of the numerical solution of the sets of equations (a), (b) and (c) for the particle density at $\kappa = 1.1$ are shown in Figure 10 in the left panel. The density $n^{(-)}(T)$ of π^- mesons is represented by a solid blue curve, which consists of several horizontal segments and one vertical segment, which reflects a phase transition of the first order. The density $n^{(+)}(T)$ of π^+ mesons is depicted as a dashed blue curve, which also consists of several horizontal segments and one vertical segment, which also reflects a first-order phase transition. It can be seen from the figure that the isospin (charge) density in the system of bosons under consideration remains constant. Indeed, for each temperature point on the graph, it can be seen that $n^{(-)}(T) - n^{(+)}(T) = 0.1\ \text{fm}^{-3}$.

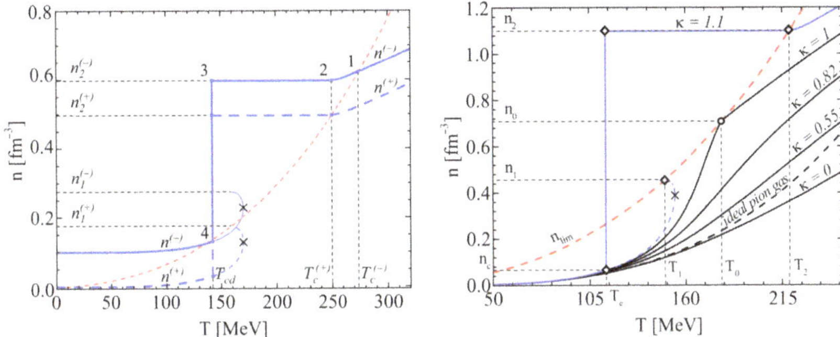

Figure 10. Interacting particle-antiparticle boson system in the thermodynamic mean-field model. **Left panel:** Particle densities vs. temperature at conserved isospin (charge) density $n_I = 0.1$ fm^{-3} as the solid blue line consisting of several segments (π^- mesons) and the dashed blue line consisting of several segments (π^+ mesons). The vertical segment for both dependencies indicates a phase transition of the first order with the creation of the condensate. In the condensate phase, $\mu_I = 0$. A dashed red line is the critical curve $n_{\lim}(T)$, see Equation (63). **Right panel:** Particle-number densities vs. temperature at $n_I = 0$ (or at $\mu_I = 0$): (1) the supercritical attraction $\kappa = 1.1$ is shown as a solid blue line consisting of several segments; the vertical segment (solid blue line) indicates a phase transition of the first order with the creation of the condensate; (2) particle densities at "weak" attraction $\kappa \leq 1$ are shown as solid black lines in the thermal phase. A dashed red line is the critical curve. Crosses on both panels separate metastable and non-physical states.

For a clearer comparison in the right panel in Figure 10, we took the liberty of depicting the right panel of Figure 3 once again. We would like to emphasize that in the condensate phase, both systems are represented by a zero chemical potential regardless of whether the particle–antiparticle system described in the left panel has a finite charge density, i.e., $n_I = 0.1$ fm^{-3}, while the particle–antiparticle system described in the right panel is characterized by zero charge density, i.e., $n_I = 0$. Therefore, if one intends to study both systems, one system with a finite charge density and another with zero charge density within the grand canonical ensemble, then the canonical variables should be $(T, \mu_I = 0)$ when describing the condensate phase in both systems.

We seem to be coming to a kind of contradiction, since the textbooks say that the chemical potential should reflect charge conservation or particle-number density conservation, as we saw in the previous Section 7.1. The resolution of this contradiction occurs according to the statement that *the grand canonical ensemble with canonical variables (T, μ) is suitable for describing only the thermal phase or for describing particles that are in kinetic states, but not in condensed states.* We verified that this is true for particle-number conservation in the case of an ideal gas of bosons, where with the canonical variables (T, μ) in the condensate phase we were able to describe only the kinetic particles, see Section 7.1.

This is also the case in our particular consideration of the relativistic particle–antiparticle boson system with conserved isospin charge n_I. Indeed, it can be seen in Figure 10 in the left panel in the temperature interval that corresponds to the condensate phase, i.e., between points 2 and 3 on the graph, that for each temperature from this interval, the thermal density of π^- mesons is equal to the thermal density of π^+ mesons, since these two densities are equal to $n_{\lim}(T)$. In other words, these kinetic densities are equal to the critical curve density. Therefore, the charge density, which is determined only by thermal particles and antiparticles, is zero. Respectively, the chemical potential, which corresponds to the charge of the system, which is determined only by thermal particles, is also zero. In addition, we see that the chemical potential μ_I is really useful for describing only thermal or kinetic particles. Actually, this can be understood from the very beginning, because the chemical potential "works" in the integral (in the distribution function), which determines only the density of kinetic particles.

The results of solving the self-consistent equations for $\kappa = 1.1$ are shown in Figure 8 in the right panel (the axis indicating the value of the chemical potential is on the right side of the graph). One can see the behavior of the chemical potential, the value of which is actually zero in the phase where the particles and antiparticles are in a condensed state. It can be seen that the chemical potential drops to zero at the temperature T_{cd}, which indicates a phase transition of the first order. We can also see in Figure 8 in the right panel that condensate forms in two different temperature intervals, at low temperatures the presence of condensate is exclusively due to charge conservation, but at higher temperatures, the formation of condensate is caused by supercritical attraction between particles.

7.3. Other Examples

Consider the thermodynamic mean-field model, where the mean field also depends on the isospin density. As shown in Ref. [35], since n and n_I are independent thermodynamic variables, the form of this mean field is as follows: $U^{(\mp)}(n, n_I) = U(n) \mp U_I(n_I)$, where $U_I(n_I)$ is an odd function, for example, $U_I(n_I) \propto n_I$, and the field $U^{(-)}$ acts on π^- mesons, while $U^{(+)}$ acts on π^+ mesons. Then, if π^- and π^+ mesons are in the condensate phase, two necessary conditions must be fulfilled: $m + U(n) - U_I(n_I) - \mu_I = 0$ and $m + U(n) + U_I(n_I) + \mu_I = 0$. From here, we obtain the equivalent equations: $m + U(n) = 0$ and $\mu_I = -U_I(n_I)$. Therefore, the chemical potential is fixed by the condition of condensate formation and is determined by the isospin density, which remains constant. Hence, when the mean interaction in the system depends on the isospin (charge) density, we again conclude that μ_I cannot be a free variable in the presence of a condensate, and hence, the grand canonical ensemble is not applicable in the condensate phase.

When describing the interacting particle–antiparticle bosonic system at a finite isospin (charge) density $n_I \neq 0$ in the field-theoretic approach formulated in Section 3, we encounter exactly the same paradox. Indeed, for development of the condensate by both particles and antiparticles, two conditions must be met: $M^2 - \mu_I = 0$ and $M^2 + \mu_I = 0$, where M is the effective mass of quasi-particles. By complete analogy with case (c) discussed above, these conditions lead to two equations: $M^2 = 0$ and $\mu_I = 0$. Therefore, it turns out that the system with a finite charge density $n_I \neq 0$ is characterized by zero value of the chemical potential. On the other hand, we see that in the presence of condensate, the density of thermal particles is the same in the negatively and positively charged components of the system, i.e., $n_{th}^{(-)}(T) = n_{th}^{(+)}(T)$. Hence, the problem can be resolved by accepting that the chemical potential is responsible only for thermal (kinetic) particles.

8. Conclusions

Therefore, in the present study, we have investigated the relativistic interacting system of Bose particles and antiparticles, which we conventionally named "pions" due to zero spin and mass $m = 139$ MeV/c^2. The repulsion between particles was fixed (hard-core repulsion), but attraction between particles, which was parameterized by the dimensionless parameter κ, changes from zero ($\kappa = 0$) to some supercritical value ($\kappa > 1$).

We proved, and by this we confirmed the conclusion obtained in [8], that at "weak" attraction ($\kappa \leq 1$), the π^- component of the system only can develop the Bose–Einstein condensate, the π^+ component is in the thermal phase for all temperatures. We have shown that for $0.93 \leq \kappa \leq 1$, in addition to the condensate of π^- mesons at low temperatures, it can appear again in some interval at higher temperatures.

- The intersections of the particle density curves with the critical curve indicate second-order phase transitions in the system.
- At the point where the particle density of π^- mesons touches the critical curve, the virtual phase transition of second order, i.e., a phase transition without setting the order parameter, appears.
- The meson system develops a first-order phase transition for sufficiently strong attractive interactions via forming a Bose condensate, thus releasing the latent heat. The

model predicts that the condensed phase is characterized by a constant total density of particles.

- The grand canonical ensemble cannot describe the state of the condensate since the chemical potential μ_I is significantly affected by the conditions of condensate formation, so it cannot be used as a free variable if the system is in the condensed phase. That is why the grand canonical ensemble is not suitable for describing a multi-component system in the condensate phase, even if only one of the components is in the condensate.

Author Contributions: writing—original draft preparation and editing, D.A.; writing—review and editing, V.G., D.Z., V.K., I.M. and H.S. All authors have read and agreed to the published version of the manuscript.

Funding: This research received no external funding.

Data Availability Statement: Not applicable.

Acknowledgments: D.A. is very grateful to J. Steinheimer and O. Philipsen for useful discussions and comments and greatly appreciates the warm hospitality and support provided by FIAS administration and the scientific community. The work of D.Zh. and D.A. was supported by the Simons Foundation and by the Program "The structure and dynamics of statistical and quantum-field systems" of the Department of Physics and Astronomy of the NAS of Ukraine. I.M. thanks FIAS for the support and hospitality. H.St. thanks for the support from the J. M. Eisenberg Professor Laureatus of the Fachbereich Physik.

Conflicts of Interest: The authors declare no conflict of interest.

Abbreviations

The following abbreviations are used in this manuscript:

SMF   Scalar mean-field model
TMF   Thermodynamic mean-field model
FED   Free energy density

Appendix A. Thermodynamically Consistent Mean-Field Model for the Interacting Particle–Antiparticle System

The consideration in this section is based on the thermodynamic mean-field model developed in Ref. [33], where a multi-component system consisting of any number of species was studied. Here, we consider specific equations of the thermodynamic mean-field model for the system of particles and antiparticles.

We limit our study to the case where at a fixed temperature, the interacting boson particles and boson antiparticles are in dynamical equilibrium with respect to annihilation and pair-creation processes. To take into account the interaction between the bosons, we introduce a phenomenological Skyrme-like mean field $U(n)$, which depends only on the total density of mesons n.

To start with, let us consider a thermodynamic system consisting of two sorts of particles. The free energy of the system and its differential can be written as

$$F(N_1, N_2, T, V) = \mu_1 N_1 + \mu_2 N_2 - pV, \tag{A1}$$

$$dF = \mu_1 dN_1 + \mu_2 dN_2 - SdT - pdV, \tag{A2}$$

where $N_{1,2}$ is the number of particles of the first and second sorts, $\mu_{1,2}$ are their chemical potentials, p is the pressure in the system and S and V are its entropy and volume. The differential of the free energy density (FED), which, for a homogeneous system, is defined as $\Phi = F/V$, reads

$$d\Phi(n_1, n_2, T) = \mu_1 dn_1 + \mu_2 dn_2 - sdT, \tag{A3}$$

where $s = S/V$, $n_{1,2} = N_{1,2}/V$ are the entropy density and the particle number density, respectively. The chemical potentials are expressed as

$$\mu_1 = \left(\frac{\partial \Phi}{\partial n_1}\right)_T, \tag{A4}$$

$$\mu_2 = \left(\frac{\partial \Phi}{\partial n_2}\right)_T. \tag{A5}$$

We assume that the FED of a system of interacting particles can be represented as a sum of FEDs of the system without interaction $\Phi_1^{(0)} + \Phi_2^{(0)}$ and the term Φ_{int} responsible for interaction, which, in turn, depends on the total density of particles $n = n_1 + n_2$,

$$\Phi(n_1, n_2, T) = \Phi_1^{(0)}(n_1, T) + \Phi_2^{(0)}(n_2, T) + \Phi_{\mathrm{int}}(n_1 + n_2, T). \tag{A6}$$

Then, in accordance with Equations (A4) and (A5), we obtain

$$\mu_1 = \frac{\partial \Phi_1^{(0)}}{\partial n_1} + \frac{\partial \Phi_{\mathrm{int}}}{\partial n} = \mu_1^{(0)} + \frac{\partial \Phi_{\mathrm{int}}}{\partial n}, \tag{A7}$$

$$\mu_2 = \frac{\partial \Phi_2^{(0)}}{\partial n_2} + \frac{\partial \Phi_{\mathrm{int}}}{\partial n} = \mu_2^{(0)} + \frac{\partial \Phi_{\mathrm{int}}}{\partial n}. \tag{A8}$$

The pressure can be written as

$$\begin{aligned} p(n_1, n_2, T) &= \mu_1 n_1 + \mu_2 n_2 - \Phi(n_1, n_2, T) \\ &= \left\{\mu_1^{(0)} n_1 - \Phi_1^{(0)}\right\} + \left\{\mu_2^{(0)} n_2 - \Phi_2^{(0)}\right\} + \left\{n\frac{\partial \Phi_{\mathrm{int}}}{\partial n} - \Phi_{\mathrm{int}}\right\}. \end{aligned} \tag{A9}$$

We introduce the following notations

$$U(n, T) = \left[\frac{\partial \Phi_{\mathrm{int}}(n, T)}{\partial n}\right]_T, \tag{A10}$$

$$P(n, T) = n\left[\frac{\partial \Phi_{\mathrm{int}}(n, T)}{\partial n}\right]_T - \Phi_{\mathrm{int}}(n, T). \tag{A11}$$

From these definitions, one immediately obtains a relation that connects these two quantities

$$n\frac{\partial U(n, T)}{\partial n} = \frac{\partial P(n, T)}{\partial n}. \tag{A12}$$

Next, in Equation (A9), we use expressions for the pressure in the single-particle ideal gas

$$p_1^{(0)} = \mu_1^{(0)} n_1 - \Phi_1^{(0)} = \frac{g}{3} \int \frac{d^3k}{(2\pi)^3} \frac{k^2}{\omega_k} f(\omega_k; \mu_1^{(0)}), \tag{A13}$$

$$p_2^{(0)} = \mu_2^{(0)} n_2 - \Phi_2^{(0)} = \frac{g}{3} \int \frac{d^3k}{(2\pi)^3} \frac{k^2}{\omega_k} f(\omega_k; \mu_2^{(0)}), \tag{A14}$$

where $f(\omega_k; T, \mu^{(0)})$ is the Bose–Einstein distribution function of ideal gas

$$f\left(\omega_k; \mu^{(0)}\right) = \left\{\exp\left[\frac{\omega_k - \mu^{(0)}}{T}\right] - 1\right\}^{-1} \quad \text{with} \quad \omega_k = \sqrt{m^2 + k^2}. \tag{A15}$$

Using relations (A7) and (A8) in the form

$$\mu_1^{(0)} = \mu_1 - U(n), \tag{A16}$$

$$\mu_2^{(0)} = \mu_2 - U(n), \tag{A17}$$

one can insert these expressions into Equations (A13) and (A14) and then rewrite the total pressure (A9) as

$$p(T, n_1, n_2) = \frac{g}{3} \int \frac{d^3k}{(2\pi)^3} \frac{k^2}{\omega_k} \left[f(E(k,n); \mu_1) + f(E(k,n); \mu_2) \right] + P(T,n), \tag{A18}$$

where g is the degeneracy factor, $E(k,n) = \sqrt{m^2 + k^2} + U(T,n)$ is the effective single-particle energy and $P(T,n)$ can be treated now as the excess pressure.

To obtain a self-consistent equation for the total particle-number density n, it is convenient to pass from the variables (T, n_1, n_2) to the variables (T, μ_1, μ_2). In this case, the total number of particles n in the system is also a function of new variables (T, μ_1, μ_2). Then, for the total number of particles n, we obtain

$$\begin{aligned}
n &= n_1 + n_2 = \left(\frac{\partial p}{\partial \mu_1} \right)_T + \left(\frac{\partial p}{\partial \mu_2} \right)_T \\
&= g \int \frac{d^3k}{(2\pi)^3} \left[f(E(k,n); \mu_1) + f(E(k,n); \mu_2) \right].
\end{aligned} \tag{A19}$$

For free energy density, one has (see Equation (A1)) an expression

$$\Phi = \mu_1 n_1 + \mu_2 n_2 - p. \tag{A20}$$

The System of Particles and Antiparticles

The chemical potential μ includes components related to different quantum numbers

$$\mu = B\mu_B + S\mu_S + Q\mu_Q + I\mu_I + \ldots, \tag{A21}$$

where B, S, Q and I correspond to the baryon quantum number, strangeness, electric charge and isospin, respectively. It is clear that the chemical potentials of boson particles μ_1 and boson antiparticles μ_2 have opposite signs [33]

$$\mu_1 = -\mu_2 \equiv \mu_I. \tag{A22}$$

Requiring the conservation of the isotopic spin n_I in the system, we obtain the set of equations

$$n = g \int \frac{d^3k}{(2\pi)^3} \left[f(E(k,n); \mu_I) + f(E(k,n); -\mu_I) \right], \tag{A23}$$

$$n_I = g \int \frac{d^3k}{(2\pi)^3} \left[f(E(k,n); \mu_I) - f(E(k,n); -\mu_I) \right], \tag{A24}$$

where the Bose–Einstein distribution function reads

$$f(E; \mu) = \left\{ \exp\left[\frac{E - \mu}{T} \right] - 1 \right\}^{-1}. \tag{A25}$$

The set of Equations (A23) and (A24) can be solved with respect to the thermodynamic quantities n and μ_I for given canonical variables T and n_I. As a result, we obtain the functions

$$n = n(T, n_I), \quad \mu_I = \mu_I(T, n_I). \tag{A26}$$

In our case, the interaction between particles is described by the Skyrme-like mean field

$$U(n) = -An + Bn^2,\tag{A27}$$

where n is the total particle-number density. Using the self-consistent solutions $n(T, n_I)$ and $\mu_I(T, n_I)$ of the set of Equations (A23) and (A24), one can obtain the expressions for the pressure and free energy density in the boson interacting system in the following form

$$p = \frac{g}{3} \int \frac{d^3k}{(2\pi)^3} \frac{k^2}{\omega_k} \left[f\big(E(k,n); \mu_I\big) + f\big(E(k,n); -\mu_I\big) \right] + P(n),\tag{A28}$$

$$\Phi = n_I \mu_I(T, n_I) - p(T, n_I),\tag{A29}$$

where $E(k, n) = \sqrt{m^2 + k^2} + U(n)$. [7] Here, the excess pressure $P(n)$ is known, and it can be calculated with the help of integration of relation (A12) using the given mean field (A27) and a natural initial condition $P(n = 0) = 0$. Hence, after integration, one obtains

$$P(n) = -\frac{A}{2}\,n^2 + \frac{2B}{3}\,n^3.\tag{A30}$$

With the help of free energy density, it is easy to calculate the volumetric heat capacity c_V

$$c_V = -T\frac{\partial^2 \Phi}{\partial T^2}.\tag{A31}$$

Notes

[1] In the nonrelativistic case, where $\mu_{\mathrm{nonrel}} = \mu - m$, the maximum value of the thermal-particle density is achieved at zero chemical potential.

[2] It should be noted that we just conventionally say "condensate phase". In fact, it is the thermodynamic state of a system that contains thermal particles and condensed particles at the same time.

[3] In fact, the name "condensate phase" is just a conventional one because this phase is a mixture of the thermal (kinetic) particles and the condensed particles.

[4] For our choice of the total charge of the system, it is the π^- mesons.

[5] Because we named it as the virtual phase transition of the second order.

[6] For our choice of the total charge of the system, these are π^- mesons.

[7] It should be noted that Equations (A23), (A24) and (A28) take place only in the absence of condensate in the system.

References

1. Bogolubov, N. On the theory of superfluidity. *Sov. J. Phys.* **1947**, *11*, 23.
2. Bogolyubov N.N. Energy levels of the imperfect Bose-Einstein gas. *Bull. Moscow State Univ.* **1947**, *7*, 43.
3. Bogolyubov, M.M. *Lectures on Quantum Statistics*; Gordon and Breach: Kyiv, Ukraine; New York, NY, USA, 1947.
4. Ginzburg, V.L.; Pitaevskii, L.P. On the theory of Superfluidity. *Sov. Phys. JETP* **1958**, *7*, 858.
5. Gross, E.P. Hydrodynamics of a Superfluid Condensate . *J. Math. Phys.* **1963**, *4*, 195–207.
6. Huang, K. *Statistical Mechanics*; John Wiley and Sons: Hoboken, NJ, USA, 1987; p. 294.
7. Haber, H.E.; Weldon, H.A. Thermodynamics of an Ultrarelativistic Ideal Bose Gas. *Phys. Rev. Lett.* **1981**, *46*, 1497. [CrossRef]
8. Anchishkin, D.; Gnatovskyy, V.; Zhuravel, D.; Karpenko, V. Self-interacting particle-antiparticle system of bosons. *Phys. Rev. C* **2022**, *105*, 045205. [CrossRef]
9. Bzdak, A.; Esumi, S.; Koch, V.; Liao, J.; Stephanov, M.; Xu, N. Mapping the phases of quantum chromodynamics with beam energy scan. *Phys. Rep.* **2020**, *853*, 1–87. [CrossRef]
10. Anchishkin, D.; Mishustin, I.; Stoecker, H. Phase transition in an interacting boson system at finite temperatures. *J. Phys. G* **2019**, *46*, 035002. [CrossRef]
11. Anselm, A.; Ryskin, M. Production of classical pion field in heavy ion high energy collisions. *Phys. Lett. B* **1991**, *226*, 482. [CrossRef]
12. Blaizot, J.-P.; Krzwitski, A. Soft-pion emission in high-energy heavy-ion collisions. *Phys. Rev. D* **1992**, *46*, 246. [CrossRef]
13. Bjorken, J.D. A full-acceptance detector for SSC physics at low and intermediate mass scales: An expression of interest to the SSC. *Int. J. Mod. Phys. A* **1992**, *7*, 4189. [CrossRef]
14. Mishustin, I.N.; Greiner, W. Multipion droplets. *J. Phys. G Nucl. Part. Phys.* **1993**, *19*, L101. [CrossRef]

15. Son, D.T.; Stephanov, M.A. QCD at Finite Isospin Density. *Phys. Rev. Lett.* **2001**, *86*, 592. [CrossRef] [PubMed]
16. Kogut, J.; Toublan, D. QCD at small nonzero quark chemical potentials. *Phys. Rev. D* **2001**, *64*, 034007. [CrossRef]
17. Toublan, D.; Kogut, J. Isospin chemical potential and the QCD phase diagram at nonzero temperature and baryon chemical potential. *Phys. Lett. B* **2001**, *564*, 212. [CrossRef]
18. Mammarella, A.; Mannarelli, M. Intriguing aspects of meson condensation. *Phys. Rev. D* **2015**, *92*, 085025. [CrossRef]
19. Carignano, S. ; Lepori, L.; Mammarella, A.; Mannarelli, M.; Pagliaroli, G. Scrutinizing the pion condensed phase. *Eur. Phys. J. A* **2017**, *53*, 35. [CrossRef]
20. Kapusta, J. Bose-Einstein condensation, spontaneous symmetry breaking, and gauge theories. *Phys. Rev. D* **1981**, *24*, 426. [CrossRef]
21. Haber, H.E.; Weldon, H.A. Finite-temperature symmetry breaking as Bose-Einstein condensation. *Phys. Rev. D* **1982**, *25*, 502. [CrossRef]
22. Bernstein, J.; Dodelson, S. Relativistic Bose gas. *Phys. Rev. Lett.* **1991**, *66*, 683. [CrossRef]
23. Shiokawa, K.; Hu, B.L. Finite number and finite size effects in relativistic Bose-Einstein condensation. *Phys. Rev. D* **1999**, *60*, 105016. [CrossRef]
24. Salasnich, L. Particles and Anti-Particles in a Relativistic Bose Condensate. *Il Nuovo C. B* **2002**, *117*, 637. [CrossRef]
25. Begun, V.V.; Gorenstein, M.I. Bose-Einstein condensation in the relativistic pion gas: Thermodynamic limit and finite size effects. *Phys. Rev. C* **2008**, *77*, 064903. [CrossRef]
26. Markó, G.; Reinosa, U.; Szép, Z. Bose-Einstein condensation and Silver Blaze property from the two-loop Φ-derivable approximation. *Phys. Rev. D* **2014**, *90*, 25021. [CrossRef]
27. Brandt, B.B.; Endrodi, G. QCD phase diagram with isospin chemical potential. *arXiv* **2016**, arXiv:1611.06758.
28. Brandt, B.B.; Endrodi, G.; Schmalzbauer, S. QCD at finite isospin chemical potential. *arXiv* **2017**, arXiv:1709.10487.
29. London, F. The λ-Phenomenon of Liquid Helium and the Bose-Einstein Degeneracy. *Nature* **1938**, *141*, 643. [CrossRef]
30. Ehrenfest, P. *Phasenumwandlungen im ueblichen und erweiterten Sinn, Classifiziert nach dem Entsprechenden Singularitaeten des Thermodynamischen Potentiales*; Communications from the Physical Laboratory of the University of Leiden, Supplement No. 75b; Kamerlingh Onnes-Institut: Leiden, The Netherlands, 1933.
31. Jaeger, G. The Ehrenfest Classification of Phase Transitions: Introduction and Evolution. *Arch. Hist. Exact Sci.* **1998**, *53*, 51. [CrossRef]
32. Mishustin, I.N.; Anchishkin, D.V.; Satarov, L.M.; Stashko, O.S.; Stoecker, H. Condensation of interacting scalar bosons at finite temperatures. *Phys. Rev. C* **2019**, *100*, 022201.
33. Anchishkin, D.; Vovchenko, V. Mean-field approach in the multi-component gas of interacting particles applied to relativistic heavy-ion collisions. *J. Phys. G Nucl. Part. Phys.* **2015**, *42*, 105102. . [CrossRef]
34. Anchishkin, D.; Mishustin, I.; Stashko, O.; Zhuravel, D.; Stoecker, H. Relativistic selfinteracting particle-antiparticle system of bosons. *Ukranian J. Phys.* **2019**, *64*, 1110–1116. [CrossRef] [PubMed]
35. Anchishkin, D.; Suhonen, E. Generalization of mean-field models to account for effects of excluded volume. *Nucl. Phys. A* **1995**, *586*, 734–754. [CrossRef]

Article

Collisional Broadening within a Hadronic Transport Approach

Branislav Balinovic [1,2], **Renan Hirayama** [2,3,*] **and Hannah Elfner** [1,2,3,4]

1   Institut für Theoretische Physik, Goethe Universität, Max-von-Laue-Strasse 1,
     60438 Frankfurt am Main, Germany; balinovic@fias.uni-frankfurt.de (B.B.); elfner@itp.uni-frankfurt.de (H.E.)
2   Frankfurt Institute for Advanced Studies, Ruth-Moufang-Strasse 1, 60438 Frankfurt am Main, Germany
3   Helmholtz Forschungsakademie Hessen für FAIR (HFHF), GSI Helmholtzzentrum für
     Schwerionenforschung, Campus Frankfurt, Max-von-Laue-Str. 12, 60438 Frankfurt am Main, Germany
4   GSI Helmholtzzentrum für Schwerionenforschung, Planckstr. 1, 64291 Darmstadt, Germany
*   Correspondence: hirayama@fias.uni-frankfurt.de; Tel.: +49-69-798-47678

Abstract: We explore the emergence of the collisional broadening of hadrons under the influence of different media using the hadronic transport approach SMASH (Simulating Many Accelerated Strongly interacting Hadrons), which employs vacuum properties and contains no a priori information about in-medium effects. In this context, we define collisional broadening as a decrease in the lifetime of hadrons, and it arises from an interplay between the cross-sections for inelastic processes and the available phase space. We quantify this effect for various hadron species, in both a thermal gas in equilibrium and in nuclear collisions. Furthermore, we distinguish the individual contribution of each process and verify the medium response to different vacuum assumptions; we see that the decay width that depends on the resonance mass leads to a larger broadening than a mass-independent scenario.

Keywords: hadronic transport; resonance properties; collisional broadening

Citation: Balinovic, B.; Hirayama, R.; Elfner, H. Collisional Broadening within a Hadronic Transport Approach. *Universe* **2023**, *9*, 414. https://doi.org/10.3390/universe9090414

Academic Editors: Péter Kovács, Sándor Lökös and Dániel Kincses

Received: 17 July 2023
Revised: 23 August 2023
Accepted: 1 September 2023
Published: 9 September 2023

1. Introduction

The properties of Quantum Chromodynamics (QCD) at finite temperature and finite baryochemical potential are not well understood, as they comprise a region of the phase diagram difficult to access both in experiments and in first-principle theories. In the confined phase, the degrees of freedom consist of colorless hadrons, and the system evolution is well described by hadronic transport. As a baseline for any further medium modifications, it is of interest to understand how hadron properties are modified when embedded in a hadronic medium.

One natural change caused by the presence of a medium is the reduction in hadron lifetimes due to absorption processes. The usual prescription in hadronic transport approaches is to associate the lifetime (τ^{vac}) of a resonance in vacuum with the inverse of its width, such that its decay in a small time interval $\Delta\tau$ happens as a Bernoulli trial with probability $P_{\text{decay}}(\Delta\tau) = \Gamma^{\text{vac}}\Delta\tau$. Here, Γ^{vac} is called *vacuum decay width*, and the medium-induced shortening of lifetimes can be thought of as an effective increase in the width. This effect is known as *collisional broadening*.

From a historical and experimental perspective, medium modifications are studied by comparing elementary collisions and heavy-ion collisions (HICs), scaling the observable appropriately. Specifically, the NA60 Collaboration revealed that the softening of the invariant mass spectra of dileptons [1] around the ρ meson pole mass is consistent with the in-medium broadening of vector mesons proposed by Rapp et al. [2,3]. This was later covered in off-shell hadronic transport approaches—where the hadron spectral function can change dynamically during propagation—by including a collisional width explicitly parameterized as a linear function of the local density [4,5]. On the other hand, on-shell hadronic transport approaches use the coarse-graining method: the average of several collisions gives local values for thermodynamic quantities, with which the corresponding rates from the in-medium model are computed [6–8].

In [8], dilepton emission was included and found to be in agreement with experimental yields in elementary collisions at HADES energies but not in the heavy-ion yields around the ρ meson pole mass, showing that the collisional broadening intrinsic to hadronic transport is not sufficient to account for the full effect of a medium. In [9], we studied the collisional broadening of ρ in an equilibrated hadron gas and in nuclear collisions. The thermal gas exhibited a spectral function similar to the full in-medium model but expectedly less broadened.

In this work, we extend that previous study to some resonances of particular interest, determine the processes that contribute the most to ρ collisional broadening, and investigate the effect of the two different model assumptions usually chosen for the vacuum properties of resonances in hadronic transport. This paper is organized as follows: Section 2 describes the aspects of the SMASH approach relevant to this work. In Sections 3.1 and 3.2, we display the behavior of the dynamically generated collisional broadening of different particles in a thermal scenario and in nuclear collisions, respectively. In Section 3.3, we compare the collisional broadening of ρ and ω mesons under different vacuum assumptions. A brief summary of the results is given in Section 4, along with a discussion of their interpretation. Appendix A shows the inelastic cross-sections of some relevant interactions.

2. SMASH Transport Approach

In this study, we used the hadronic transport approach SMASH-2.2 (Simulating Many Accelerated Strongly interacting Hadrons) to simulate different states of nuclear matter, such as a thermal gas in equilibrium and nuclear collisions [10]. In this microscopic transport approach, the complete information of the phase space is accessible at all times according to effective solutions of the relativistic Boltzmann equation.

We employ the geometric collision criterion of SMASH to determine possible scatterings, in which an interaction can happen if

$$d_{\text{trans}} < d_{\text{int}} = \sqrt{\frac{\sigma_{\text{tot}}}{\pi}},$$ (1)

where σ_{tot} is the total cross-section and d_{trans} is the distance between two particles in a given time interval in the center of the mass frame. With this criterion, the only possible processes are binary: resonance formation ($2 \to 1$), its corresponding resonance decay ($1 \to 2$), as well as elastic and inelastic scatterings ($2 \to 2$). To account for multi-particle interactions, intermediate resonances are produced or decay in a chain. The available species, their vacuum mass M_0 and pole width Γ_0, possible decay channels, and associated branching ratios are taken from Particle Data Group 2016 [11].

The mass of a new resonance is constant during propagation and randomly chosen at production using the normalized vacuum spectral function

$$\mathcal{A}^{\text{vac}}(m) = \frac{2\mathcal{N}}{\pi} \frac{m^2 \Gamma^{\text{dec}}(m)}{(m^2 - M_0^2)^2 + [m\Gamma^{\text{dec}}(m)]^2}$$ (2)

as a probability distribution, where normalization factor $\mathcal{N}$ is defined by the relation

$$1 = \int_{m_{\text{min}}}^{\infty} \mathcal{A}^{\text{vac}}(m)\, dm,$$ (3)

with threshold mass m_{min} being equal to the sum of masses from its lightest decay channel. The *mass-dependent* decay width is based on the Manley formalism [12], given by

$$\Gamma^{\text{dec}}_{R \to ab}(m) = \Gamma^0_{R \to ab} \frac{\rho_{ab}(m)}{\rho_{ab}(M_0)},$$ (4)

where $\Gamma^0_{R \to ab}$ is the partial width at the pole mass and $\rho_{ab}(\sqrt{s})$ is a mass-dependent function evaluating the available phase space for the creation of particles a and b from energy $\sqrt{s}$. The total decay width of R in (2) is the sum of the partial widths in (4) over all possible final states ab. If decay happens, the particle decays in a randomly chosen decay channel according to the corresponding branching ratios. Stable hadrons[1] have $\Gamma^{\mathrm{dec}} \approx 0$, so $\mathcal{A}^{\mathrm{vac}}$ reduces to a δ-distribution, and the particle always has pole mass. We also employ a second assumption for vacuum decays to investigate the impact on collisional broadening. Here, $\Gamma^{\mathrm{vac}} = \Gamma_0$ independently of the resonance mass. This is referred to as the *mass-independent* decay assumption.

The SMASH approach uses vacuum properties, so the average lifetime of a particle corresponds to its inverse width only in vacuum. When it is surrounded by other hadrons with which it can interact inelastically, the average lifetime naturally decreases, as illustrated in Figure 1 for a ρ embedded in a medium. We describe this setup in Section 3.1. We notice that the medium suppresses the lifetimes of low-mass particles more, while higher masses are not very affected. This reflects the overall inelastic cross-section between the particle and the rest of the medium.

We remark that this prescription of sampling lifetimes from the (inverse) vacuum width breaks down close to the threshold. The available phase space, ρ_{ab}, approaches 0, so Equation (4) leads to $1/\Gamma^{\mathrm{dec}} \to \infty$, and the resonance can live forever. A more grounded definition was introduced in [13] from the fundaments of quantal scattering theory, associating the lifetime of a resonance with the time delay equal to the derivative of the phase shift, which can be computed analytically from the resonance shape. However, there is no consensus on how to implement this prescription appropriately in real transport model calculations, as it can generate negative time delays or require cross-sections for the "forward-going" part of the resonant wave packet, which are unmeasurable and must be parametrized [14,15]. Therefore, we stick to the usual prescription and investigate the consequences by comparing it to the aforementioned mass-independent assumption, which does not lead to infinite-lasting resonances.

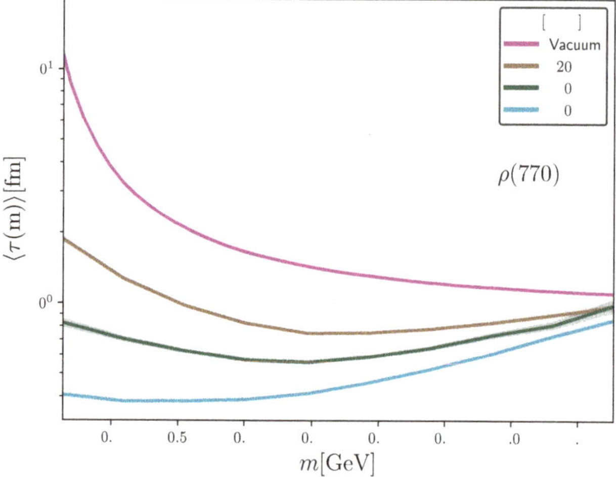

Figure 1. Proper lifetime of the ρ meson for a gas in equilibrium at different temperatures and baryochemical potential $\mu_B = 400$ MeV.

To probe the effect of these inelastic interactions, we follow [9] and define the effective width as

$$\Gamma^{\mathrm{eff}} = \langle \tau \rangle^{-1} = \left\langle \frac{t_f - t_i}{\gamma} \right\rangle^{-1},$$
(5)

where t_i and t_f are the initial time ("birth") and final time ("death") of the particle, respectively, and γ is its Lorentz factor in the computational frame. This definition follows naturally from the prescription we use for the resonance lifetimes in *vacuum*; since they are sampled from the *vacuum* width, the *effective* width should be defined in terms of the *effective* lifetime. In this work, we only consider the dependence of Γ^{eff} on the resonance (on-shell) mass. Because the decays are randomly selected at each time step, it may happen that $\Gamma^{\mathrm{eff}} < \Gamma^{\mathrm{dec}}$ if there is little to no broadening and the statistics are insufficient.

We consider the "death" of a particle when it either decays or scatters inelastically, and we compute γ with its initial momentum. This means that if it goes through elastic scatterings at times $t_1, ..., t_N$ before its death, the exact lifetime is

$$\tilde{\tau} = \frac{t_f - t_N}{\gamma_N} + \frac{t_N - t_{N-1}}{\gamma_{N-1}} + ... + \frac{t_1 - t_i}{\gamma_i}. \tag{6}$$

We checked that this leads to the same effective width as simply computing $\tau = \dfrac{t_f - t_i}{\gamma_i}$ in the analyzed systems. We believe that this happens because the momentum change in elastic collisions is small enough and can be either positive or negative, such that $\langle \tau \rangle \approx \langle \tilde{\tau} \rangle$.

To further quantify collisional broadening, we also define the collisional width as

$$\Gamma^{\mathrm{coll}}(m) = \Gamma^{\mathrm{eff}}(m) - \Gamma^{\mathrm{dec}}(m). \tag{7}$$

Subtracting the contribution of the vacuum from equation (5) results in the contributions solely caused by absorption processes. The medium effects can be further reframed in terms of the *dynamical* spectral function,

$$\mathcal{A}^{\mathrm{dyn}}(m) = \frac{2\tilde{\mathcal{N}}}{\pi} \frac{m^2 \Gamma^{\mathrm{eff}}(m)}{(m^2 - M_0^2)^2 + m^2 \Gamma^{\mathrm{eff}}(m)^2}, \tag{8}$$

which amounts to replacing the vacuum width in (2) with the effective width (5). Since the system is restricted to finite phase space, the support of this spectral function is not infinite. Therefore, we normalize it with respect to the vacuum spectral function (2):

$$\tilde{\mathcal{N}} = \frac{\int_{m_{\min}}^{m_{\max}} \mathcal{A}^{\mathrm{vac}}(m)dm}{\int_{m_{\min}}^{m_{\max}} \mathcal{A}^{\mathrm{dyn}}(m)dm}, \tag{9}$$

allowing for a proper comparison between the spectral functions for different assumptions. Usually, the expression "spectral function" is used interchangeably with "invariant mass spectra". We find it important to highlight that this is not the case here. The former refers to $\mathcal{A}^{\mathrm{dyn}}$, which encompasses the modifications to the propagator of the resonance, while the latter is denoted by dN/dm and describes the production of resonances; it was also studied in [9,16].

3. Results

3.1. Hadron Gas in Equilibrium

In this section, we demonstrate how particle interactions with a thermal medium affect the effective width (5). To do so, we employ an equilibrated hadron gas with different temperatures T and baryochemical potential $\mu_B = 400$ MeV. The system is simulated with a large box with periodic boundary conditions, initialized with thermal multiplicities according to the given (T, μ_B), and momenta are assigned according to the Boltzmann distribution. To ensure thermalization, we allow the gas to relax and only include particles with $t_i > 1000$ fm, which modifies the nominal (T, μ_B) values. The systems starting at $T = 120/140/160$ MeV fall to $106/128/149$ MeV, respectively. In the following, legends denote the initial temperature, and error bands are statistical.

Since collisional broadening originates from absorption processes, it is also possible to probe it for stable particles. For example, the $\pi\rho \to \omega$ process contributes to the broadening of both π and ρ.

Using (5), we extract the effective width for nucleons and pions. Table 1 shows that similar to ρ, they display an increasing effective width for rising temperatures, and we observe it to be significantly stronger for π than for N. The reason for this is the different number of inelastic channels; many mesonic and baryonic resonances decay into pions, while nucleons only participate in baryonic processes. As the temperature rises, more and more of these channels are opened, so the difference between the effective widths increases.

Table 1. Effective width of stable particles in thermal equilibrium. Errors are statistical.

	Γ^{eff} [GeV]		
T [MeV]	120	140	160
N	0.063 (1)	0.1082 (8)	0.1789 (7)
π	0.0802 (4)	0.2033 (5)	0.4376 (6)

Next, we investigate the collisional broadening of some selected resonances of particular interest. Along with the ρ meson, ω and $\Delta(1232)$ are relevant for dilepton yield, and the latter is also important for pion production, which is famously too high in most transport models. The reconstructability of $K^*(892)$ in HICs was investigated in [17]. $a_1(1260)$ is the chiral partner to ρ, so whether it also broadens is a natural question. Since the average lifetime of the ρ meson is shown in Figure 1, we do not repeat its inverse plot here. The effective width for ω in Figure 2a shows that Γ^{eff} grows with the medium temperature and that the difference from the vacuum is higher at lower masses. The curves converge to the vacuum decay width at large masses, so the decay probability dominates over the absorption probability, meaning that collisional broadening decreases. This happens because the processes that absorb the resonance become less likely, as detailed in Appendix A. We also see a non-monotonic structure in the effective width, caused by the shape of $\Gamma^{\text{dec}}(m)$, which increases relatively sharply while the medium effect decreases.

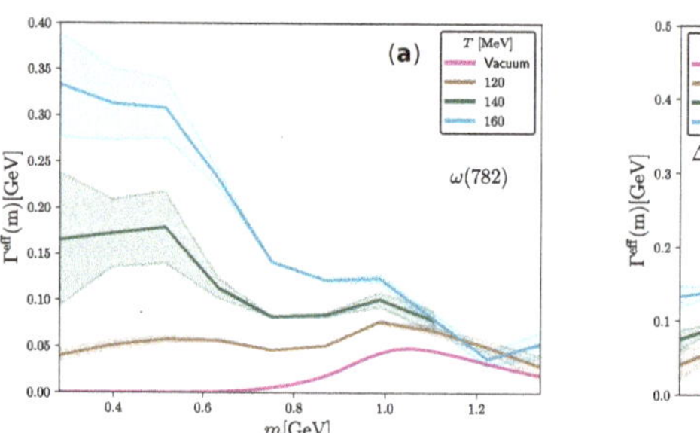
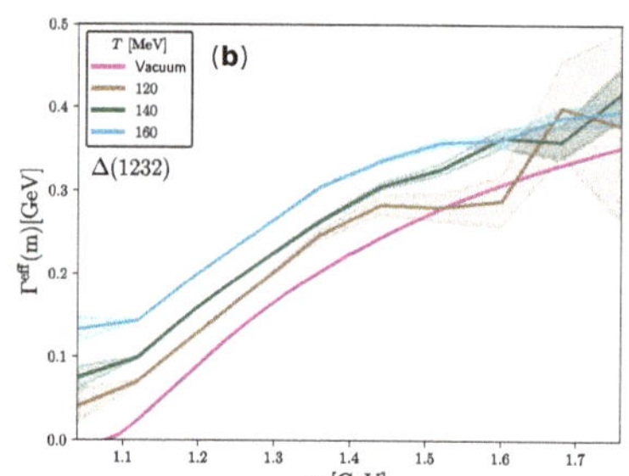

Figure 2. Effective width of (**a**) ω mesons and (**b**) $\Delta(1232)$ baryons in thermal equilibrium.

This convergence at high enough masses is generally shared between the resonances we probe, as shown in Figure 2b for the $\Delta(1232)$ baryon and in Figure 3a for the $K^*(892)$ meson, respectively. Another possible effect coming into play here is that if the vacuum width is large, the particle decays more quickly, so it has less time to scatter inelastically. Unlike the other resonances, low-mass $\Delta(1232)$ do not seem more sensitive to the medium temperature.

Moreover, $a_1(1260)$ does not develop any collisional broadening, as we show in Figure 3b. This is because there is no known process in which it emerges as a decay product; hence, there is no absorption process for it. One possible interesting consequence is that if the a_1 spectral function does broaden in a real QCD medium, as expected if chiral symmetry is restored [2], it has no contribution from collisional broadening, unlike the broadening of the ρ meson, its chiral partner.

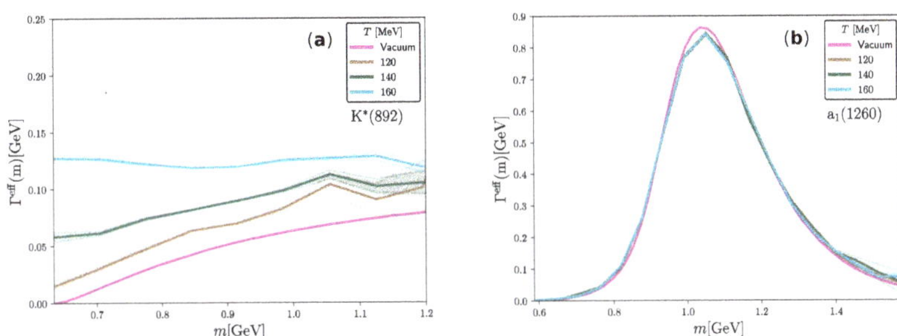

Figure 3. Effective width of (**a**) $K^*(892)$ and (**b**) $a_1(1260)$ mesons in thermal equilibrium.

3.2. Heavy-Ion Collisions

In this chapter, we study the collisional broadening of different particles in the off-equilibrium systems created after HICs. In SMASH, the nucleons in each nucleus are sampled from a Woods–Saxon distribution and move along the z-axis with the input beam energy. One key difference from the thermal hadron gas setup is the presence of strings, representing $2 \to N$ processes. They are formed in an interaction when the incoming particles have sufficient energy and are handled by PYTHIA 8.2 [18]. Another consequence of this initial state is in the system chemistry; in the thermal gas, all possible hadrons are initialized, using the full support of $\mathcal{A}^{\text{vac}}$. On the other hand, HICs start with only nucleons, and the available energy limits the phase space for particle production.

We restrict the analysis to central Au+Au collisions at $1.23A$ GeV kinetic energy and C+C collisions at $1A$ GeV, which are setups run by the HADES experiment at GSI [19,20]. A larger set of systems was investigated in [9], but these two provide a good grasp of the effect of medium size.

Figure 4a shows the effective width of the ω meson, where the broadening increases with system size. Since the vacuum width is small and close to the pole mass, the spectral function (2) is sharp; therefore, particles with masses far from the pole value are rare. For the C+C system, which contains little energy in total, this means that large-mass ω mesons are not produced. Compared with Figure 2a, lower-mass particles have a small broadening. We understand this in light of the medium expansion: as the system expands, the energy available to produce resonances decreases; consequently, lower masses become more likely. At the same time, the medium is diluted; therefore, the broadening of these particles is suppressed.

The $\Delta(1232)$ baryon also broadens more in a collision between heavier nuclei, as we show in Figure 4b. In the C+C collision, lower masses are not produced. The behavior is similar to the thermal gas in Figure 2b, where the collisional broadening displays weak dependence on the mass. This suggests that the Δ baryons behave thermally, with most being created via the first $NN \to N\Delta$ interactions.

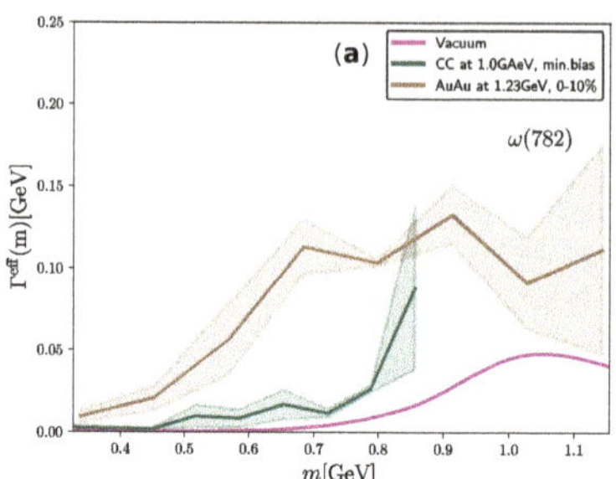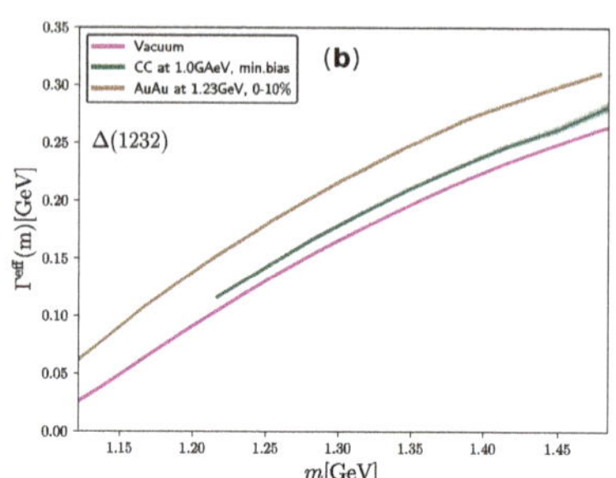

Figure 4. Effective width of the (**a**) $\omega(782)$ meson and (**b**) $\Delta(1232)$ baryon for different nuclear collision systems.

In Figure 5a, we see a small broadening of $K^*(892)$, which increases at high masses in the Au+Au system. In our implementation of low-energy nuclear collisions, the only[2] channel for strange hadron production is the decay of heavier resonances into a strange–antistrange pair. The first NN interactions rarely produce states able to decay into $K^*(892)$ (see Appendix A). Then, at least three interactions must happen to produce it, when the medium may have already become dilute. This is consistent with a previous study that used reconstructable K^* [17]; since they leave the medium before being absorbed, their decay products are also unaffected.

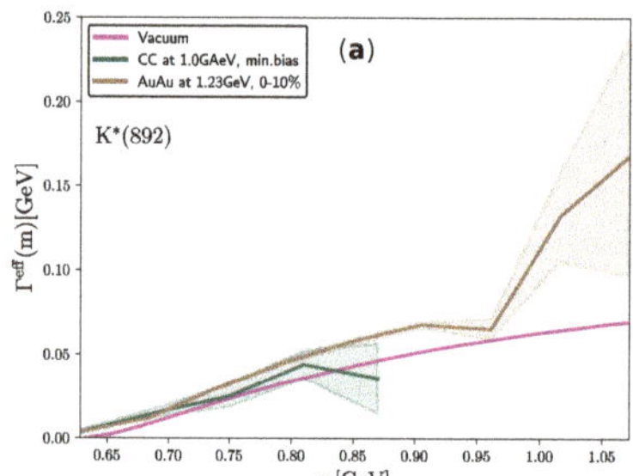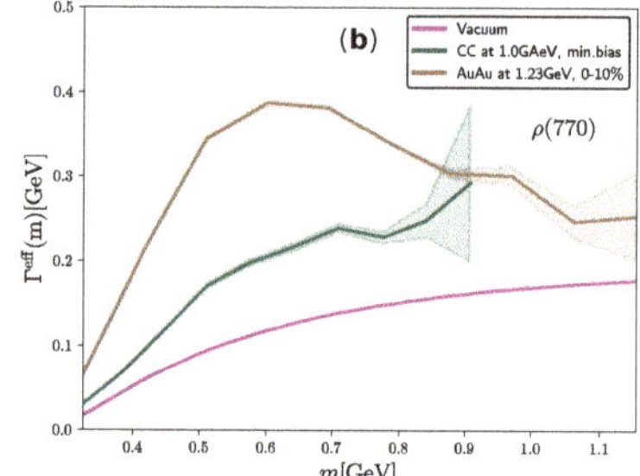

Figure 5. Effective width of the (**a**) $K^*(892)$ and (**b**) ρ mesons for different nuclear collision systems.

Much like in Figure 3b, we do not observe the collisional broadening of the $a_1(1260)$ meson, since there is no process where it is a decay product; therefore, we do not plot the result. We show the effective width of the ρ meson in these collision systems in Figure 5b. Similar to the ω meson, the difference from the thermal gas behavior in lower masses happens because of the medium dilution, since they are mostly produced in the late stage [9].

To discriminate the processes that cause this broadening, we weigh the collisional width (7) with the fraction of each process at a given mass. The dominant contributions are shown in Figure 6 in the Au+Au system. Out of the five most important processes, four are baryonic, similar to the results of Rapp's full in-medium model [2], where ρ couples to nucleons more. As previously suggested in [21], we see a significant contribution from the $\rho N \to N(1520)$ channel around $m \approx 0.5$ GeV. The biggest mesonic contribution is from chiral partner $a_1(1260)$, as was the case in [22] in the same mass range.

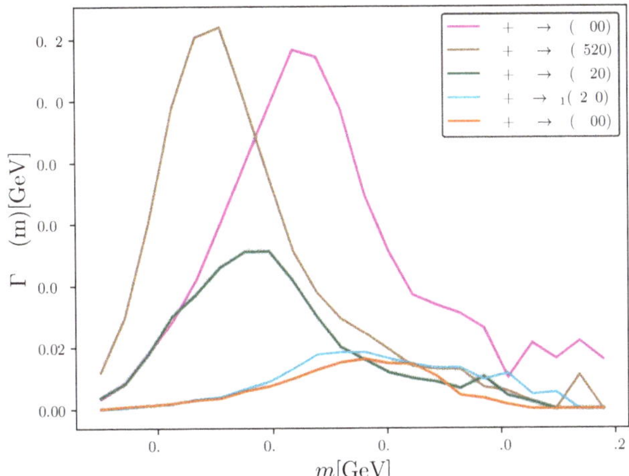

Figure 6. Contributions of the 5 most significant absorption channels to the collisional width of ρ in central Au+Au collisions at 1.23 GeV.

3.3. Collisional Broadening under Different Vacuum Assumptions

In this section, we discuss the effects of the different vacuum decay assumptions (described in Section 2) on collisional broadening. We evaluate the collisional width (7) for both assumptions in the framework of a hadronic thermal gas and show it for ρ and ω in Figure 7, using $\Gamma_\rho^0 = 149$ MeV and $\Gamma_\omega^0 = 8.5$ MeV. ρ is more broadened in the mass-dependent case, while ω displays a more complicated structure. This arises from different effects:

1. Particles that decay cannot be absorbed, so a larger vacuum width suppresses collisional broadening. At low masses, the vacuum decay width tapers down to 0 in the mass-dependent assumption. This makes the particles more prone to be absorbed by the medium in comparison with the mass-independent case.

2. Inelastic cross-section σ_{ab} affects the broadening of both a and b, since it determines how much one absorbs the other. It has peaks around the pole mass (M_R^0) of possible resonances $ab \to R$ [10]. The masses of the incoming particles control the off-shell mass of the outgoing resonance ($m_R = \sqrt{s_{ab}}$), so such peaks lead to structures in the collisional width of a and b, as exemplified by Figure 6; the contribution of the process $\rho N \to N(1520)$ is higher and close[3] to $M_{N(1520)}^0 - m_N = 0.57$ GeV, and heavier resonances lead to peaks in larger m_ρ. This effect is not relevant for very small masses, when $\mathcal{A}_R^{\text{vac}} \to 0$.

3. Absorption cross-section $\sigma_{ab \to R}$ is also proportional to Γ_R^{vac}, so that different mass assumptions give different weights to the resonance peaks.

4. At high enough masses, the absorption cross-section decreases so much that particles stop undergoing collisional broadening, as detailed in Appendix A, such that the vacuum assumption has no effect.

The interplay among the aforementioned effects causes a mass-dependent ρ to always be more absorbed than a mass-independent one. In the case of ω, both cases lead to the same broadening in the range $m_\omega = 0.6 - 0.75$ GeV, but a peak is more pronounced in the mass-dependent assumption in the range $m_\omega = 0.75 - 1.0$ GeV.

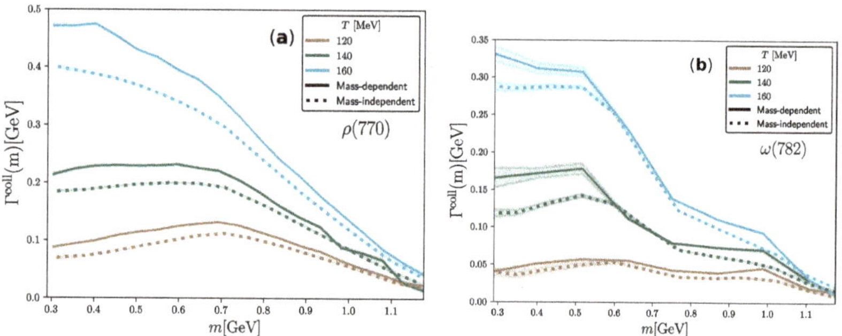

Figure 7. Collisional widths of (**a**) ρ and (**b**) ω in thermal equilibrium under different vacuum decay assumptions.

In order to assess the net effect of both assumptions, we show the corresponding broadened spectral functions ($\mathcal{A}^{\mathrm{dyn}}$) in Figure 8, using definition (8) with the same normalization (9) for both assumptions. As expected, the mass-dependent assumption leads to a broader spectral function, but this is only significant around the pole mass.

As mentioned in Section 2, the prescription of setting resonance lifetimes to their inverse widths is not consistent with scattering theory derivations, as they approach infinity close to the threshold. In the mass-independent assumption, this does not happen; subsequently, the medium effect in that region is reduced. In the phase-shift prescription, the time delay (that is, the vacuum lifetime) approaches 0 at the threshold, so we predict that if it is used appropriately, the collisional broadening at low masses is suppressed, as resonances decay immediately after formation.

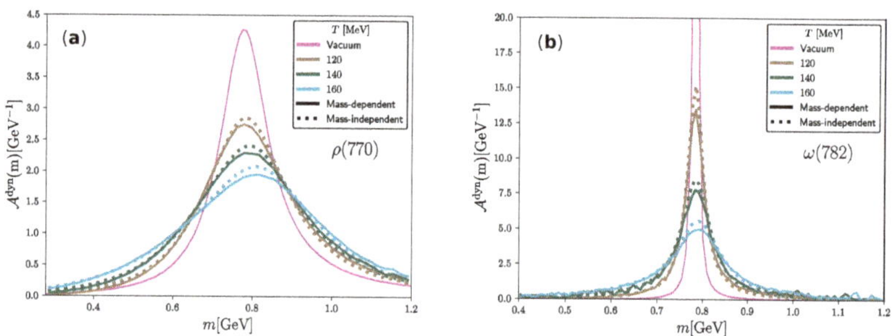

Figure 8. Dynamical spectral function of the (**a**) ρ and (**b**) ω mesons at different temperatures and baryochemical potential $\mu_B = 400$ MeV in thermal equilibrium under different vacuum decay assumptions.

4. Conclusions and Discussion

In this work, we investigate the collisional broadening of different particles by computing their effective width in the framework of a hadronic transport approach. First, we evolve a hadron gas to equilibrium at different temperatures and baryochemical potential $\mu_B = 400$ MeV, allowing us to establish the thermal behavior of such particles. The effective width shows dependence on the system temperature, where large temperatures enhance the collisional broadening of all hadron species except the $a_1(1260)$ meson, which does not have any absorption channel available. The particles that can broaden are generally more affected at lower masses, because the absorption cross-sections decrease at high masses.

Furthermore, we study the effect of collisional broadening in non-equilibrium systems created in HICs. In this framework, the effective width shows dependence on the system size, as collisional broadening is enhanced by a larger system. We observe that each resonance behaves differently, depending on when it is produced the most. $\Delta(1232)$

baryons show a behavior similar in both the thermal gas and HICs, since they are mostly created in the first NN interactions and move across a relatively thermalized medium [8]. On the other hand, $K^*(892)$ mesons, being strange particles, tend to appear after the third interactions, when the medium has already dissipated.

We also investigate the processes that cause a collisional broadening of ρ mesons. Particularly, the contribution of the $\rho N \to N(1520)$ absorption channel shows a significant effect around $m_\rho \approx 0.5$ GeV. In agreement with the full in-medium model [3,22], the coupling to nucleons is the most important, with π coupling being a distant second. We also find that the pole mass of the outgoing particle in each absorption channel determines the peaks in their contribution to collisional broadening, with some differences to account for the final kinetic energy.

Lastly, we compare two assumptions for the decay probability in vacuum in the thermal gas framework. We observe that a mass-dependent description of Γ^{dec} slightly enhances collisional broadening close to the pole mass.

Author Contributions: Conceptualization, R.H. and H.E.; Methodology, R.H.; Data curation, B.B.; Writing—original draft, B.B.; Writing—review and editing, R.H.; Supervision and Project administration, H.E. All authors have read and agreed to the published version of the manuscript.

Funding: This work was supported by Helmholtz Forschungsakademie Hessen für FAIR (HFHF). The authors also acknowledge the support by the State of Hesse within the Research Cluster ELEMENTS (Project ID 500/10.006) and by Deutsche Forschungsgemeinschaft (DFG; German Research Foundation)—Project number 315477589—TRR 211.

Data Availability Statement: The data presented in this study are openly available from FigShare at https://figshare.com/articles/dataset/Data-Balinovic_zip/23695755. Accessed on 6 September 2023.

Acknowledgments: Computational resources were provided by GreenCube at GSI.

Conflicts of Interest: The authors declare no conflicts of interest.

Appendix A. Interaction Cross-Sections

As discussed in Section 3, the collisional broadening of a particle results from the processes where it is absorbed, which are determined by inelastic cross-sections. These are functions of the center-of-mass energy between the incoming particles and thus depend on the off-shell mass of each, as well as on the relative momenta. We exemplify this in Figure A1, which shows the inelastic cross-sections for $\omega + p$ and $\omega + \pi$ scatterings and how they depend on the excess energy for different values of m_ω.

In the thermal hadron gas, the inelastic processes consist of $2 \to 1$ and $2 \to 2$ interactions. The peaks of each cross-section are always around the same $\sqrt{s}$, so the peaks in *excess* energy trivially shift with the increase in resonance mass. This happens until the peak cannot shift any further, since the center-of-mass energy is bounded from below by the sum of incoming masses. After this, heavier resonances have a progressively smaller σ_{inel}, and consequently a smaller collisional width, as seen in Figure 2a. The incoming particles still interact, but mostly through elastic scattering, which does not cause collisional broadening.

Figure A2 shows the cross-sections of $p + p$ scatterings that happen in the first moments of an HIC. For energies below $\sqrt{s} = 3.5$ GeV, the largest inelastic contribution is the excitation of Δ via $NN \to N\Delta$, followed by double Δ production. This is why the Δ baryons in the nuclear collisions of Section 3.2 behave similarly to the those in the thermal gas of Section 3.1. Other $2 \to 2$ channels are possible but very unlikely, with branching ratio $\sigma/\sigma_{\mathrm{tot}} \leq 2\%$.

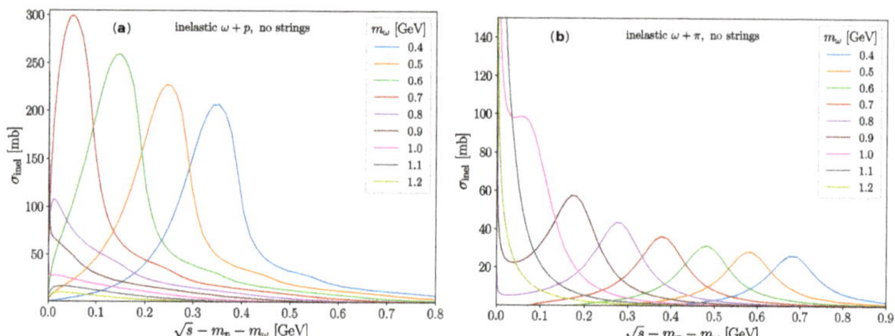

Figure A1. Inelastic cross-sections of (**a**) $\omega + p$ and (**b**) $\omega + \pi$ scatterings in a thermal gas for different off-shell masses m_ω.

Between $\sqrt{s} = 3.5$ and 4.5 GeV, we use a transition from the resonance to the string picture, with non-diffractive string fragmentations quickly dominating; above that, only strings are produced.

Concerning the $K^*(892)$ meson, the lowest mass states that can decay into it are $N(1875)$ and $\Delta(1900)$, both of which are rarely produced in these interactions. The systems analyzed in Section 3.2 have $\sqrt{s_{NN}} = 2.32$–2.41 GeV, so most $K^*(892)$ will be produced from a tertiary or later interaction.

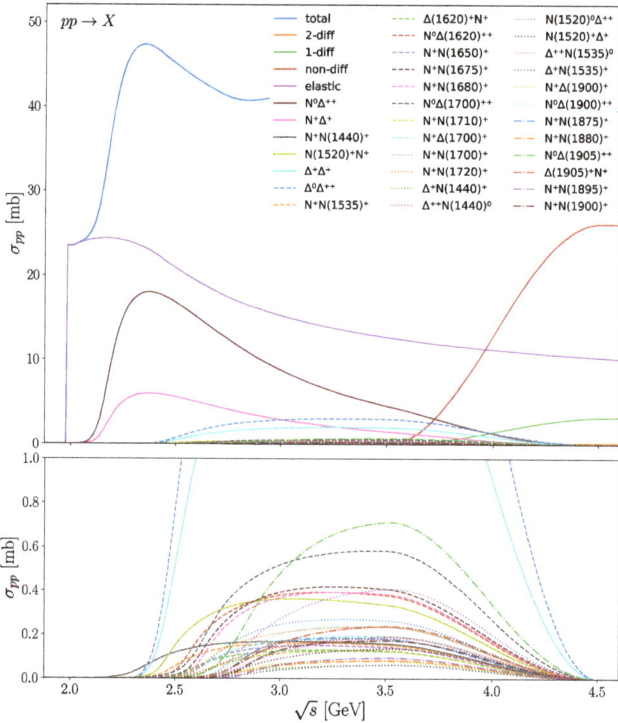

Figure A2. Cross-sections of $p + p$ scatterings we use in a nuclear collision, including string fragmentation ("2-diff", "1-diff", and "non-diff"). Below is an enlargement of smaller contributions.

Notes

[1]   We consider stable the hadrons with $\Gamma_0 \leq 10$ keV.

[2]   In collisions with energies higher than $\sqrt{s_{NN}} \approx 3.5$ GeV, strange hadrons also come from string fragmentation (see Appendix A).

Universe **2023**, *9*, 414

3   The difference from the actual peak is due to the kinetic energy given to the created resonance.

References

1. Larionov, A.B.; Effenberger, M.; Leupold, S.; Mosel, U. Resonance lifetime in BUU: Observable consequences. *Phys. Rev. C* **2002**, *66*, 054604. [CrossRef]
2. Rapp, R.; Wambach, J. Low mass dileptons at the CERN SPS: Evidence for chiral restoration? *Eur. Phys. J. A* **1999**, *6*, 415–420. [CrossRef]
3. van Hees, H.; Rapp, R. Dilepton Radiation at the CERN Super Proton Synchrotron. *Nucl. Phys. A* **2008**, *806*, 339–387. [CrossRef]
4. Buss, O.; Gaitanos, T.; Gallmeister, K.; van Hees, H.; Kaskulov, M.; Lalakulich, O.; Larionov, A.B.; Leitner, T.; Weil, J.; Mosel, U. Transport-theoretical Description of Nuclear Reactions. *Phys. Rept.* **2012**, *512*, 1–124. [CrossRef]
5. Linnyk, O.; Bratkovskaya, E.L.; Ozvenchuk, V.; Cassing, W.; Ko, C.M. Dilepton production in nucleus-nucleus collisions at top SPS energy within the Parton-Hadron-String Dynamics (PHSD) transport approach. *Phys. Rev. C* **2011**, *84*, 054917. [CrossRef]
6. Endres, S.; van Hees, H.; Weil, J.; Bleicher, M. Coarse-graining approach for dilepton production at energies available at the CERN Super Proton Synchrotron. *Phys. Rev. C* **2015**, *91*, 054911. [CrossRef]
7. Endres, S.; van Hees, H.; Weil, J.; Bleicher, M. Dilepton production and reaction dynamics in heavy-ion collisions at SIS energies from coarse-grained transport simulations. *Phys. Rev. C* **2015**, *92*, 014911. [CrossRef]
8. Staudenmaier, J.; Weil, J.; Steinberg, V.; Endres, S.; Petersen, H. Dilepton production and resonance properties within a new hadronic transport approach in the context of the GSI-HADES experimental data. *Phys. Rev. C* **2018**, *98*, 054908. [CrossRef]
9. Hirayama, R.; Staudenmaier, J.; Elfner, H. Effective spectral function of vector mesons via lifetime analysis. *Phys. Rev. C* **2023**, *107*, 025208. [CrossRef]
10. Weil, J.; Steinberg, V.; Staudenmaier, J.; Pang, L.G.; Oliinychenko, D.; Mohs, J.; Kretz, M.; Kehrenberg, T.; Goldschmidt, A.; Bauchle, B.; et al. Particle production and equilibrium properties within a new hadron transport approach for heavy-ion collisions. *Phys. Rev. C* **2016**, *94*, 054905. [CrossRef]
11. Olive, K.A. et al. [Particle Data Group] Review of Particle Physics. *Chin. Phys. C* **2014**, *38*, 090001. [CrossRef]
12. Manley, D.M.; Saleski, E.M. Multichannel resonance parametrization of pi N scattering amplitudes. *Phys. Rev. D* **1992**, *45*, 4002–4033. [CrossRef] [PubMed]
13. Danielewicz, P.; Pratt, S. Delays associated with elementary processes in nuclear reaction simulations. *Phys. Rev. C* **1996**, *53*, 249–266. [CrossRef]
14. Bass, S.A.; Belkacem, M.; Bleicher, M.; Brandstetter, M.; Bravina, L.; Ernst, C.; Gerland, L.; Hofmann, M.; Hofmann, S.; Konopka, J.; et al. Microscopic models for ultrarelativistic heavy ion collisions. *Prog. Part. Nucl. Phys.* **1998**, *41*, 255–369. [CrossRef]
15. Li, Q.; Bleicher, M. A Model comparison of resonance lifetime modifications, a soft equation of state and non-Gaussian effects on pi-pi correlations at FAIR/AGS energies. *J. Phys. G* **2009**, *36*, 015111. [CrossRef]
16. Schumacher, D.; Vogel, S.; Bleicher, M. Theoretical analysis of dilepton spectra in heavy ion collisions at GSI-FAIR energies. *Acta Phys. Hung. A* **2006**, *27*, 451–458. [CrossRef]
17. Reichert, T.; Bleicher, M. Kinetic mass shifts of $\rho(770)$ and $K^*(892)$ in Au+Au reactions at $E_{beam} = 1.23$ AGeV. *Nucl. Phys. A* **2022**, *1028*, 122544. [CrossRef]
18. Sjöstrand, T.; Ask, S.; Christiansen, J.R.; Corke, R.; Desai, N.; Ilten, P.; Mrenna, S.; Prestel, S.; Rasmussen, C.O.; Skands, P.Z. An introduction to PYTHIA 8.2. *Comput. Phys. Commun.* **2015**, *191*, 159–177. [CrossRef]
19. Agakishiev, G. et al. [HADES Collaboration] Study of dielectron production in C+C collisions at 1-A-GeV. *Phys. Lett. B* **2008**, *663*, 43–48. [CrossRef]
20. Adamczewski-Musch, J. et al. [HADES Collaboration] Centrality determination of Au + Au collisions at 1.23A GeV with HADES. *Eur. Phys. J. A* **2018**, *54*, 85, [CrossRef]
21. Vogel, S.; Bleicher, M. Reconstructing $\rho 0$ and ω mesons from nonleptonic decays in C+ C collisions at 2 GeV/nucleon in transport model calculations. *Phys. Rev. C* **2006**, *74*, 014902. [CrossRef]
22. Rapp, R.; Gale, C. ρ properties in a hot meson gas. *Phys. Rev. C* **1999**, *60*, 024903. [CrossRef]

Article

The Formulation of Scaling Expansion in an Euler-Poisson Dark-Fluid Model

Balázs Endre Szigeti [1,2,*], Imre Ferenc Barna [1] and Gergely Gábor Barnaföldi [1]

[1] Wigner Research Centre for Physics, Institute for Particle and Nuclear Physics, 1121 Budapest, Hungary
[2] Faculty of Science, Department of Atomic Physics, Eötvös Loránd University, 1117 Budapest, Hungary
* Correspondence: szigeti.balazs@wigner.hu

Abstract: We present a dark fluid model described as a non-viscous, non-relativistic, rotating, and self-gravitating fluid. We assume that the system has spherical symmetry and that the matter can be described by the polytropic equation of state. The induced coupled nonlinear partial differential system of equations was solved using a self-similar time-dependent ansatz introduced by L. Sedov and G.I. Taylor. These kinds of solutions were successfully used to describe blast waves induced by an explosion following the Guderley–Landau–Stanyukovich problem. We show that the result of our quasi-analytic solutions are fully consistent with the Newtonian cosmological framework. We analyzed relevant quantities from the model, namely, the evolution of the Hubble parameter and the density parameter ratio, finding that our solutions can be applied to describe normal-to-dark energy on the cosmological scale.

Keywords: dark fluid; Sedov–Taylor Ansatz; self-similarity

Citation: Szigeti, B.E.; Barna, I.F.; Barnaföldi, G.G. The Formulation of Scaling Expansion in an Euler-Poisson Dark-Fluid Model. *Universe* **2023**, *9*, 431. https://doi.org/10.3390/universe9100431

Academic Editor: Máté Csanád

Received: 21 June 2023
Revised: 30 August 2023
Accepted: 13 September 2023
Published: 27 September 2023

1. Introduction

In the second half of the 20th century, various self-similar solutions have been found following Gottfried Gurderley's famous discovery of spherically symmetric self-similar solutions that describe an imploding gas that collapses to the center [1]. In this paper, we use the kinds of self-similar solutions that were independently found by Leonid Ivanovich Sedov and Sir Geoffrey Ingram Taylor during the 1940s [2,3]. Despite the fact that such models have been well-known for decades, they have recently received attention again. This *ansatz* has been already applied successfully in several hydrodynamical systems, such as in the context of the three-dimensional Navier–Stokes and Euler equations [4], heat equations [5,6], and star formation [7]. In addition, the concept of self-similarity has a wide range of applications in general relativity. Homothetic solutions were first introduced by Cahill and Taub [8], and have been studied extensively for such different topics as asymptotic solutions in cosmology [9] and the gravitational collapse of black holes [10].

The existence of the dark matter was first proposed by the Dutch astronomer Jacobus Cornelius Kapteyn [11], and became widely known through Zwicky's famous work from 1933 [12]. During the second half of the century, solid experimental evidence was provided by Vera Rubin, Ken Ford, and others [13,14]. However, the general existence and specified properties of dark matter remain one of the most disputed topics in theoretical astrophysics. Dark fluid is a theoretical attempt to describe the properties of dark matter and its unification with dark energy into one hypothesized substance [15].

Our goal is to use the Sedov–Taylor *ansatz* to describe the time evolution of a dark fluid-like material characterized by a coupled nonlinear partial differential equation system. In our model, we study one of the simplest dark fluid materials, described by a polytropic (linear) equation of state. The dynamical evolution of the dark fluid is governed by the Euler equation and the gravitational field is described by the corresponding Poisson equation. We find time-dependent scaling solutions of the velocity flow, density flow, and gravitational fields, which can be good candidates to describe the evolution of the gravitationally coupled dust-like dark matter in the universe. We show that these kinds of solutions are consistent with the Newtonian Friedmann equations. They satisfy the mass conservation and acceleration equations and provide a similar expansion rate of the universe as the traditional solutions. The aim of this study is to broaden the knowledge of time-dependent self-similar solutions in these dark fluid models by improving upon and extending our previous model [16]. We tested our model on cosmological scales; moreover, we studied how the obtained evolution of the Hubble parameter and the ratio of the density parameters resemble other physical models.

2. The Model

We consider a set of coupled nonlinear partial differential equations which describe the non-relativistic dynamics of a compressible fluid with zero thermal conductivity and zero viscosity [17],

$$\partial_t \rho + \mathrm{div}(\rho \boldsymbol{u}) = 0 \,, \tag{1a}$$

$$\partial_t (\rho \boldsymbol{u}) + \mathrm{div}(\rho \boldsymbol{u} \otimes \boldsymbol{u}) = -\nabla P(\rho) + \rho \boldsymbol{g} \,, \tag{1b}$$

$$P = P(\rho) \,. \tag{1c}$$

These equations are the continuity, Euler equation, and equation of state (EoS), respectively. We assume that the system has spherical symmetry, and we are interested in solving it in one dimension. If we assume the fluid is ideal and the system has spherical symmetry, we can reduce the multi-dimensional partial differential equation (PDE) system into the one-dimensional radial-dependent one:

$$\partial_t \rho + (\partial_r \rho)u + (\partial_r u)\rho + \frac{2u\rho}{r} = 0 \,, \tag{2a}$$

$$\partial_t u + (u\partial_r)u = -\frac{1}{\rho}\partial_r P + g \,, \tag{2b}$$

$$P = P(\rho) \,. \tag{2c}$$

Here, the dynamical variables are $\rho = \rho(r, t)$, $u = u(r, t)$, and $P = P(r, t)$, respectively referring to the density, radial velocity flow, and pressure field distributions, with g being the radial component of an exterior force density. As we noted briefly in the introduction, we use the following general linear equation of state:

$$P(\rho) = w\rho^n, \qquad n = 1 \,. \tag{3}$$

Several forms of the EOS are available in astrophysics, and polytropic ones have been successfully used in the past; see Emden's famous book [18]. A great variety of applications can be found [19]. In Equation (3), the w parameter can vary depending on the type of matter that governs the system's evolution. Traditionally, $w = 0$ is used; this value corresponds to the EoS for ordinary non-relativistic matter or cold dust. For our case, we choose a negative value for w which leads to different kinds of dark-fluid scenarios, as was presented in detail by Perkovic [20]. In this paper, we choose $w = -1$ which represents the simplest case of an expanding universe governed by dark matter. Smaller values could cause a "big rip". The adiabatic speed of sound can be evaluated from Equation (3), and it is easy to show that it is constant:

$$\frac{\mathrm{d}P(\rho)}{\mathrm{d}\rho} = c_s^2 = w \,, \tag{4}$$

which is a necessary physical condition. Furthermore, let us assume that we have an additional self-gravitating term in Equation (2b). In this case, the exterior force density g can be expressed in the following way:

$$g = -\partial_r \Phi ,$$
(5)

where $\Phi = \Phi(r, t)$ is the Newtonian gravitational potential and satisfies the Poisson equation, which is coupled to the previously proposed PDE system [21]

$$\nabla^2 \Phi = 4\pi G \rho ,$$
(6)

where G is the universal gravitational constant, which is set to unity in our further calculations. Note that we can add an additional constant term Λ to Equation (1b) which plays a similar role to the cosmological constant in Einstein's equations:

$$\partial_t u + (u\partial_r)u = -\frac{1}{\rho}\partial_r p - \partial_r\Phi(r) + \Lambda .$$
(7)

Below, we show that this constant cannot be used, as it does not lead to a consistent self-similar solution, which is what we are looking for here. Note that this observation in our model can be an *indirect proof of the nonexistence of the static Universe picture*. We can extend the exterior force density further with a rotating term. In this case, we add a phenomenological rotation term to Equation (2b), meaning that the equation takes the following modified form:

$$\partial_t u + (u\partial_r)u = -w\frac{1}{\rho}\partial_r p - \partial_r\Phi(r) + \frac{\sin\theta\,\omega^2 r}{t^2} ,$$
(8)

where ω is a dimensionless parameter that describes the strength of the rotational effect and θ is the polar angle. We assume that the rotation is slow; therefore, we can expect that the spherical symmetry is not broken. This statement is satisfied if the ω parameter is sufficiently small, implying that the rotational energy is negligible compared to the gravitational energy. The self-similar analysis of various rotating and stratified incompressible ideal fluids was investigated in two Cartesian coordinates in [22]. Note that the geometrized unit system ($c = 1$, $G = 1$) was applied for the calculations below, which can be converted to other units as well. See Appendix A for more details and unit conversations.

3. Scaling Solution and Sedov–Taylor Ansatz

We would like to find and study analytic solutions of equations by applying the long-established self-similar *ansatz* by Sedov and Taylor [2,3], which can be expressed in the following form:

$$u(r,t) = t^{-\alpha} f\left(\frac{r}{t^\beta}\right) ,$$
(9a)

$$\rho(r,t) = t^{-\gamma} g\left(\frac{r}{t^\beta}\right) ,$$
(9b)

$$\Phi(r,t) = t^{-\delta} h\left(\frac{r}{t^\beta}\right),$$
(9c)

where r means radial and t means time dependence. Note that the so-called shape functions (f, g, h) only depend on $rt^{-\beta}$; thus, we introduce a new variable

$$\zeta = rt^{-\beta}.$$
(10)

where ζ is a dimensionless quantity in geometrized units. The (not yet determined) exponents are called similarity exponents (α, β, γ, and δ), and have physical relevance; β

describes the rate of spread of the spatial distribution during the time evolution (if the exponent is positive) or contraction (if $\beta < 0$). In addition, the other exponents describe the rate of decay of the intensity of the corresponding field. Solutions with integer exponents are called self-similar solutions of the first kind, while the second kind denotes the non-integer ones. Self-similarity is based on the concept that physical quantities preserve their shape during time evolution. A general description of the properties of these types of scaling solutions can be found in our previous publication [16].

We assume that the shape functions are sufficiently smooth and continuously differentiable at least twice in ζ over the entire domain. Thus, we calculate the relevant time and space derivatives of the shape functions and substitute them into Equation (2). As a consequence, we usually obtain an overdetermined algebraic equation system for the similarity exponents. Other possible scenarios may play out as well; these have been presented in detail in [16]. We obtain the following numerical value for the exponents for both the non-rotating and rotating cases: $\alpha = 0$, $\beta = 1$, $\gamma = 2$, and $\delta = 0$. If we add the Λ constant to the Euler equation, such a solution for the similarity equation cannot be found. From these results, it is evident that the dynamical variables such as the velocity, gravitational potential, and density flow have spreading properties. Our physical intuition says that spreading is somehow similar to expansion, which is a basic property of the universe at astronomical or cosmological scales.

By substituting the obtained numerical values of the similarity exponents, we can reduce the induced PDE system into an ordinary differential equation (ODE) system that depends only on the ζ independent variable. We find that the obtained equation system has the following form:

$$-\zeta g'(\zeta) + f'(\zeta)g(\zeta) + f(\zeta)g'(\zeta) + \frac{2f(\zeta)g(\zeta)}{\zeta} = 0, \tag{11a}$$

$$-\zeta^2 f'(\zeta) + \zeta f'(\zeta)f(\zeta) = -\frac{wg'(\zeta)}{g(\zeta)} - h'(\zeta)\zeta + \omega^2 \sin\theta\zeta^2, \tag{11b}$$

$$h'(\zeta) + h''(\zeta)\zeta = g(\zeta)4\pi G\zeta . \tag{11c}$$

It can easily be noticed that the ordinary differential equation system presented in Equation (11) cannot be solved analytically. For linearized nonautonomous ordinary differential equation systems, the stationary point of the phase space can be found, and we can say something about the general asymptotic behavior of the solutions as well [23]. Nonetheless, there is no generally known method for nonlinearized nonautonomous differential equation systems. Moreover, the existence and uniqueness of smooth solutions has not yet been proven in multiple dimensions. Therefore, it is a reasonable approach to solve the obtained ordinary differential equation system in Equation (11) numerically for a large number of parameter sets (based on physical considerations) in order to explore the behavior of the solution of the system with different boundary and initial conditions. One example of such a numerical solution can be seen in Figure 1.

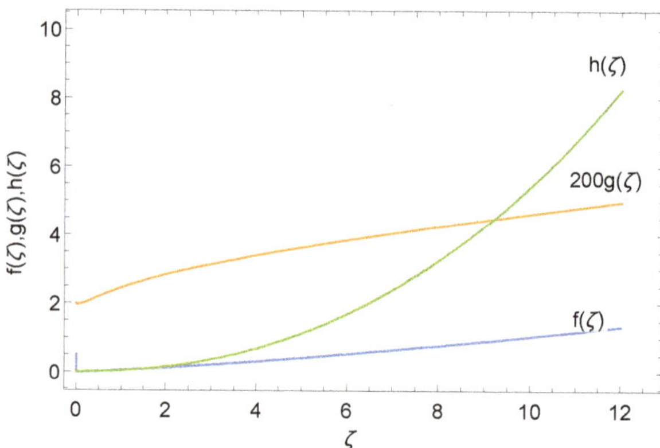

Figure 1. Numerical solutions of the shape functions; integration was started at $\zeta_0 = 0.001$ and initial conditions of $f(\zeta_0) = 0.5$, $g(\zeta_0) = 0.01$, $h(\zeta_0) = 0$, and $h'(\zeta_0) = 1$ were used. For better visibility, the $g(\zeta)$ function was upscaled by a factor of 200. The values are provided in geometrized units.

As an example, at a specific parameter and initial condition set, the shape function of the velocity $f(\zeta)$ is almost linear and increasing after a short decrease, providing a hint of Hubble expansion-like behavior. The shape function $g(\zeta)$ is asymptotically flat after a quick ramp-up, corresponding to the conservation of matter. The last shape function $h(\zeta)$ has an increasing polynomial trend with a slight positive exponent, which is connected to the gravitational potential. To obtain a sufficiently smooth numerical solution, we solved the ODE system using an adaptive numerical integration provided by *Wolfram Mathematica 13.1* [24]. For all of our calculations, the integration limits were $\zeta_0 = 0.001$ and $\zeta_{max} = 40$, the same as in [16]. As has been said before, we established initial conditions to obtain the numerical solution; for this reason, we use ranges $\mathcal{R}$ of $f(\zeta_0) = 0.005 - 0.5$ and $g(\zeta_0) = 0.001 - 0.1$, while for the second-order differential equation we have $h(\zeta_0) = 0$ and $h'(\zeta_0) = 1$.

This choice of initial conditions reflects that, first it is physically reasonable that the density flow range is $\mathcal{R}(g) \subset \mathbb{R}^+$ and finite. Recent results suggest that dark fluid could possibly have negative mass [25]; however, in the present model this leads to singular solutions. Second, our choice of initial velocity flow $\mathcal{R}(f) \subset \mathbb{R}^+$ makes for an initially radially expanding fluid. We have seen that if the initial value for $f(\zeta)$ and $g(\zeta)$ is set outside of the previously given range, the solution of the differential equation becomes singular. Moreover, we have seen that the variation in the initial condition corresponding to the shape function of the gravitational potential does not affect the trend of the time evolution of the system, as it only causes vertical shifts. Therefore, we set the initial numerical value equal to zero.

We are interested in finding the solution of the ODE system as a function of the spatial and time coordinates. We transform our single-variable numerical solutions into two-variable functions, for which we use the inverted form of Equation (10). It can easily be noticed that if we look at the shape of the *ansatz*, the solution has a singularity at $t = 0$. Thus, we use the $0.001 \leq t \leq 25$ and $0.001 \leq r \leq 25$ domains to obtain the space- and time-dependent initial dynamical functions $u(r,t)$, $\rho(r,t)$, and $\Phi(r,t)$.

4. Results

Here, we present the solutions of the self-gravitating non-relativistic dark fluid. First, we provide a detailed introduction to the global properties of the solutions in a non-rotating system. Second, we show the effect of slow rotation on the solutions. In addition, we compare the results from the two cases with each other and with the previous results in [16]. Note that the spherical symmetry of the system was kept conserved for all the cases.

4.1. Non-Rotating System

In the first case, we set the ω parameter to zero and used the obtained numerical values for the similarity exponents (α, β, γ, and δ) to obtain the exact ordinary differential equation. For the numerical integration, we used the same initial conditions $f(\zeta_0) = 0.5$ and $g(\zeta_0) = 0.01$ applied in our previous paper for the velocity and density flows, respectively. First, we used the time and radial projection of the unknown functions to obtain a better understanding. Figure 2 illustrates the spatial and time projections of the obtained velocity, density, and gravitational potential. These results for the velocity and density flows are consistent with our initial statement that these kinds of solutions of dark fluids can be used as a model to describe the exploding system (e.g., the universe). Similar behaviors can be seen for the radial velocity and the density, which both have fast decays in time at all distances. In addition, they both have a real singularity at $t = 0$ due to the shape of the *ansatz*. However, the radial distribution has a different nature; the density increases excessively at distances near the center of the explosion, and becomes linear at larger distances. On the contrary, the velocity grows polynomially with the radial distance. It can be seen that the gravitational potential decreases hyperbolically over time and becomes asymptotically flat. The gravitational potential has a natural singularity at the origin. The local existence of global weak solutions on a domain outside of the origin in spite of the existence of a singularity has been shown by Tsunge and others.

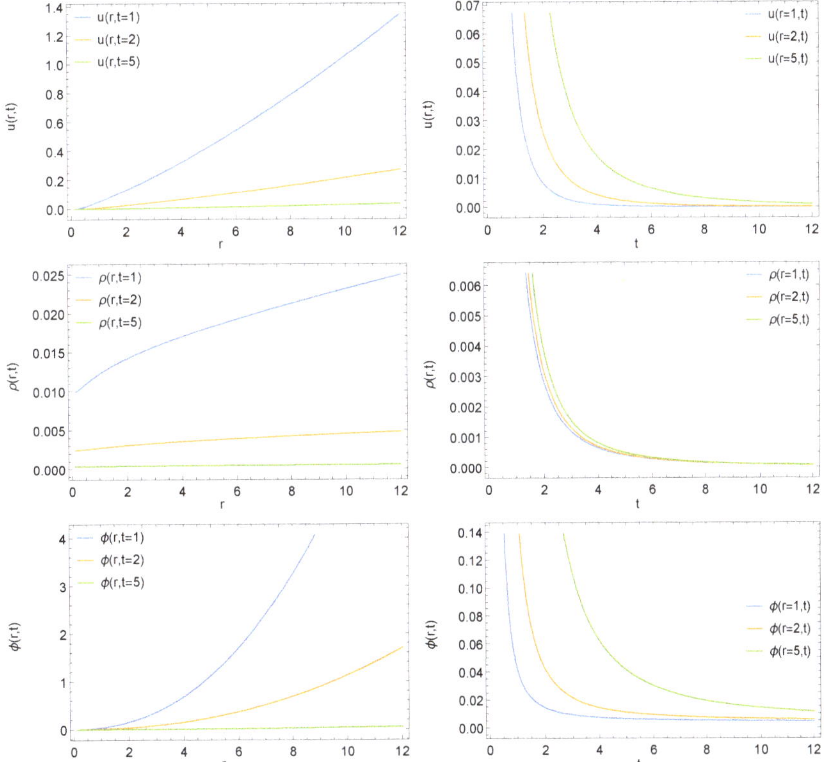

Figure 2. Different radial (left) and time (right) projections of the velocity flow (first row), density (second row), and gravitational potential (third row), respectively, for the non-rotating case. A detailed explanation is provided in the main text. The domain range is provided in geometrized units.

Here, we find a different radial velocity profile than in the previous non-rotating model with two equations presented in [16]. Furthermore, the similarity exponents are different ($\alpha = 0$, $\beta = 1$, and $\gamma = -1$ in the two-equation model). This is most likely due to the new smoother solution forming as a consequence of a result of the second derivative appearing in the Poisson equation. It can be seen that the solution depicted above in Figure 2 is numerically stable in the specified initial and boundary condition range. It is more relevant to investigate the dynamics of the complete fluid in time and space in order to understand certain general trends or physical phenomena as the function of the initial conditions. For this reason, we now evaluate the related energy densities, which are as follows:

$$\epsilon_{kin}(r,t) = \frac{1}{2}\rho(r,t)u^2(r,t), \qquad \Phi(r,t) = h(r,t), \qquad \epsilon_{tot}(r,t) = \epsilon_{kin}(r,t) + \Phi(r,t). \quad (12)$$

Figure 3 further illustrates the fact that the kinetic energy density has a singularity at $t = 0$, as we have already seen in the case of the radial velocity. It has linearly enhancing maxima at larger distances, and has a fast decay in time for all radial distances. As mentioned above, the gravitational potential is a negative polynomial in time; therefore, we can obtain the total energy density of the system. From the total energy density distribution, it is apparent that the short-time behavior of the system is dominated by the initial explosion, while the long-range structure is regulated by the gravitational potential.

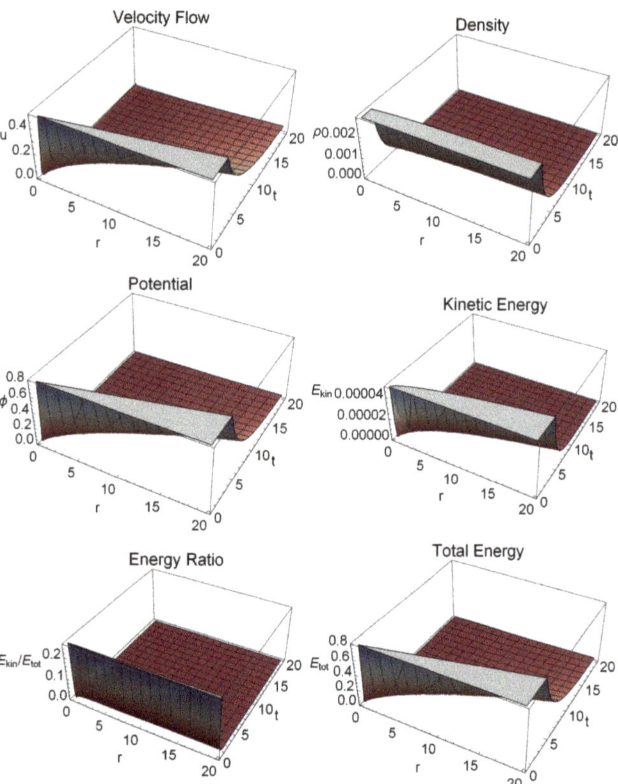

Figure 3. Numerical solutions of the velocity flow $u(r,t)$, density flow $\rho(r,t)$, and gravitational potential $\Phi(r,t)$ as a function of the spatial and time coordinates in the case of a non-rotating system, additionally showing the distribution of the total and kinetic energy densities. We used $\zeta_0 = 0.001$ for numerical integration and initial conditions $f(\zeta_0) = 0.5$, $g(\zeta_0) = 0.01$, $h(\zeta_0) = 0$, and $h'(\zeta_0) = 1$.

4.2. Rotating System

In this section, we analyzed the effect of slow rotation compared to the non-rotating case. First, we studied the effect of the variation of the maximal angular velocity (the ω parameter). We chose the polar angle (the θ parameter at the equatorial), as it provides the largest effect. As previously stated, we assume in our further analysis that the spherical symmetry is not broken. According to this assumption, we fix the gravitational force density to be at least one magnitude larger than the centrifugal force density at every time and space ($\|f_{grav}\| \gg \|f_{centr}\|$). The numerical results show that we can find a range of ω within which the constraint is fulfilled as long as the previously specified initial condition set is valid. We demonstrate that the asymptotic behavior of the numerical solution includes a significant dependence of ω on the acceptable domain ($0 < \omega < 0.3$) of the parameters and initial conditions.

Comparing the results shown in Figure 4 with the non-rotating case in Figure 2, it is evident that slow and constant rotation does not affect the time or spatial distribution of the gravitational potential. Moreover, it can be seen that the radial density profile of the system is nearly uniform, and is identical to the previous case in that it decreases rapidly over time. Thus, we can conclude that the rotation accelerates the even distribution of the material in space and speeds up inflation.

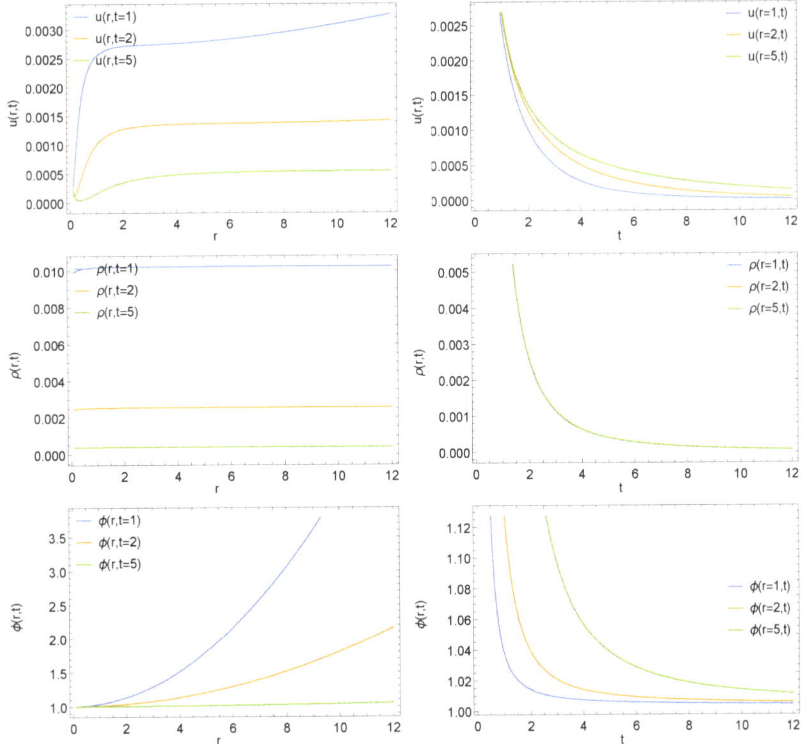

Figure 4. Time and radial projections of the velocity flow (first row), density (second row), and gravitational potential (third row) for the rotating system ($\omega = 0.2535$). We used $\zeta_0 = 0.001$ for numerical integration and initial conditions $f(\zeta_0) = 0.5$, $g(\zeta_0) = 0.01$, $h(\zeta_0) = 0$, and $h'(\zeta_0) = 1$.

The singular behavior close to $t = 0$ is not affected by the rotation, as was expected. However, a significant difference can be seen in the first graph (the top left panel of Figure 4).

It can be seen that the radial profile of the velocity flow starts from zero in the origin and shows exponential growth for the short-range behavior.

Figure 5 illustrates the critical influence of the ω on the long-range asymptotic behavior of the time evolution of the velocity flow. Moreover, an increase in the ω value causes significant modifications in the radial profile for both the velocity and density flows while leaving the time evolution unaltered. In our analysis of the behavior of the obtained numerical solutions, we found that the inspected initial value range shows similar behavior; $\omega < 1$ can be found for every initial and boundary condition in which the long-range asymptotic structure alternates. An example of this can be seen in Figure 5. Likewise, we studied the properties of the relevant dynamic variables for the previous case. The energy density associated with the rotation and the total energy is

$$\epsilon_{rot}(r,t) = \frac{1}{2}\rho(r,t)\omega^2 r \qquad \epsilon_{tot}(r,t) = \Phi(r,t) + \epsilon_{kin}(r,t) + \epsilon_{rot}(r,t). \tag{13}$$

these quantities are depicted in Figure 6.

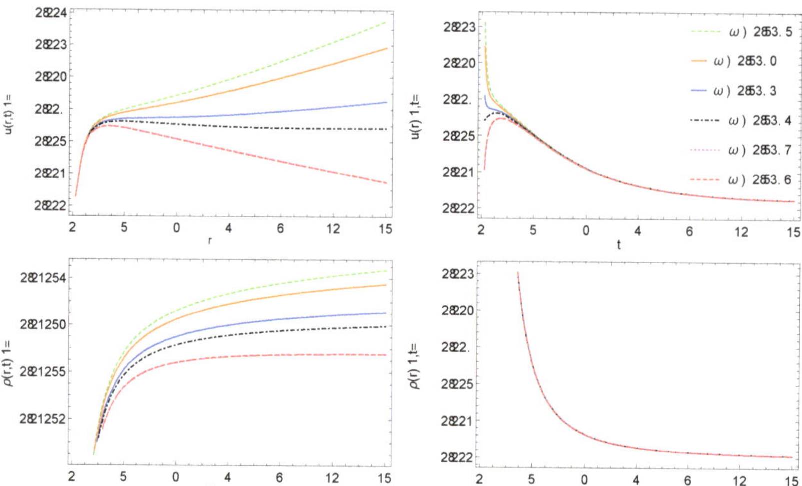

Figure 5. Dependence of the space and time evolution on the maximal angular velocity ω; the different lines correspond to different values of the angular velocity ω. The curves were evaluated at a particular time (left), with radial coordinates provided on the vertical axis (right). A detailed explanation is provided in the main text.

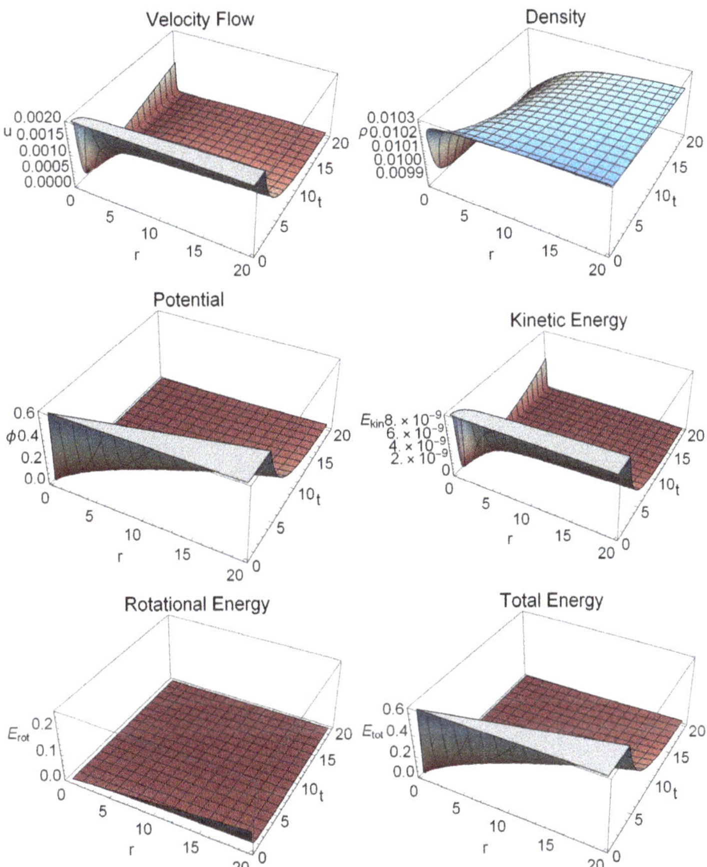

Figure 6. Numerical solutions of the velocity flow $u(r,t)$, density flow $\rho(r,t)$, and gravitational potential $\Phi(r,t)$ as a function of the spatial and time coordinates for the rotating case. In addition, we present the distribution of the total and kinetic energy densities. We used $\zeta_0 = 0.001$ for numerical integration, $\omega = 0.2535$, and initial conditions $f(\zeta_0) = 0.5$, $g(\zeta_0) = 0.01$, $h(\zeta_0) = 0$, and $h'(\zeta_0) = 1$.

5. Connection to Newtonian Friedmann Equation

Next, we apply the obtained scaling solution to cosmology in order to explain the evolution of the universe. Our strategy is to describe the concept of a cosmic fluid in terms of Newtonian cosmology, which we use because relativistic effects are not of great significance in this context. In this case, there is no need for a global reference point; thus, a *scale factor* $a(t)$ can be introduced which contains the entire time evolution that affects our chosen reference frame. Henceforth, the relative distances in time can be denoted in this term:

$$R(t) = a(t)\, l \tag{14}$$

where $R(t)$ is a continuous function and l is real number. In this chapter, we use classical conservation equations to show the connection between the obtained self-similar solution and the traditional Friedmann equations.

5.1. Connection to the Expansion Rate

Here, we introduce the total mass inside a radius r:

$$M(t) = \int_{\mathcal{V}(t)} \rho(R(t), t)\, \mathrm{d}V = 4\pi \int_0^r \rho(R(t), t)\, R(t)^2\, \mathrm{d}R(t), \tag{15}$$

where $\mathcal{V}(t) \subset \mathbb{R}^3$ is a sphere with radius $R(t)$ and $r \in (0, R(t))$. Therefore, within the classical Newtonian framework, the energy and mass conservation principle are fulfilled separately:

$$\frac{\mathrm{d}}{\mathrm{d}t} M(t) = 4\pi \frac{\mathrm{d}}{\mathrm{d}t} \int \rho(a(t)l, t)\, a^3(t)\, l^2\, \mathrm{d}l \overset{!}{=} 0. \tag{16}$$

Replacing the derivation with the integral formula, we can assume that equality holds for every l:

$$\rho(a(t)l, t) \frac{\mathrm{d}}{\mathrm{d}t} \left[\rho(a(t)l, t) \right] = -3 \frac{\dot{a}(t)}{a(t)}. \tag{17}$$

In addition, we impose the kinematic condition that

$$\frac{\mathrm{d}}{\mathrm{d}t} R(t) = u(R(t), t) \;\Rightarrow\; \frac{1}{g(R(t), t)} \frac{\mathrm{d}}{\mathrm{d}t} \left[t^{-\gamma} g(R(t), t) \right] = -3 \frac{t^{-\alpha} f(R(t), t)}{R(t)}. \tag{18}$$

We impose $\rho(r, t) > 0 \quad$ for $\forall r \geq R(t)$ and assume that the velocity $u(r, t)$ and density $\rho(r, t)$ are obtained from the self-similar *ansatz* in Equation (9). We expand the density and velocity flow into the following respective power series:

$$\rho(r, t) \sim t^{-\gamma} \sum_{m=0}^{\infty} \rho_m \zeta^m \quad \text{and} \quad u(r, t) \sim t^{-\alpha} \sum_{n=0}^{\infty} u_n \zeta^n. \tag{19}$$

After substituting $\zeta = R(t) t^{-\beta}$, Equation (18) becomes

$$-\gamma t^{-(\gamma+1)} + t^{-\gamma} \sum_{m=1}^{\infty} \rho_m m \left(R(t) t^{-\beta} \right)^{m-1} \left(\dot{R}(t) t^{-\beta} - \beta R(t) t^{-(\beta+1)} \right) =$$
$$= -\frac{3}{R(t)} \left[t^{-\alpha} \sum_{n=0}^{\infty} u_n \left(R(t) t^{-\beta} \right)^n \right] \left[t^{-\gamma} \sum_{m=0}^{\infty} \rho_m \left(R(t) t^{-\beta} \right)^m \right]. \tag{20}$$

On the relevant space and time scale, we only consider the contributions of the finite term. The discussion of the rotating case in Section 4 requires that we use a high-order polynomial to approximate the $f(\eta)$ function. Based on the analysis of the numerical results in Figure 4, we restrict ourselves to using polynomes up to the eighth order, which are well able to approximate the curves:

$$u(r, t) \sim t^{-\alpha} \sum_{n=0}^{8} \tilde{u}_n \zeta^n \quad \text{and} \quad \rho(r, t) \sim t^{-\gamma} \sum_{m=0}^{8} \tilde{\rho}_m \zeta^m. \tag{21}$$

Moreover, a further simplification can be applied within the domain of interest for ρ by describing it as a polynomial with a rational exponent $\rho(\zeta) \sim A\zeta^\kappa$, where $\kappa = 6/7$. Hence, Equation (20) can be rewritten as

$$\kappa \dot{R}(t) - \frac{1}{t}[\gamma + \kappa \beta] R(t) + 3 t^{-\alpha} \sum_{n=0}^{8} \tilde{u}_k \left(R(t) t^{-\beta} \right)^n = 0. \tag{22}$$

Consequently, the mass conservation equation in (15) becomes a nonautonomous first-order differential equation that cannot be solved analytically due to its highly nonlinear terms.

We can find further simplifications in the non-rotating limit ($\omega \to 0$), as the higher-order terms are less and less significant; therefore, the velocity flow simply becomes $u(r,t) \sim u_1\zeta^1 + u_2\zeta^2$ and the mass conservation equation is

$$\kappa\dot{R}(t) + 3u_2 t^{-(\alpha+2\beta)}[R(t)]^2 - \frac{1}{t}[\gamma + \kappa\beta]R(t) + 3u_1 R(t)t^{-(\alpha+\beta)} = 0. \tag{23}$$

The above equation can be solved analytically; the solution is

$$R(t) = u_1 t^{\beta+\frac{\gamma}{\kappa}} \exp\left[-\frac{3u_1 t^\mu}{\mu\kappa}\right] \times \left[3^{-\frac{\gamma}{\mu\kappa}} u_2 t^{\gamma/\kappa}\left(\frac{u_1 t^\mu}{\nu}\right)^{-\frac{\gamma}{\mu\kappa}} \Gamma\left(1+\frac{\gamma}{\nu'},\frac{3u_1 t^\mu}{\nu}\right) - \mathcal{H}_1 u_1\right]^{-1}, \tag{24}$$

where $\mu = 1 - (\alpha+\beta)$, $\nu = \kappa - \beta\kappa$, α, β, γ are the similarity exponents and A, u_1, u_2 the positive constants, the κ positive exponent is obtained from the solution of Equation (11), $\mathcal{H}_1$ is an integration constant, and $\Gamma()$ is the upper incomplete Gamma function [26]. Equation (24) can be simplified by substituting the similarity exponents and constants from the non-rotating case:

$$R(t) = \frac{t}{\mathcal{H}_1 t^{\frac{3u_1-2}{\kappa}} + \frac{3u_2}{2-3u_1}}, \quad \text{where } \kappa = \frac{6}{7}. \tag{25}$$

We can now use Hubble's law of expansion to determine the $\mathcal{H}_1$ integration constant and the numerical integration for the rotating case:

$$\left.\frac{\dot{a}(t)}{a(t)}\right|_{t=t_0} = H_0, \text{ if } a(t_0) = 1, \tag{26}$$

where $H_0 = 66.6^{+4.1}_{-3.3}$ km/s/Mpc is the experimental value of the Hubble constant [27]. The expansion rates for the rotating (yellow line) and non-rotating (blue line) Universe are drawn in Figure 7 in the usual units of standard cosmology. The dashed line represents the present measured value of the Hubble parameter.

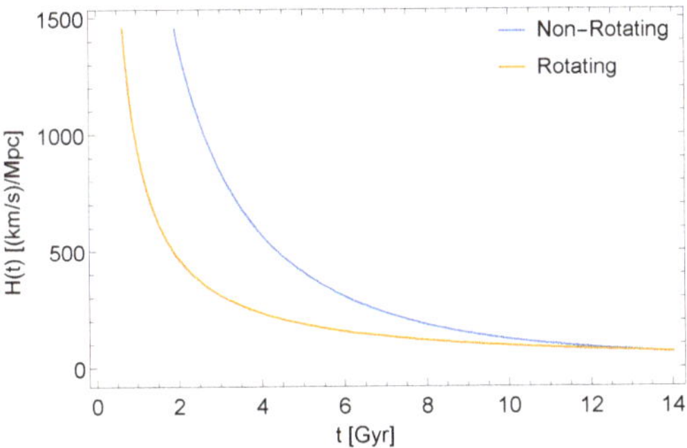

Figure 7. Analytical (non-rotating) and numerical (rotating) solutions of the expansion rate of the universe; integration started at $\zeta_0 = 0.001$, and the initial conditions were $f(\zeta_0) = 0.5$, $g(\zeta_0) = 0.008$, $h(\zeta_0) = 0$, and $h'(\zeta_0) = 1$. These results match well with the data from [28].

5.2. Connection to the Critical Density

Energy conservation has to be investigated as well, as it leads to the critical density parameter of the universe. We can apply the Newton equation to obtain the so-called acceleration equation [29]:

$$\frac{\ddot{a}(t)}{a(t)} = -\frac{4\pi G}{3}\rho(R(t),t) \quad \Rightarrow \quad \ddot{R}(t) = -\frac{4\pi G}{3}\rho(R(t),t)R(t) . \tag{27}$$

Next, we multiply the equation by $\dot{R}(t)$ and integrate over time; then, Equation (27) becomes

$$\dot{R}(t)^2 = \frac{8\pi G}{3}\rho(t)R^2(t) + U_0, \tag{28}$$

which is the energy conservation equation. In general relativity, U_0 is related to the curvature of spacetime. However, because we are working in the framework of Newtonian cosmology it is instead associated with a kind of mechanical energy. Equation (28) can be transformed into the following form using the definition $H(t) := \dot{R}(t)/R(t)$:

$$H^2(t) = \frac{8\pi G}{3}\rho(R(t),t) + U_t, \tag{29}$$

where $U_t = U_0/R(t)^2$ is a dynamical constant, $U_0 = H_0^2(1 - \Omega_{0,k})$, and we use the following definition for the critical density:

$$\rho_0 = \Omega_0\rho_c, \quad \text{where} \quad \rho_c = \frac{3H_0^2}{8\pi G} . \tag{30}$$

The value of the dynamical constant U_t determines the evolution of the system. If $U_t > 0$, then the universe eventually re-collapses. On the other hand, if $U_t < 0$ then the system expands to infinity. Thus, $U_t = 0$ is a special case in which the universe shows similar expanding behavior but reaches zero velocity only in infinite time [29]. Here, we are searching for an isentropic solution; therefore, we introduce the entropy conservation equation $dE + pdV = 0$ from [30]. Using the non-steady flow internal energy formula $E = V\rho(u + 1)$ in the $c = 1$ unit system, and using the equation of state from Equation (3) in [31], the entropy conservation [32] can be expressed in terms of

$$\dot{E} + p\dot{V} = 0 \quad \Rightarrow \quad \frac{\dot{\rho}}{\rho} = -3\left[\frac{\dot{R}}{R} + \frac{\ddot{R}}{\dot{R}}\right], \tag{31}$$

with the solution

$$\frac{\rho}{\rho_0} = H_0^3(\dot{a}(t)a(t))^{-3}. \tag{32}$$

Thus, in this model the relationship between the time-dependent Hubble parameter and the density parameter is

$$H(t) = H_0^2\sqrt{\Omega_{0,CDM}\left(\frac{H_0}{H(t)}\right)^3\frac{1}{a^3(t)} + \Omega_{DE,0}\frac{1}{a^2(t)}}, \tag{33}$$

which cannot be solved explicitly. Here, $\Omega_{0,CDM}$ or $\Omega_{DE,0}$ are density parameters related to the cold dark matter (CDM) and dark energy (DE), respectively. The relationship between the density parameters and the dynamical parameters of the self-similar solution is defined and explained in the next section.

6. Discussion

According to current scientific understanding, dark matter and dark energy make up about 95% of the total energy density of the observable universe today. The dark fluid theory suggests that a single substance may explain both dark matter and dark energy. The behaviour of the hypothetical dark fluid is believed to resemble that of cold dark matter on galactic scales while exhibiting similar characteristics to dark energy at larger scales [25]. Predictions can be obtained from our Sedov–von Neumann–Taylor blast wave-inspired nonrelativistic dark fluid model on galactic and cosmological scales. A useful feature of this model is that the initial value problem of the reduced ordinary differential equation system is easier to handle than the boundary and initial condition problem of the original partial differential equations. To provide a reliable practical basis for our dark fluid model, we have tested the theoretical results on astrophysical scales in order to demonstrate the similar nature of the solutions on the cosmological scale.

Our solution was developed based on cosmological observations, using the Hubble law to scale the expansion of the universe. Our model includes various scaling mechanisms through the use of a Sedov-type self-similar *ansatz*, which allows for describing different time decay scenarios [33]. In the case of a non-rotating system, we can conclude that the radial velocity profile of the solution provides Hubble-like expansion, while the non-rotating model provides inflation-like behavior $u(r,t) > 1$ on a long-range timescale, which cannot be physical (causal). We have seen that the high initial velocity of the dark fluid relaxes to a small, constant, and non-relativistic value on long timescales ($t > 6$ billion years). Another interesting feature of the model is that the gravitation potential is a negative polynomial in time, which is consistent with the distribution of the density of the dark fluid.

One notable aspect in the case of the rotating model is that it does not show superluminous behavior in the expected time range, contrary to the non-rotating model. Simultaneously, we have found that the radial profile of the density becomes saturated and close to flat at far distances from the initial point; see the top right panel of Figure 6. In addition, it is possible to set the initial conditions such that the universe today is observed as flat Euclidean with the density parameter

$$\Omega(t) = \frac{\sum_i \rho_i}{\rho_c} \approx 1 \tag{34}$$

being the critical density corresponding to the Hubble parameter H_0. The flatness of the universe is indicated by the recent measurements of WMAP [34]. The sum index uses the baryonic energy (B), dark energy (DE), and cold dark matter (CDM). We can define the matter part of $\Omega_M = \Omega_{CDM} + \Omega_B$ along with the full $\Omega = \sum_i \Omega_i$ and $i \in \{B, CDM, DE\}$ [35]. If the $\Omega_{CDM} \gg \Omega_B$ relation is correct, then we can assume the following identity:

$$\left.\frac{\Omega_M(t)}{\Omega(t)}\right|_{t=t_0} \sim \left.\frac{E_{kin}(R(t),t)}{E_{tot}(R(t),t)}\right|_{t=t_0} = 0.26. \tag{35}$$

Accordingly, we can determine the relevant time and radial coordinates from the obtained results, which correspond to this specific energy relation evaluated at $r = R(t_0)$ and $t = t_0$. In the absence of dark fluid, the universe continues to expand indefinitely, though at a gradually slowing rate that eventually approaches zero. This causes an open-topology universe. In this case, the ultimate fate of the universe is that the temperature asymptotically approaches absolute zero in a so-called "big freeze". At the same time, it is essential to mention that a weakness of our non-relativistic model is that its results are not as precise as those of relativistic Friedmann equation-based models [25].

7. Conclusions and Outlook

In this paper, we have studied the behaviour of self-similar time-dependent solutions in a coupled nonlinear partial differential equation system describing a non-viscous, non-relativistic, and self-gravitating fluid (Euler–Poisson system). The reason behind the applied self-similar solutions is that they provide a very efficient method to analyze various kinds of physical systems, particularly as concerns the hydrodynamical description of systems that involve collapse and explosion.

The analysis presented in this paper is a Euler–Poisson extension of our previous model [16]. We found that a Sedov–Taylor type solutions exists and that the algebraic equation obtained for the similarity exponents has only one unique solution. We have used the obtained solution to describe the behaviour of a non-relativistic dark fluid on cosmological scales, and have presented the relevant kinematical and dynamical quantities.

In addition, we have shown that our model based on self-similarity is in agreement with the Newtonian Friedmann equations.

It can easily be noticed that this model has certain limitations due to its classical nature, although it does provide relatively adequate results on the cosmological scale. We have shown that the obtained quasi-analytical solution for the evolution of the Hubble parameter is in agreement with the standard cosmological model, e.g., [28]. Moreover, in the previous section we have shown that the energy ratio from our model closely resembles the ratio of $\Omega_M(t)/\Omega(t)$, i.e., the relevant density parameters as of today.

This model has the additional practical benefit that the calculation does not require high computing performance and resources. Therefore, it can be used to estimate the physical value of the initial and boundary values when more sophisticated theoretical or numerical simulations are used. Moreover, it can provide a reliable basis for comparing for two- and three-dimensional hydrodynamical simulations. In future research, through reasonable effort it would be possible to improve this model in order to describe relativistic matter [36], Chaplygin gas [37], and two-fluid models.

Author Contributions: Original idea, I.F.B.; formal analysis, B.E.S.; software, B.E.S.; visualization, B.E.S.; writing—original draft, B.E.S.; correction, I.F.B. and G.G.B. All authors have read and agreed to the published version of the manuscript.

Funding: The authors gratefully acknowledge the financial support of the Hungarian National Research, Development, and Innovation Office (NKFIH) under Contracts No. OTKA K135515, No. NKFIH 2019-2.1.11-TET-2019-00078, and No. 2019-2.1.11-TET-2019-00050, and of the Wigner Scientific Computing Laboratory (WSCLAB, the former Wigner GPU Laboratory).

Data Availability Statement: This work is based on analytic calculation of the given formulae. All data are included in the plots. The data underlying this article can be shared upon reasonable request to the corresponding author.

Acknowledgments: The authors gratefully acknowledge their useful discussions with N. Barankai. Author I.F.B. offers this study in memory of the astronomer György Paál (1934–1992), who taught him physics and sailing in the summer of 1990 at Lake Balaton.

Conflicts of Interest: The authors declare no conflict of interest.

Appendix A

Table A1 summarizes the macroscopic and astronomical scales, quantities, and couplings in different units (SI) and geometrized) along with the factors to convert them. Table A2 shows the astronomical units used in the Discussion section 6.

Table A1. Relevant physical quantities in SI and geometrized units. Geometrized units can be converted into SI units using the factors provided.

Variable	SI Unit	Geometrized Unit	Factor
mass	kg	m	$c^2 G^{-1}$
length	m	m	1
time	s	m	c^{-1}
density	$\mathrm{kg\,m^{-1}}$	$\mathrm{m^{-2}}$	$c^2 G$
velocity	$\mathrm{m\,s^{-1}}$	1	c
acceleration	$\mathrm{m\,s^{-2}}$	$\mathrm{m^{-1}}$	c^2
force	$\mathrm{kg\,m\,s^{-2}}$	1	$c^4 G^{-1}$
energy	$\mathrm{kg\,m^2\,s^{-2}}$	$\mathrm{m^{-1}}$	$c^4 G^{-1}$
energy density	$\mathrm{kg\,m^{-1}\,s^{-2}}$	$\mathrm{m^{-2}}$	$c^4 G^{-1}$

Table A2. Relevant astronomical quantities and their corresponding values in SI units.

Variable	Astronomical Unit	SI Unit
length	ly	$9.46073047258 \cdot 10^{15}$ m
length	Gly	$9.46073047258 \cdot 10^{24}$ m
length	kPc	$3.08567758128 \cdot 10^{19}$ m
time	Gy	$3.1556926 \cdot 10^{16}$ s

References

1. Guderley, K.G. Starke kugelige und zylindrische verdichtungsstosse in der nahe des kugelmitterpunktes bnw. der zylinderachse Luftfahrtforschung. *Luftfahrtforschung* **1942**, *19*, 302.
2. Sedov, L.I. Propagation of strong shock waves. *J. Appl. Math. Mech.* **1946**, *10*, 241–250.
3. Taylor, G.I. The Formation of a Blast Wave by a Very Intense Explosion. I. Theoretical Discussion. *Proc. R. Soc. Lond. Ser. A Math. Phys. Sci.* **1950**, *201*, 159–174. [CrossRef]
4. Barna, I.F.; Mátyás, L. Analytic solutions of a two-fluid hydrodynamic model. *AJR2P* **2021**, *4*, 14–26. [CrossRef]
5. Bressan, A.; Murray, R. On self-similar solutions to the incompressible Euler equations. *J. Differ. Equ.* **2020**, *269*, 5142–5203. [CrossRef]
6. Barna I.F.; Kersner, R. Heat conduction: A telegraph-type model with self-similar behavior of solutions. *Phys. Math. Theor.* **2010**, *43*, 375210. [CrossRef]
7. Guo, Y.; Hadžić, M.; Jang, J.; Schreckerk, M.A. Gravitational Collapse for Polytropic Gaseous Stars: Self-Similar Solutions. *Arch. Ration. Mech. Anal.* **2022**, *246*, 957–1066. [CrossRef]
8. Cahill, M.E.; Taub, A.H. Spherically symmetric similarity solutions of the Einstein field equations for a perfect fluid. *Commun. Math. Phys.* **1971**, *21*, 1. [CrossRef]
9. Eardley, D.M. Self-similar spacetimes: Geometry and dynamics. *Commun. Math. Phys.* **1974**, *37*, 287. [CrossRef]
10. Gundlach, C.; Martín-García, J.M. Critical phenomena in gravitational collapse. *Living Rev. Relativ.* **2007**, *10*, 4. [CrossRef]
11. Jacobus Cornelius, K. First Attempt at a Theory of the Arrangement and Motion of the Sidereal System. *Astrophys. J.* **1922**, *55*, 302–327. [CrossRef]
12. Zwicky, F. Die Rotverschiebung von extragalaktischen Nebeln. *Helv. Phys. Acta* **1933**, *6*, 110–127. (In English) [CrossRef]
13. Rubin, V.; Ford, C.; Kent, W., Jr. Rotation of the Andromeda Nebula from a Spectroscopic Survey of Emission Region. *Astrophys. J.* **1970**, *159*, 379–403. [CrossRef]
14. Einasto, J.; Kaasik, A.; Saar, E. Dynamic evidence on massive coronas of galaxies. *Nature* **1974**, *250*, 309–310. [CrossRef]
15. Arbey, A. Cosmological constraints on unifying Dark Fluid models. *Open Astron. J.* **2008**, *1*, 27–38. [CrossRef]
16. Barna, I.F.; Pocsai, M.A.; Barnaföldi, G.G. Self-Similar Solutions of a Gravitating Dark Fluid. *Mathematics* **2022**, *10*, 3220. [CrossRef]
17. Euler, L. Principes généraux du mouvement des fluides. *Mém. Acad. Sci. Berl.* **1757**, *11*, 274–315. (In English) [CrossRef]
18. Emden, R. *Gaskugeln: Anwendungen der Mechanischen Wärmetheorie auf Kosmologische und Meteorologische Probleme*; B. G. Teubner: Wiesbaden, Germany, 1907.
19. Horedt, G.P. *Polytropes Applications in Astrophysics and Related Fields*; Kluver Academic Publishers: Amsterdam, The Netherlands, 2004. [CrossRef]
20. Perkovic, D.; Stefancic, H. Dark sector unifications: Dark matter-phantom energy, dark matter - constant w dark energy, dark matter-dark energy-dark matter. *Phys. Lett. B* **2019**, *797*, 134816. [CrossRef]
21. Poisson, S.D. Mémoire sur la théorie du magnétisme en movement. *Mém. Acad. R. Sci. Inst. Fr.* **1823**, *6*, 441–570. (In French)
22. Barna I.F.; Mátyas, L. General self-similar solutions of diffusion equation and related constructions. *Asian J. Res. Rev. Phys.* **2021**, *4*, 14–26. [CrossRef]
23. Kenneth B. *Howell: Ordinary Differential Equations: An Introduction to the Fundamentals*, 1st ed.; CRC Press: Boca Raton, FL, USA, 2015. [CrossRef]
24. Wolfram Research, Inc. *Mathematica*, Version 13.1; Wolfram Research, Inc.: Champaign, IL, USA, 2023.
25. Farnes, J.S. A unifying theory of dark energy and dark matter: Negative masses and matter creation within a modified ΛCDM framework. *Astron. Astrophys.* **2018**, *620*, A92. [CrossRef]
26. Olver, F.W.J.; Lozier, D.W.; Boisvert, R.F.; Clark, C.W. *NIST Handbook of Mathematical Functions*; Cambridge University Press: Cambridge, UK, 2011; ISBN 978-0-521-14063-8.
27. Kelly, P.L.; Rodney, S.; Treu, T.; Oguri, M.; Chen, W.; Zitrin, A.; Birrer, S.; Bonvin, V.; Dessart, L.; Diego, J.M.; et al. Constraints on the Hubble constant from Supernova Refsdal's reappearance. *Science* **2023**, *380*, eabh1322. [CrossRef] [PubMed]
28. Mohan, N.J.; Sasidharan, A.; Mathew, T.K. Bulk viscous matter and recent acceleration of the universe based on causal viscous theory. *Eur. Phys. J. C* **2017**, *77*, 1–13. [CrossRef]
29. Deruelle, N.; Uzan, J.P. *Relativity in Modern Physics*; Oxford University Press: Oxford, UK, 2018. [CrossRef]
30. Ryden, B. *Introduction to Cosmology*, 2nd ed.; Cambridge University Press: Cambridge, UK, 2016. [CrossRef]
31. Landau, L.; Lifshitz, E. *Course of Theoretical Physics*, 2nd ed.; Butterworth-Heinemann: Oxford, UK, 1987; Volume 6.

32. Coles, P.; Lucchin, F. *Cosmology. The Origin and Evolution of Cosmic Structure*, 2nd ed.; John Wiley and Sons: Hoboken, NJ, USA, 2002.
33. Hubble, E. A relation between distance and radial velocity among extra-galactic nebulae. *Proc. Natl. Acad. Sci. USA* **1929**, *15*, 168–173. [CrossRef] [PubMed]
34. Bennett, C.L.; Larson, D.; Weil, J.L.; Jarosik, N.; Hinshaw, G.; Odegard, N.; Smith, K.M.; Hill, R.S.; Gold, B.; Halpern, M.; et al. Nine-year Wilkinson Microwave Anisotropy Probe (WMAP) observations: Final maps and results. *Astrophys. J. Suppl. Ser.* **2013**, *208*, 20. [CrossRef]
35. Dimitar, V. Estimations of total mass and energy of the universe. *Phys. Int.* **2014**, *5*, 15–20. [CrossRef]
36. Rezzolla, L. *Relativistic Hydrodynamics*; Oxford University Press: Oxford, UK, 2018. [CrossRef]
37. Popov, V.A. Dark energy and dark matter unification via superfluid Chaplygin gas. *Phys. Lett. B* **2010**, *686*, 4–5. [CrossRef]

 universe

Article

Even and Odd Self-Similar Solutions of the Diffusion Equation for Infinite Horizon

László Mátyás [1],* and Imre Ferenc Barna [2]

[1] Department of Bioengineering, Faculty of Economics, Socio-Human Sciences and Engineering, Sapientia Hungarian University of Transylvania, Libertătii sq. 1, 530104 Miercurea Ciuc, Romania
[2] Wigner Research Center for Physics, Konkoly-Thege Miklós út 29-33, 1121 Budapest, Hungary
* Correspondence: matyaslaszlo@uni.sapientia.ro

Abstract: In the description of transport phenomena, diffusion represents an important aspect. In certain cases, the diffusion may appear together with convection. In this paper, we study the diffusion equation with the self-similar Ansatz. With an appropriate change of variables, we have found an original new type of solution of the diffusion equation for infinite horizon. We derive novel even solutions of diffusion equation for the boundary conditions presented. For completeness, the odd solutions are also mentioned as well, as part of the previous works. We have found a countable set of even and odd solutions, of which linear combinations also fulfill the diffusion equation. Finally, the diffusion equation with a constant source term is discussed, which also has even and odd solutions.

Keywords: partial differential equations; diffusion and thermal diffusion; self-similarity

MSC: 60G18; 76R50

Citation: Mátyás, L.; Barna, I.F. Even and Odd Self-Similar Solutions of the Diffusion Equation for Infinite Horizon. *Universe* 2023, 9, 264. https://doi.org/10.3390/universe9060264

Academic Editors: Máté Csanád, Sándor Lökös and Dániel Kincses

Received: 23 February 2023
Revised: 25 May 2023
Accepted: 26 May 2023
Published: 31 May 2023

1. Introduction

It is evident that mass diffusion or heat conduction is a fundamental physical process which attracted enormous intellectual interest from mathematicians, physicists and engineers over the last two centuries. The existing literature about mass and heat diffusion is immense; we only mention some fundamental textbooks [1–5].

Regular diffusion is the cornerstone of many scientific disciplines, such as surface growth [6–8], reactions diffusion [9] or even flow problems in porous media. In our last two papers, we gave an exhaustive summary of such processes with numerous relevant reviews [10,11].

In connection with thermal diffusion [12,13], the simultaneous presence of heat and mass transfer is also possible, which may lead to cross effects [14]. One may find relevant applications related to general issues of heat transfer or engineering in [15]. Important diffusive phenomena occur in the universe [16], which is another field of interest.

The study of population dynamics or biological processes [17–19] also involves diffusive processes, especially in spatial extended systems. In environmental sciences, the effects of spreading, distribution and adsorption of particulate matter or pollutants are also relevant [20–23]. Furthermore, diffusion coefficients have been measured for practical purposes in food sciences as well [24].

New applications of diffusion have gained ground in social sciences in the last decades as well. As examples, we can mention diffusion of innovations [25,26], diffusion of technologies and social behavior [27] or even diffusion of cultures, humans or ideas [28,29]. One may also find aspects related to diffusion in the theory of pricing [30,31]. The structure of the network has also a crucial role which influences the spread of innovations, ideas or even computer viruses [32]. Parallel to such diffusion activities, generalization of heat-transport equations was done by Ván and coauthors [33], e.g., fourth-order partial differential equations (PDE)s were formulated to elaborate the problem of non-regular heat conduction phenomena. Finally, we should not forget the continuously developing numerical methods of PDEs; it is worth mentioning the new results obtained by Kovács and coworkers [34,35]. Such spirit of the times clearly shows that investigation of diffusion (and heat conduction) is still an important task.

Having in mind that diffusion can be a general, three-dimensional process beyond Cartesian symmetry, here we investigate the one-dimensional diffusion equation. The change in time of variable $C(x,t)$ is influenced by the presence of it in the neighbors:

$$\frac{\partial C(x,t)}{\partial t} = D\frac{\partial^2 C(x,t)}{\partial x^2},\tag{1}$$

where D is the diffusion coefficient which should have positive real values. Usually, it can be considered constant for given temperature and pressure in gases. A counter-example is the heat diffusion process in large-temperature-gradient semiconductor crystals where heat conduction coefficients have a complicated temperature dependence [36].

One assumes that $C(x,t)$ is a sufficiently smooth function together with existing derivatives regarding both variables. In this general form, one may observe that if $C(x,t)$ is a solution, then $C(x,t) + C_0$ is also a solution, where C_0 is a constant.

In this study, we consider a constant diffusion coefficient. From a practical point of view, a typical case is the diffusion in gases, where at constant temperature and constant pressure, the diffusion coefficient is constant as is described in [37].

For a finite horizon or interval, in case the concentration is fixed at the two ends $C(x = 0, t) = C_0$ and $C(x = L, t) = C_0$, the solutions are

$$C_k(x,t) = C_0 + e^{-D\frac{\pi^2 k^2 t}{L^2}} \cdot \sin\left(\frac{k\pi}{L}x\right),\tag{2}$$

where $k = 1,2,3\ldots$; it can be any positive integer number. In general, beyond C_0, any linear combination of the product of the exponent and sine for different k is a solution. For finite horizon, in the case when the density is fixed to zero on both ends, the solutions are changed to

$$C_k(x,t) = C_0 + e^{-D\frac{\pi^2 n^2 t}{L^2}} \cdot \cos\left(\frac{n\pi}{L}x\right),\tag{3}$$

where $n = 1,2,3\ldots$ can be any positive integer number. Thanks to the Fourier theorem, with the help of Equations (2) and (3) arbitrary diffusion profile can be approximated on a closed interval. These are well-known analytic results and can be found in any usual physics textbooks such as [1,2].

In the present study—with the help of the self-similar Ansatz—we are going to present generic symmetric solutions for infinite horizon. These solutions have their roots at the very beginning of the theory, in the form of the Gaussian [1,2]:

$$C(x,t) = \text{Const.} \cdot \frac{1}{\sqrt{t}}e^{-\frac{x^2}{4Dt}}.\tag{4}$$

For infinite horizon, there are also certain works which present a given aspect of the diffusion, and it may arrive to a slightly more general aspect than the classical solution presented above [38].

In the following, we will go much beyond that point and will present and analyze completely new type of solutions. For finite horizon, an even initial condition can be expressed as a linear combination of the countable solutions of Equation (3), for $t = 0$. For positive times, the linear combination gives the dynamics of $C(x,t)$. In a similar way, we expect in the following that if we can find a countable set of even solutions, for infinite horizon, then by linear combinations of these functions, we can give the dynamics in time of a certain number of even functions. The initial conditions for infinite horizon can be set more easily after the change of variables, which will be discussed in more detail later.

2. Theory and Results

In the case of infinite horizon, when we want to derive the corresponding solutions, we make the following self-similar transformation:

$$C(x,t) = t^{-\alpha} f\left(\frac{x}{t^{\beta}}\right) = t^{-\alpha} f(\eta). \tag{5}$$

Note that the spatial coordinate x now runs along the whole real axis. The role of of α and β is illustrated on Figure 1. As is indicated in the figure, β shows the spreading and α the decay in time.

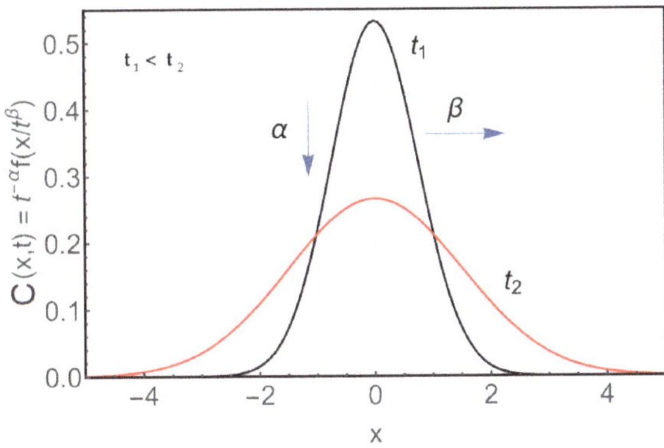

Figure 1. The importance of α and β in case of the change of variables of Equation (5).

This kind of Ansatz has been applied by Sedov [39], and was later also used by Raizer and Zel'dowich [40] For certain systems, Barenblatt applied it successfully [41] as well. We have also used it for linear or non-linear partial differential equation (PDE) systems, which are from fluid mechanics [42–44] or quantum mechanical systems [45]. In certain cases, the equation of state of the fluid also plays a role [46,47]. Diffusion-related applications of the self-similar analysis method can be found in relatively recent works as well [48–50].

The transformation takes into account the (4) formula, and before the function f, instead of $1/\sqrt{t}$ there is a generalized function $1/t^{\alpha}$, and in the argument of f, the fraction x/t^{β} is possible, with a β which should be determined later.

We evaluate the first and second derivative of relation (5), and insert it in the equation of diffusion (1). This yields the following ordinary differential equation (ODE)

$$-\alpha t^{-\alpha-1} f(\eta) - \beta t^{-\alpha-1} \eta \frac{df(\eta)}{d\eta} = D t^{-\alpha-2\beta} \frac{d^2 f(\eta)}{d\eta^2}. \tag{6}$$

The reasoning is self-consistent if all three terms have the same decay in time. This is possible if

$$\alpha = \text{arbitrary real number}, \quad \beta = 1/2, \tag{7}$$

and yields the following ODE

$$-\alpha f - \frac{1}{2}\eta f' = D f''. \tag{8}$$

This ODE is a kind of characteristic equation, with the above-presented change of variable. One can observe that for $\alpha = 1/2$, this equation can be written as

$$-\frac{1}{2}(\eta f)'' = D f''. \tag{9}$$

If this equation is integrated once

$$\mathrm{Const}_0 - \frac{1}{2}\eta f = D f',$$ (10)

where Const_0 is an arbitrary constant, which may depend on certain conditions related to the problem. If we take this $\mathrm{Const}_0 = 0$, then one arrives to the generic solution

$$f = f_0 e^{-\frac{\eta^2}{4D}},$$ (11)

where f_0 is a constant. Inserting this form of f in form of $C(x,t)$ given by Equation (5)—for $\alpha = 1/2$ as it was mentioned earlier—one obtains an even solution for the space variable:

$$C(x,t) = f_0 \frac{1}{t^{\frac{1}{2}}} e^{-\frac{x^2}{4Dt}}.$$ (12)

By this, we have recovered the generic Gaussian solution, which can be seen in Figure 2a.

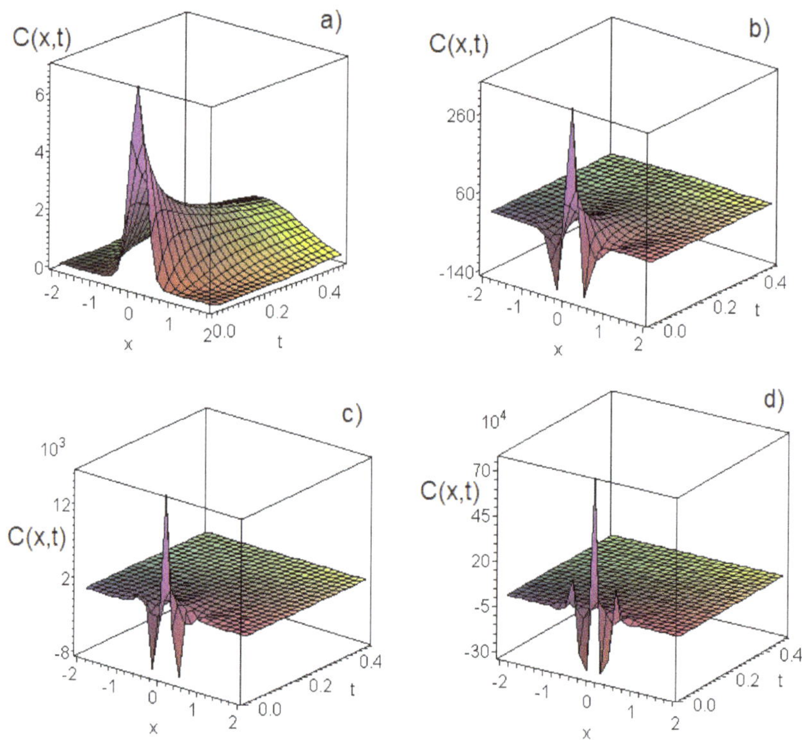

Figure 2. The solution $C(x,t)$ for (**a**) $\alpha = 1/2$, (**b**) $\alpha = 3/2$ (**c**) $\alpha = 5/2$ and (**d**) $\alpha = 7/2$, respectively.

If we want to find further solutions, the Equation (8) has to be solved for general α. The general solution for infinite horizon of (8) can be written as:

$$f(\eta) = \eta \cdot e^{-\frac{\eta^2}{4D}} \left(c_1 M\left[1 - \alpha, \frac{3}{2}, \frac{\eta^2}{4D}\right] + c_2 U\left[1 - \alpha, \frac{3}{2}, \frac{\eta^2}{4D}\right] \right),$$ (13)

where c_1 and c_2 are real integration constants, which are fixed by the initial conditions, and $M(,,)$ and $U(,,)$ are the Kummer's functions. For exhaustive details, consult [51].

If α are positive integer numbers, then both special functions M and U are finite polynomials in terms of the third argument $\frac{\eta^2}{4D}$

$$f(\eta) = \eta \cdot e^{-\frac{\eta^2}{4D}}\left(\kappa_0 + \kappa_1\frac{\eta^2}{4D} + \ldots + \kappa_{n-1}\cdot\left[\frac{\eta^2}{4D}\right]^{n-1}\right). \tag{14}$$

These give the *odd solutions* of the diffusion equation for $\alpha = n$, (where n positive integer), in terms of the space variable. It follows for the complete solution $C(x, t)$

$$C(x,t) = \frac{1}{t^n}f(\eta) = \frac{1}{t^n}\frac{x}{\sqrt{t}}e^{-\frac{x^2}{4Dt}}\cdot\left(\kappa_0 + \kappa_1\frac{x^2}{4Dt} + \ldots + \kappa_{n-1}\cdot\left[\frac{x^2}{4Dt}\right]^{n-1}\right). \tag{15}$$

These odd solutions have been studied thoroughly by Mátyás and Barna in previous works [10,11] and for completeness, we present these solutions in Appendix A.

For the *even solutions*, we denote by $g(\eta)$ the following function

$$f(\eta) = \eta \cdot e^{-\frac{\eta^2}{4D}}g(\eta), \tag{16}$$

Inserting this equation into Equation (8), we have

$$\eta g'' + 2g' - \frac{\eta^2}{2D}g' + (\alpha - 1)\frac{\eta}{D}g = 0. \tag{17}$$

In concordance with Equation (13), we get the general solution

$$g(\eta) = \left(c_1 M\left[1 - \alpha, \frac{3}{2}, \frac{\eta^2}{4D}\right] + c_2 U\left[1 - \alpha, \frac{3}{2}, \frac{\eta^2}{4D}\right]\right). \tag{18}$$

At this point, we make the conjecture from the forms of U and M, that if we had the classical spatially even solution for $\alpha = 1/2$, than the next spatially even solution would be for $\alpha = 3/2$, with the form of g

$$g(\eta) = K_0\frac{1}{\eta} + K_1\eta, \tag{19}$$

where K_0 and K_1 are arbitrary constants, which should be determined later. We insert this form of g in (17); we find that the form (19) fulfills the Equation (17) if

$$K_1 = -\frac{1}{2D}K_0. \tag{20}$$

We obtain the same result if we insert the form

$$f(\eta) = \eta \cdot e^{-\frac{\eta^2}{4D}}\left(K_0\frac{1}{\eta} + K_1\eta\right), \tag{21}$$

directly into the Equation (8). By this, for $\alpha = 3/2$, we get for the function f

$$f(\eta) = K_0 \cdot \eta \cdot e^{-\frac{\eta^2}{4D}}\left(\frac{1}{\eta} - \frac{1}{2D}\eta\right) = K_0 \cdot e^{-\frac{\eta^2}{4D}}\left(1 - \frac{1}{2D}\eta^2\right). \tag{22}$$

Substituting this form into (5), one gets

$$C(x,t) = K_0\frac{1}{t^{\frac{3}{2}}}e^{-\frac{x^2}{4Dt}}\left(1 - \frac{1}{2D}\frac{x^2}{t}\right). \tag{23}$$

This result is visualized in Figure 2b.

If we follow the case $\alpha = 5/2 = 2.5$, then the following form for the function $g(\eta)$ can be considered

$$g(\eta) = K_0 \cdot \frac{1}{\eta} + K_1 \cdot \eta + K_2 \cdot \eta^3. \tag{24}$$

If we insert this form in the Equation (17), the following relations for the constants K_0, K_1 and K_2 can be derived

$$K_1 = \frac{K_0}{D}, \tag{25}$$

and

$$K_2 = -\frac{K_1}{12D} = \frac{K_0}{12D^2}. \tag{26}$$

By this, we get for the $g(\eta)$

$$g(\eta) = K_0 \left(\frac{1}{\eta} - \frac{1}{D}\eta + \frac{1}{12D^2}\eta^3 \right). \tag{27}$$

Correspondingly the final form for $f(\eta)$ for $\alpha = 2.5$ is

$$f(\eta) = K_0 \cdot e^{-\frac{\eta^2}{4D}} \left(1 - \frac{1}{D}\eta^2 + \frac{1}{12D^2}\eta^4 \right). \tag{28}$$

Inserting this form into (5), one gets

$$C(x,t) = K_0 \frac{1}{t^{\frac{5}{2}}} e^{-\frac{x^2}{4Dt}} \left(1 - \frac{1}{D}\frac{x^2}{t} + \frac{1}{12D^2}\frac{x^4}{t^2} \right). \tag{29}$$

This result can be seen in Figure 2c.

If we follow the case $\alpha = 7/2 = 3.5$, then the following form for the function $g(\eta)$ can be considered:

$$g(\eta) = K_0 \cdot \frac{1}{\eta} + K_1 \cdot \eta + K_2 \cdot \eta^3 + K_3 \cdot \eta^5. \tag{30}$$

If we replace this form into Equation (17), the next relations among the constants K_0, K_1, K_2 and K_3 can be derived:

$$K_1 = -\frac{3}{2}\frac{K_0}{D}, \tag{31}$$

for the next coefficient

$$K_2 = -\frac{K_1}{6D} = \frac{K_0}{4D^2}. \tag{32}$$

Finally, for the third coefficient one obtains

$$K_3 = -\frac{K_2}{30D} = -\frac{K_0}{120D^3}. \tag{33}$$

Inserting these coefficients into the Formula (30), one obtains the following expression

$$g(\eta) = K_0 \left(\frac{1}{\eta} - \frac{3}{2D} \cdot \eta + \frac{1}{4D^2} \cdot \eta^3 - \frac{1}{120D^3} \cdot \eta^5 \right). \tag{34}$$

This form of g yields, by Equation (16), for the function f

$$f(\eta) = K_0 \cdot e^{-\frac{\eta^2}{4D}} \left(1 - \frac{3}{2D}\eta^2 + \frac{1}{4D^2}\eta^4 - \frac{1}{120D^3}\eta^6 \right). \tag{35}$$

Inserting this form into (5), one obtains

$$C(x,t) = K_0 \frac{1}{t^{\frac{7}{2}}} e^{-\frac{x^2}{4Dt}} \left(1 - \frac{3}{2D}\frac{x^2}{t} + \frac{1}{4D^2}\frac{x^4}{t^2} - \frac{1}{120D^3}\frac{x^6}{t^3} \right). \tag{36}$$

This result is clearly visualized in Figure 2d.

It is evident that by including higher terms in the finite series of Equation (30), the solutions for $\alpha = 9/2, 11/2$, etc. can be evaluated in a direct way.

For completeness, we present the shape functions $f(\eta)$s on Figure 3. Note that solutions with higher α values have more oscillations and quicker decay. The same features appear for odd solutions as well.

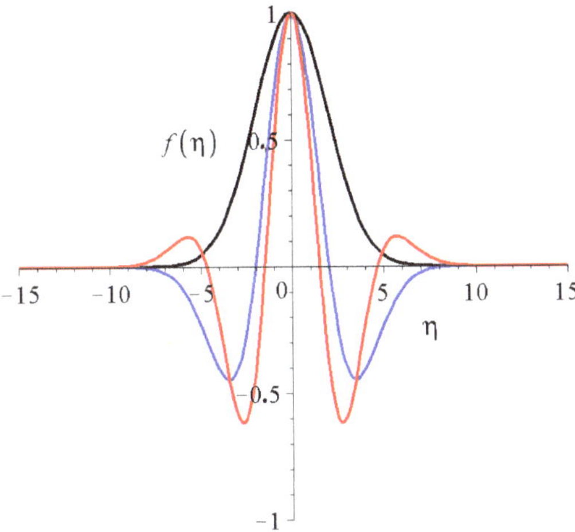

Figure 3. Even shape functions $f(\eta)$ of Equation (16) for three different self-similar α exponents. The black, blue and red curves are for $\alpha = 1/2, 3/2$ and $5/2$ numerical values, with the same diffusion constant ($D = 2$), respectively. Note that shape functions with larger αs have more zero transitions. We will show that for $\alpha > 0$ integer values, the integral of the shape functions give zero on the whole and the half-axis as well.

As we can see, at this point, the solutions fulfills the boundary condition $C() \to 0$ if $x \to \pm\infty$, for positive α values.

The general initial value problem can be solved with the usage of the Green's functions formalism. According to the standard theory of the Green's functions, the solution of the diffusion Equation (1) can be obtained via the next convolution integral:

$$C(x,t) = \frac{1}{2\sqrt{\pi t}} \int_{-\infty}^{+\infty} w(x_0)G(x - x_0)dx_0, \tag{37}$$

where $w(x_0)$ defines the initial condition of the problem, $C|_{t=0} = w(x_0)$. The Green's function for diffusion is well defined and can be found in many mathematical textbooks e.g., [52–55]:

$$G(x - x_0) = exp\left[-\frac{(x - x_0)^2}{4tD}\right]. \tag{38}$$

On the other side, the Gaussian function is a fundamental solution of diffusion.

We will see in the following that for some special forms of the initial conditions, such as polynomials, Gaussian, Sinus or Cosines, the convolution integral can be done analytically.

In the following, we evaluate the convolution integral for $\alpha = 1/2$.

As an example for the initial condition problem, we may consider the following smooth function with a compact support:

$$w(x_0) = \frac{\text{Heaviside}(3 - x_0) \cdot \text{Heaviside}(3 + x_0) \cdot (9 - x_0^2)}{9}. \tag{39}$$

This initial condition is a typical initial distribution for diffusion, and one can see on Figure 4a.

The convolution integral for $\alpha = 1/2$:

$$C(x,t) = \frac{1}{2\sqrt{\pi t}} \int_{-\infty}^{+\infty} \frac{\text{Heaviside}(3 - x_0) \cdot \text{Heaviside}(3 + x_0) \cdot (9 - x_0^2)}{9} \cdot e^{\frac{(x-x_0)^2}{4Dt}} dx_0. \quad (40)$$

The result of this evaluation is

$$
\begin{aligned}
C(x,t) = \frac{1}{2\sqrt{\pi t}} &\left[\sqrt{\pi t}\,\text{erf}\left(\tfrac{3+x}{2\sqrt{t}}\right) + \tfrac{2}{9}xt\,e^{-\frac{6x+x^2+9}{4t}} + \tfrac{2}{3}t\,e^{-\frac{6x+x^2+9}{4t}} - \tfrac{2}{9}t^{\frac{3}{2}}\sqrt{\pi}\,\text{erf}\left(\tfrac{3+x}{2\sqrt{t}}\right) \right. \\
&- \tfrac{1}{9}x^2\sqrt{\pi t}\,\text{erf}\left(\tfrac{3+x}{2\sqrt{t}}\right) - \sqrt{\pi t}\,\text{erf}\left(\tfrac{x-3}{2\sqrt{t}}\right) - \tfrac{2}{9}xt\,e^{-\frac{-6x+x^2+9}{4t}} + \tfrac{2}{3}t\,e^{-\frac{-6x+x^2+9}{4t}} \\
&\left. + \tfrac{2}{9}t^{\frac{3}{2}}\sqrt{\pi}\,\text{erf}\left(\tfrac{x-3}{2\sqrt{t}}\right) + \tfrac{1}{9}x^2\sqrt{\pi t}\,\text{erf}\left(\tfrac{x-3}{2\sqrt{t}}\right) \right],
\end{aligned}
\quad (41)
$$

which is presented on Figure 4b.

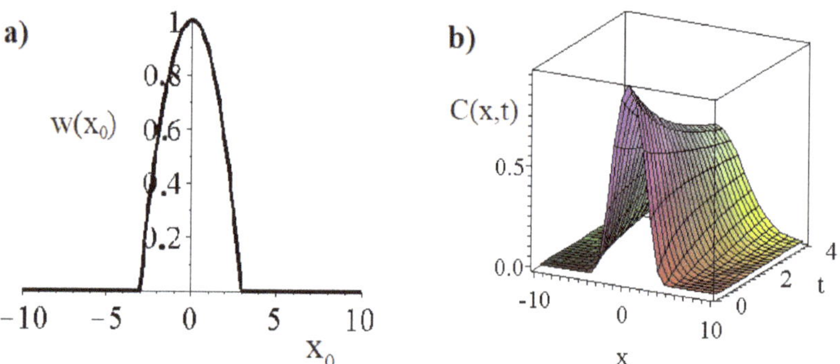

Figure 4. (**a**) The initial condition (39) (**b**) The convolution integral for $\alpha = 1/2$ of Equation (40).

3. The Properties of the Shape Functions and Solutions

In the following, we study some properties of the shape functions $f(\eta)$ and of the complete solutions $C(x,t)$. First, we consider the L^1 integral norms.

For the case $\alpha = 1/2$ the form of

$$\int_{-\infty}^{\infty} f(\eta)d\eta = \int_{-\infty}^{\infty} f_0 e^{-\frac{\eta^2}{4D}} d\eta = f_0\, 2\sqrt{\pi D}. \quad (42)$$

The constant f_0 is chosen, depending on the problem. If C stands for the density which diffuses, f_0 in the above integral is related to the total mass of the system.

Correspondingly,

$$\int_{-\infty}^{\infty} C(x,t)dx = \int_{-\infty}^{\infty} f_0 \frac{1}{\sqrt{t}} e^{-\frac{x^2}{4Dt}} dx = f_0\, 2\sqrt{\pi D}. \quad (43)$$

For the case $\alpha = 3/2$:

$$\int_{-\infty}^{\infty} f(\eta)d\eta = \int_{-\infty}^{\infty} K_0 \cdot e^{-\frac{\eta^2}{4D}} \left(1 - \frac{1}{2D}\eta^2\right) d\eta = 0. \quad (44)$$

It is interesting to see that the integral of first even shape function beyond Gaussian is zero. An even more remarkable feature is, however, that

$$\int_{-\infty}^{0} f(\eta)d\eta = \int_{0}^{\infty} f(\eta)d\eta = 0. \tag{45}$$

So the oscillations, the positions of the zero transitions, divide the function in such a way that the integral is not only on the whole real axis $(-\infty\ldots\infty)$ but on the half axis $(0\ldots\infty)$ or $(-\infty\ldots0)$ gives zero as well.

Evaluating the same type of integrals for the corresponding solution $C(x,t)$, we have

$$\int_{-\infty}^{\infty} C(x,t)dx = \int_{0}^{\infty} C(x,t)dx = \int_{-\infty}^{0} C(x,t)dx = \int_{-\infty}^{\infty} K_0 \cdot \frac{1}{t^{3/2}} e^{-\frac{x^2}{4Dt}}\left(1 - \frac{1}{2D}\frac{x^2}{t}\right)dx = 0,$$

at any time point, (and for any diffusion coefficient D).

The same property is true for all possible higher harmonic solutions if α is positive half-integer number $\alpha = (2n+1)/2$ when $(n \in \mathbb{N})$. This property has far-reaching consequences. The linearity of the regular diffusion equation and this additional property of this even series of solutions makes it possible to perturb the usual Gaussian in such a way that the total number of particles is conserved during the diffusion process; however, the initial distribution can be changed significantly. One can see from the final form of the solutions $C(x,t)_\alpha \sim \frac{1}{t^\alpha}$ that the decay of these perturbations are, however, short-lived because they have a quicker decay than the standard Gaussian solutions. For completeness, we present a $C(x,t)$ solution which is a linear combination of the first two even solutions $\alpha = 1/2, 3/2$ in the form of

$$C(x,t) = \frac{60}{t^{\frac{1}{2}}} e^{-\frac{x^2}{4t}} - \frac{0.001}{t^{\frac{3}{2}}} e^{-\frac{x^2}{4t}}\left(1 - \frac{x^2}{2t}\right), \tag{46}$$

on Figure 5. Note that coefficients with different orders of magnitude have to be applied to reach a visible effect when the sum of two functions have to be visualised with different power-law decay.

As a second property, we investigate the cosine Fourier transform of the shape functions:

$$C_\alpha(k) = \int_{-\infty}^{\infty} Cos(k \cdot \eta) f_\alpha(\eta)d\eta. \tag{47}$$

It can be shown with direct integration that the Fourier transform is

$$C_{\alpha = \frac{2N+1}{2}}(k) \propto l \cdot \sqrt{\pi} \cdot \frac{k^{2N} \cdot D^N \cdot e^{-k^2 D}}{\sqrt{\frac{1}{D}}}, \tag{48}$$

for all $N \in \mathbb{N}\backslash 0$ positive integer and l is a real constant. This means that qualitatively, the spectra for all positive half integer α are similar. They start from zero, have a global positive maximum and a quick decay to zero. It is generally known from spectral analysis that pulses of finite length have band spectra which have a minimal, a maximal and a central frequency.

In Appendix A, the corresponding normalization coefficients are given for the odd functions as well.

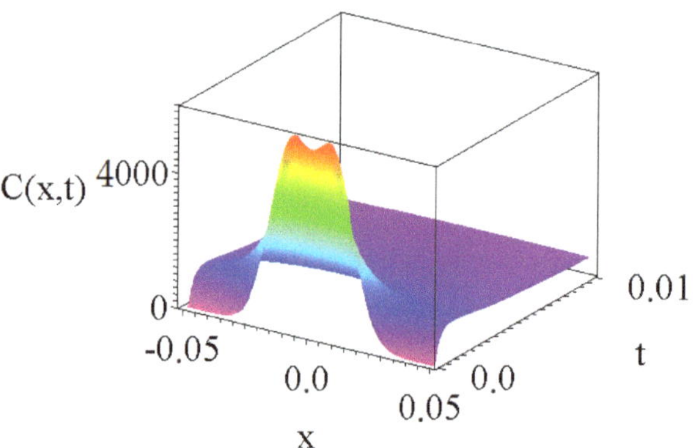

Figure 5. The function $C(x,t)$, solution of Equation (46).

4. The Diffusion Equation with Constant Source

At this point we try to find solutions of the diffusion equation, mainly with the self similar Ansatz, where on the right hand side, there is a constant source term:

$$\frac{\partial C(x,t)}{\partial t} = D\frac{\partial^2 C(x,t)}{\partial x^2} + n. \tag{49}$$

For this equation, one also apply the self-similar transformation (5), and we get a modified equation relative to the homogeneous one

$$-\alpha t^{-\alpha-1}f(\eta) - \beta t^{-\alpha-1}\eta\frac{df(\eta)}{d\eta} = Dt^{-\alpha-2\beta}\frac{d^2 f(\eta)}{d\eta^2} + n. \tag{50}$$

The free term on the r.h.s. has no explicit time decay; consequently, we expect the same from the other terms, which means

$$-\alpha - 1 \;=\; 0 \tag{51}$$

$$-\alpha - 2\beta \;=\; 0. \tag{52}$$

The two equations have to be fulfilled simultaneously. Solving these equations, we get the following values for α and β:

$$\alpha = -1 \text{ and } \beta = \frac{1}{2} \tag{53}$$

Inserting these values to the Equation (50), we get the following ODE

$$f(\eta) - \frac{1}{2}\eta\frac{df(\eta)}{d\eta} = D\frac{d^2 f(\eta)}{d\eta^2} + n. \tag{54}$$

We emphasize that we arrived to this equation by a self-similar transformation. At this point, we observe that if we shift the function f by a constant, and introduce the function h:

$$h(\eta) = f(\eta) - n \tag{55}$$

we arrive to a slightly modified equation

$$h(\eta) - \frac{1}{2}\eta\frac{dh(\eta)}{d\eta} = D\frac{d^2 h(\eta)}{d\eta^2}. \tag{56}$$

One may observe that if the transformation $\eta \to -\eta$ and $h(-\eta) = h(\eta)$ is applied, the equation still remains the same; consequently, we expect at least one even solution.

If we look for the even solution by polynomial expansion,

$$h(\eta) = A + B\eta^2 + \ldots \tag{57}$$

then, we get by direct substitution

$$A = 2 \cdot B \cdot D. \tag{58}$$

This means that the even solution reads as follows

$$h(\eta) = B(2D + \eta^2) \tag{59}$$

where B is a constant depending on initial conditions.

Furthermore, we observe that the transformation $\eta \to -\eta$ and $h(-\eta) = -h(\eta)$ also leaves Equation (56) unchanged. This means that it is worthwhile to look for an odd solution as well. The odd solution of the equation is

$$h(\eta) = 2D\,\eta\,e^{-\frac{\eta^2}{4D}} + \sqrt{\pi}\,(2D^{3/2} + \sqrt{D}\,\eta^2)\,erf\left(\frac{1}{2}\frac{\eta}{\sqrt{D}}\right) \tag{60}$$

One can see the form of this odd solution in Figure 6.

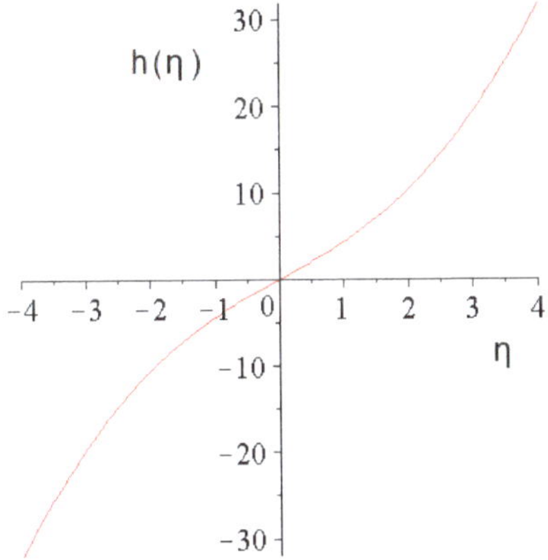

Figure 6. The shape function $h(\eta)$, described by Equation (60), the odd solution of Equation (56).

We mention, that Equation (56) may have a further solution, which eventually does not have the symmetry to be even or odd, and that may be expressed in terms of a Hermite function with a negative integer as one can see in Equation (8) of the Reference [11]. Such a solution reads as follows

$$h(\eta) = e^{-\frac{\eta^2}{4D}} \cdot Hermite\left[-3, \frac{\eta}{2\sqrt{D}}\right] \tag{61}$$

We try to find certain relevant features of this result. In an interesting way, the series expansion of the above solution (61) means a sum of an even function with second order and another odd function, which appears to be proportional to the series of solution (60). The first terms of these series are presented in Appendix B.

If n is positive in the Equation (49), then we can talk about a source in the equation, and if n is negative, than we say that there is a sink in the diffusion process. The sink can be considered physical by the time $C(x, t) \geq 0$. Diffusive systems with sinks have been studied in ref. [56], and water purification by adsorption also means a process with change of concentration in space and decrease in time [57].

A general solution for the shape function can be obtained from the linear combination of the even and odd solutions presented above

$$h(\eta) = \kappa_1 \left[2D\,\eta\,e^{-\frac{\eta^2}{4D}} + \sqrt{\pi}\,(2D^{3/2} + \sqrt{D}\,\eta^2)\,erf\left(\frac{1}{2}\frac{\eta}{\sqrt{D}}\right) \right] + \kappa_2[2D + \eta^2] \tag{62}$$

where κ_1 and κ_2 are constants depending on the initial or boundary conditions of the problem.

Inserting this shape function to the general solution (5), we get for the final form of $C(x,t)$ in the presence of a constant source

$$C(x,t) = t \cdot \left[\kappa_1\left(2D\,\frac{x}{\sqrt{t}}\,e^{-\frac{x^2}{4Dt}} + \sqrt{\pi}\left(2D^{3/2} + \sqrt{D}\,\frac{x^2}{t}\right)erf\left(\frac{1}{2}\frac{x}{\sqrt{Dt}}\right)\right) + \kappa_2\left(2D + \frac{x^2}{t}\right) + n \right] \tag{63}$$

For relatively shorter times, the general solution has interesting features depending on the weight of the even or the odd part of the solution, as one can see in Figure 7a.

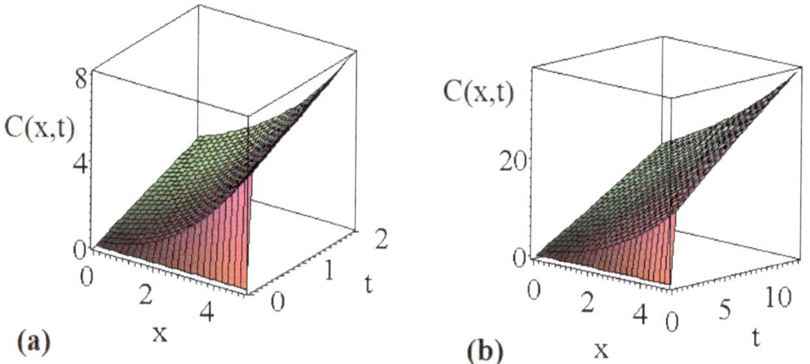

Figure 7. The shape function $C(x,t)$, solution of Equation (63), for $D = 1$ and $n = 1$, in case (a) $\kappa_1 = 0.1$ $\kappa_2 = 0.03$ and (b) $\kappa_1 = 0.2$ $\kappa_2 = 0.2$.

The long time behavior is dominated by the constant of the even solution and the source term. Correspondingly, for sufficiently long times, the relation $C(x,t) \sim (2\kappa_2 D + n) \cdot t$ characterizes the dynamics, as one can see in Figure 7b.

5. Summary and Outlook

Applying the well-known self-similar Ansatz—together with an additional change of variables—we derived symmetric solutions for the one-dimensional diffusion equations. Using the Fourier series analogy, we might say that these solutions may be considered as possible higher harmonics of the fundamental Gaussian solution. As unusual properties, we found that the integral of these solutions—beyond Gaussian—gives zero on both the half and the whole real axis as well. Thanks to the linearity of the diffusion equation, these kinds of functions can be added to the particle- (or energy-) conserving fundamental Gaussian solution; therefore, a new kind of particle diffusion process can be described. Due to the higher α self-similar exponents, these kinds of solutions give relevant contributions only at smaller time coordinates, because the corresponding solutions decay more quickly than the usual Gaussian solution. In case of a constant source or sink term in the diffusion equation, the value of α is no more arbitrary; it has a constant value $\alpha = -1$. Even for this fixed value of α, the diffusion equation with source term has even and odd solutions as well.

These kinds of solutions can also be evaluated for two- or three-dimensional, cylindrical or spherical symmetric systems as well. Work is in progress to apply this kind of analysis to more sophisticated diffusion systems as well. We hope that our new solutions have far-reaching consequences and that they will be successfully applied in other scientific disciplines such as quantum mechanics, quantum field theory, astrophysics, probability theory or in financial mathematics in the near future.

Author Contributions: Methodology, original draft and validation, L.M.; conceptualization, investigation and software, I.F.B. All authors have read and agreed to the published version of the manuscript.

Funding: There was no extra external funding.

Data Availability Statement: The data that support the findings of this study are all available within the article.

Conflicts of Interest: The authors declare no conflict of interest.

Appendix A

For completeness and for direct comparison, we show the first five odd shape functions $f(\eta)$ and the corresponding solutions $C(x,t)$:

$$
\begin{aligned}
f(\eta) &= erf\left(\frac{\eta}{2\sqrt{D}}\right), \\
f(\eta) &= \kappa_0 \cdot \eta \cdot e^{-\frac{\eta^2}{4D}}, \\
f(\eta) &= \kappa_0 \cdot \eta \cdot e^{-\frac{\eta^2}{4D}} \cdot \left(1 - \frac{1}{6D}\eta^2\right), \\
f(\eta) &= \kappa_0 \cdot \eta \cdot e^{-\frac{\eta^2}{4D}} \cdot \left(1 - \frac{1}{3D}\eta^2 + \frac{1}{60}\frac{1}{D^2}\eta^4\right), \\
f(\eta) &= \kappa_0 \cdot \eta \cdot e^{-\frac{\eta^2}{4D}} \cdot \left(1 - \frac{1}{2D}\eta^2 + \frac{1}{20}\frac{1}{D^2}\eta^4 - \frac{1}{840}\frac{1}{D^3}\eta^6\right),
\end{aligned}
\tag{A1}
$$

for $\alpha = 0,1,2,3,4\ldots\mathbb{N}$. The first case with the change of variable $x/\sqrt{t}$ with no α (or implicitly $\alpha = 0$) dates back to Boltzmann [58], as is also mentioned by [59,60].

All integrals of the functions from (A1) on the whole real axis give zero:

$$
\int_{-\infty}^{\infty} f_\alpha(\eta)d\eta = 0,
\tag{A2}
$$

However, on the half-axis:

$$
\int_{0}^{\infty} f_{\alpha=0}(\eta)d\eta = \infty,
\tag{A3}
$$

and for additional non-zero integer αs, we get:

$$
\int_{0}^{\infty} f_\alpha(\eta)d\eta = \frac{D}{\alpha - 1/2}.
\tag{A4}
$$

Integrals on the opposite half-axis $(-\infty\ldots0]$ have the same value with a negative sign, respectively. The forms for odd $C(x,t)$s are the following:

$$C(x,t) = erf\left(\frac{x}{2\sqrt{Dt}}\right),$$

$$C(x,t) = \left(\frac{\kappa_1 x}{t^{\frac{3}{2}}}\right)e^{-\frac{x^2}{4Dt}},$$

$$C(x,t) = \left(\frac{\kappa_1 x}{t^{\frac{5}{2}}}\right)e^{-\frac{x^2}{4Dt}}\left(1 - \frac{x^2}{6Dt}\right),$$

$$C(x,t) = \left(\frac{\kappa_1 x}{t^{\frac{7}{2}}}\right)e^{-\frac{x^2}{4Dt}}\left(1 - \frac{x^2}{3Dt} + \frac{x^4}{60(Dt)^2}\right),$$

$$C(x,t) = \left(\frac{\kappa_1 x}{t^{\frac{9}{2}}}\right)e^{-\frac{x^2}{4Dt}}\left(1 - \frac{x^2}{2Dt} + \frac{x^4}{20(Dt)^2} - \frac{x^6}{840(Dt)^3}\right). \tag{A5}$$

The space integrals of $\int_{-\infty}^{\infty} C_\alpha(x,t)dx = 0$ for all positive integers αs. On the positive half-axis for $\alpha = 0$, the integral of the error function in infinite, for positive αa, it is:

$$\int_{-\infty}^{\infty} C_\alpha(x,t) = \frac{Dt^{\frac{1}{2}-\alpha}}{\alpha - \frac{1}{2}}. \tag{A6}$$

which are well-defined values for finite, D, t and α. On the $(-\infty \ldots 0]$ half axis, the sign is opposite. Additional detailed analysis of the odd functions was presented in our former study [11].

Appendix B

The power series of the Equation (61) reads as follows

$$
\begin{aligned}
e^{-\frac{\eta^2}{4D}} \cdot Hermite[-3, \frac{\eta}{2\sqrt{D}}] &= \frac{\sqrt{\pi}}{8} - \frac{\eta}{4\sqrt{D}} + \frac{\sqrt{\pi}\eta^2}{16D} \\
&\quad - \frac{\eta^3}{48D^{3/2}} + \frac{\eta^5}{1920D^{5/2}} - \frac{\eta^7}{53760D^{7/2}} + o(\eta^9) \\
&= \frac{\sqrt{\pi}}{8} + \frac{\sqrt{\pi}}{16}\frac{\eta^2}{D} \\
&\quad + \frac{1}{16}\left(-4\frac{\eta}{\sqrt{D}} - \frac{1}{3}\frac{\eta^3}{D^{3/2}} + \frac{1}{120}\frac{\eta^5}{D^{5/2}} - \frac{1}{3360}\frac{\eta^7}{D^{7/2}}\right) + o(\eta^9).
\end{aligned}
\tag{A7}
$$

The series of relation (60) yields the following

$$2D\,\eta\,e^{-\frac{\eta^2}{4D}} + \sqrt{\pi}\,(2D^{3/2} + \sqrt{D}\,\eta^2)\,erf\left(\frac{1}{2}\frac{\eta}{\sqrt{D}}\right) = \tag{A8}$$

$$= D^{3/2}\left(4\frac{\eta}{\sqrt{D}} + \frac{1}{3}\frac{\eta^3}{D^{3/2}} - \frac{1}{120}\frac{\eta^5}{D^{5/2}} + \frac{1}{3360}\frac{\eta^7}{D^{7/2}} + o(\eta^9)\right).$$

As one can see—based on power expansions—the solution related to the Hermite function (A7) is still a kind of linear combination of the quadratic even solution and the odd solution (A8).

References

1. Crank, J. *The Mathematics of Diffusion*; Clarendon Press: Oxford, UK, 1956.
2. Ghez, R. *Diffusion Phenomena*; Dover Publication: Mineola, NY, USA, 2001.
3. Bennett, T.D. *Transport by Advection and Diffusion: Momentum, Heat and Mass Transfer*; John Wiley & Sons: Hoboken, NJ, USA, 2013.
4. Lienhard, J.H., IV; Lienhard, J.H., V. *A Heat Transfer Textbook*, 4th ed.; Phlogiston Press: Cambridge, MA, USA, 2017.
5. Newman, J.; Battaglia, V. *The Newman Lectures on Transport Phenomena*; Jenny Stanford Publishing: Dubai, United Arab Emirates, 2021.
6. Kardar, M.; Parisi, G.; Zhang, Y.-C. Dynamic scaling of growing interfaces. *Phys. Rev. Lett.* **1986**, *56*, 889. [CrossRef]

7. Barna, I.F.; Bognár, G.; Guedda, M.; Hriczó, K.; Mátyás, L. Analytic self-similar solutions of the Kardar-Parisi-Zhang interface growing equation with various noise terms. *Math. Model. Anal.* **2020**, *25*, 241. [CrossRef]

8. Barabási, A.L.; Stanley, E. *Fractal Concepts in Surface Growth*; Cambridge University Press: Cambridge, MA, USA, 1995.

9. Mátyás, L.; Gaspard, P. Entropy production in diffusion-reaction systems: The reactive random Lorentz gas. *Phys. Rev. E* **2005**, *71*, 036147. [CrossRef] [PubMed]

10. Mátyás, L.; Barna, I.F. General self-similar solutions of diffusion equation and related constructions. *Rom. J. Phys.* **2022**, *67*, 101.

11. Barna, I.F.; Mátyás, L. Advanced Analytic Self-Similar Solutions of Regular and Irregular Diffusion Equations. *Mathematics* **2022**, *10*, 3281. [CrossRef]

12. Cannon, J.R. *The One-Dimensional Heat Equation*; Addison-Wesley Publishing: Reading, MA, USA, 1984.

13. Cole, K.D.; Beck, J.V.; Haji-Sheikh, A.; Litkouhi, B. *Heat Conduction Using Green's Functions*; Series in Computational and Physical Processes in Mechanics and Thermal Sciences (2nd ed); CRC Press: Boca Raton, FL, USA, 2011.

14. Mátyás, L.; Tél, T.; Vollmer, J. Thermodynamic cross effects from dynamical systems. *Phys. Rev. E* **2000**, *61*, R3295. [CrossRef]

15. Thambynayagam, R.K.M. *The Diffusion Handbook: Applied Solutions for Engineers*; McGraw-Hill: New York, NY, USA, 2011.

16. Michaud, G.; Alecian, G.; Richer, G. *Atomic Diffusion in Stars*; Springer: New York, NY, USA, 2013; Volume 70.

17. Murray, J.D. *Mathematical Biology II: Spatial Models and Biomedical Applications*, 3rd ed.; Springer: New York, NY, USA, 2003.

18. Alebraheem, J. Predator interference in a predator-prey model with mixed functional and numerical responses. *Hindawi J. Math.* **2023**, *2023*, 4349573. [CrossRef]

19. Perthame, B. *Parabolic Equations in Biology*; Springer International Publishing: Berlin/Heidelberg, Germany, 2015.

20. Szép, R.; Mateescu, E.; Nechifor, A.C.; Keresztesi, Á. Chemical characteristics and source analysis on ionic composition of rainwater collected in the Charpatians "Cold Pole", Ciuc basin, Eastern Carpatians, Romania. *Environ. Sci. Pollut. Res.* **2017**, *24*, 27288. [CrossRef] [PubMed]

21. Gillespie, D.T.; Seitaridou, E. *Simple Brownian Diffusion*; Oxford University Press: Oxford, UK, 2013.

22. Tálos, K.; Páger, C.; Tonk, S.; Majdik, C.; Kocsis, B.; Kilár, F.; Pernyeszi, T. Cadmium biosorption on native Saccharomyces cerevisiae cells in aqueous suspension. *Acta Univ. Sapientiae Agric. Environ.* **2009**, *1*, 20.

23. Nechifor, G.; Voicu, S.I.; Nechifor, A.C.; Garea, S. Nanostructured hybrid membrane polysulfone-carbon nanotubes for hemodialysis. *Desalination* **2009**, *241*, 342. [CrossRef]

24. Lv, J.; Ren, K.; Chen, Y. CO_2 diffusion in various carbonated beverages: A molecular dynamic study. *Phys. Chem.* **2018**, *122*, 1655. [CrossRef]

25. Hägerstrand, T. *Innovation Diffusion as a Spatial Process*; The University of Chicago Press: Chicago, IL, USA, 1967.

26. Rogers, E.M. *Diffusion of Innovations*; The Free Press: Los Angeles, CA, USA, 1983.

27. Nakicenovic, N.; Grübler, A. *Diffusion of Technologies and Social Behavior*; Springer: Berlin/Heidelberg, Germany, 1991.

28. Bunde, A.; Kärger, J.C.; Vogl, G. *Diffusive Spreading in Nature, Technology and Society*; Springer: Berlin/Heidelberg, Germany, 2018.

29. Vogel, G. *Adventure Diffusion*; Springer: Berlin/Heidelberg, Germany, 2019.

30. Mazzoni, T. *A First Course in Quantitative Finance*; Cambridge University Press: Cambridge, MA, USA, 2018.

31. Lázár, E. Quantifying the economic value of warranties: A survey. *Acta Univ. Sapientiae Econ. Bus.* **2014**, *2*, 75. [CrossRef]

32. Albert, R.; Barabási, A.-L. Statistical mechanics of complex networks. *Rev. Mod. Phys.* **2002**, *74*, 47. [CrossRef]

33. Rogolino, P.; Kovács, R.; Ván, P.; Cimmelli, V.A. Generalized heat-transport equations: Parabolic and hyperbolic models. *Contin. Mech. Thermodyn.* **2018**, *30*, 1245. [CrossRef]

34. Jalghaf, H.K.; Kovács, E.; Majár, J.; Nagy, A.; Askar, A.H. Explicit stable finite difference methods for diffusion-reaction type equations. *Mathematics* **2021**, *9*, 3308. [CrossRef]

35. Nagy, A.; Saleh, M.; Omle, I.; Kareem, H.; Kovács, E. New stable, explicit shifted-hopscotch algorithms for the heat equation. *Math. Comput. Appl.* **2021**, *26*, 61.

36. Ezzahri, Y.; Ordonez-Miranda, J.; Joulain, K. Heat transport in semiconductor crystals under large temperature gradients. *Int. J. Heat Mass Transf.* **2017**, *108*, 1357. [CrossRef]

37. Cussler, E.L. *Diffusion: Mass Transfer in Fluid Systems*, 3rd ed.; Cambridge University Press: Cambridge, MA, USA, 2009.

38. Bluman, G.W.; Cole, J.D. The general similarity solution of the heat equation. *J. Math. Mech.* **1969**, *18*, 1025.

39. Sedov, L. *Similarity and Dimensional Methods in Mechanics*; CRC Press: Boca Raton, FL, USA, 1993.

40. Zel'dovich, Y.B.; Raizer, Y.P. *Physics of Shock Waves and High Temperature Hydrodynamic Phenomena*; Academic Press: New York, NY, USA, 1966.

41. Baraneblatt, G.I. *Similarity, Self-Similarity, and Intermediate Asymptotics*; Consultants Bureau: New York, NY, USA, 1979.

42. Barna, I.F.; Mátyás, L. Analytic solutions for the three dimensional compressible Navier-Stokes equation. *Fluid Dyn. Res.* **2014**, *46*, 055508. [CrossRef]

43. Barna, I.F.; Pocsai, M.A.; Lökös, S.; Mátyás, L. Rayleigh-Benard convection in the generalized Oberbeck-Boussinesq system. *Chaos Solitons Fractals* **2017**, *103*, 336. [CrossRef]

44. Barna, I.F.; Pocsai, M.A.; Barnaföldi, G.G. Self-similar solutions of a gravitating dark fluid. *Mathematics* **2022**, *10*, 3220. [CrossRef]

45. Barna, I.F.; Pocsai, M.A.; Mátyás, L. Analytic solutions of the Madelung equation. *J. Gen. Lie Theory Appl.* **2017**, *11*, 1000271.

46. Csanád, M.; Vargyas, M. Observables from a solution of (1+3)-dimensional relativistic hydrodynamics. *Eur. Phys. J. A* **2010**, *44*, 473. [CrossRef]

47. Barna, I.F.; Mátyás, L. Analytic solutions for the one-dimensional compressible Euler equation with heat conduction closed with different kind of equation of states. *Miskolc Math. Notes* **2013**, *14*, 785. [CrossRef]
48. Nath, G.; Singh, S. Approximate analytical solution for shock wave in rotational axisymetric perfect gas with azimutal magnetic field: Isotermal flow. *J. Astrophys. Astron.* **2019**, *40*, 50. [CrossRef]
49. Sahu, P.K. Shock wave driven out by a piston in a mixture of a non-ideal gas and small solid particles under the influence of azimuthal or axial magnetic field. *Braz. J. Phys.* **2020**, *50*, 548. [CrossRef]
50. Kanchana, C.; Su, Y.; Zhao, Y. Primary and secondary instabilities in Rayleigh-Benard convention of water-copper nanoliquid. *Commun. Nonlinear Sci. Numer. Simul.* **2020**, *83*, 105129. [CrossRef]
51. Olver, F.W.J.; Lozier, D.W.; Boisvert, R.F.; Clark, C.W. *NIST Handbook of Mathematical Functions*; Cambridge University Press: Cambridge, MA, USA, 2010.
52. Kythe, P.K. *Green's Functions and Linear Differential Equations*; Chapman & Hall/CRC Applied Mathematics and Nonliner Science; CRC Press: Boca Raton, FL, USA, 2011.
53. Rother, T. *Green's Functions in Classical Physics*; Lecture Notes in Physics; Springer International Publishing: New York, NY, USA, 2017; Volume 938.
54. Bronshtein, I.N.; Semendyayev, K.A.; Musiol, G.; Mühlig, H. *Handbook of Mathematics*; Springer: Wiesbaden, Germany, 2007.
55. Greiner, W.; Reinhardt, J. *Quantum Electrodynamics*; Springer: Berlin/Heidelberg, Germany, 2009.
56. Claus, I.; Gaspard, P. Fractals and dynamical chaos in a two-dimensional Lorentz gas with sinks. *Phys. Rev. E* **2001**, *63*, 036227. [CrossRef]
57. Rápó, E.; Tonk, S. Factors affecting synthetic dye adsorption; desorption studies: A review of results from the last five years (2017–2021). *Molecules* **2021**, *26*, 5419 [CrossRef]
58. Boltzmann, L. Zur Intergration der Diffusionsgleichung bei variabeln Diffusionscoefficienten. *Ann. Phys.* **1894**, *53*, 959. [CrossRef]
59. Lonngren, K.E. Self similar solution of plasma equations. *Proc. Indian Acad. Sci.* **1977**, *86*, 125. [CrossRef]
60. Mehrer, H.; Stolwijk, N.A. Heroes and highlights in the history of diffusion. *Diffus. Fundam.* **2009**, *11*, 1.

Communication

With Nanoplasmonics towards Fusion

Tamás Sándor Biró [1,2,3,*,†], Norbert Kroó [1], László Pál Csernai [1,4,5], Miklós Veres [1], Márk Aladi [1], István Papp [1], Miklós Ákos Kedves [1], Judit Kámán [1], Ágnes Nagyné Szokol [1], Roman Holomb [1], István Rigó [1], Attila Bonyár [6], Alexandra Borók [1,6], Shireen Zangana [6], Rebeka Kovács [6], Nóra Tarpataki [6], Mária Csete [7], András Szenes [7], Dávid Vass [7], Emese Tóth [7], Gábor Galbács [8] and Melinda Szalóki [9]

[1]   Wigner Research Centre for Physics, 1121 Budapest, Hungary
[2]   Institute for Physics, Babeş Bolyai University, 3400 Cluj-Napoca, Romania
[3]   Complexity Science Hub, 1080 Vienna, Austria
[4]   Department of Physics and Technology, Bergen University, 5020 Bergen, Norway
[5]   Csernai Consult, 5119 Bergen, Norway
[6]   Department of Electronics Technology, Faculty of Electrical Engineering and Informatics, Budapest University of Technology and Economics, 1111 Budapest, Hungary
[7]   Department for Quantum Optics, University of Szeged, 6720 Szeged, Hungary
[8]   Department of Inorganic and Analytical Chemistry, University of Szeged, 6720 Szeged, Hungary
[9]   Department of Biomaterials and Prosthetic Dentistry, Faculty of Dentistry, University of Debrecen, 4032 Debrecen, Hungary
*   Correspondence: biro.tamas@wigner.hu
†   Based on a talk given by T.S.Biró at the Zimanyi School, 5–9 December 2022.

Abstract: A status report is presented about the Nanoplasmonic Laser Induced Fusion Experiment (NAPLIFE). The goal is to investigate and verify plasmonically enhanced phenomena on the surfaces of nanoantennas embedded in a polymer target at laser intensities up to a few times 10^{16} W/cm^2 and pulse durations of 40–120 fs. The first results on enhanced crater formation for Au-doped polymer targets are shown, and SERS signals typical for CD_2 and ND bound vibrations are cited. Trials to detect D/H ratio by means of LIBS measurments are reported. Plasmonics has the potential to work at these intensities, enhancing the energy and deuterium production, due to thus far unknown mechanisms.

Keywords: fusion; plasmonics; nanotechnology; energy production

Citation: Biró, T.S.; Kroó, N.; Csernai, L.P.; Veres, M.; Aladi, M.; Papp, I.; Kedves, M.A.; Kámán, J.; Nagyné Szokol, A.; Holomb, R.; et al. With Nanoplasmonics towards Fusion. *Universe* **2023**, *9*, 233. https://doi.org/10.3390/universe9050233

Academic Editor: Máté Csanád

Received: 29 March 2023
Revised: 28 April 2023
Accepted: 15 May 2023
Published: 17 May 2023

1. Laser Fusion Ignition Improvement by Nanoantennas

Taming nuclear fusion is a long-term dream [1]. Since the 1950s, several approaches have been developed towards this goal. To date, the most characteristic experiments belong to two groups of approaches: (i) magnetic confinement fusion (MCF) techniques and (ii) laser inertial heating and compression (ICF). The former approach has developed different techniques for producing the magnetic field which confines the hot plasma, such as tokamak and stellarator, while the latter uses more and more powerful lasers to increase the compression and to reach the ignition temperature necessary for starting elementary fusion reactions in thermal equilibrium. ITER in Cadarash, France is building the largest tokamak ever seen, and NIF in Los Alamos, Nevada collects the energy of 192 huge lasers onto a miniscule target (hohlraum). In both cases, magnetic confinement or inertial compressed fusion, there are plasma instabilities to fight.

The third type of approach, which generally do not aim for thermal equilibrium, have been discredited over the years by the "cold" fusion fallacy which presented high claims but provided no evidence. Yet, some fusion as a by-product of mechanical manipulations, such as sonoluminescence or cracking of metals enriched with hydrogen, have been reported in the passing decades. Generally, the chemical type of spectral or other evidence hunted down the presence of deuterium (atoms, not nuclei) in such cases. The real goal is to reach tamed fusion, even as nanofusion in a controllable quantity is preferred for slow but rich energy production.

Nanoplasmonics is one of the efficient means of squeezing electromagnetic energy into nanosized volumes. It may result in hotspots around nanoparticles with high electron density, and thus high electromagnetic fields with characteristic lifetimes in the few times ten femtosecond range. The near field of these localized plasmons has a screening effect around positively charged particles (e.g., protons) and the momentum of the correlated motion of the plasmonic electrons may be transferred to these positively charged particles, resulting in their acceleration to high momentum and energy. These positive effects may play significant role in increasing the probability of fusion of these positive (e.g., proton, deuterium, boron, etc.) ions.

Our Nanoplasmonic Laser-Induced Fusion Experiment (NAPLIFE) investigates the extent to which nanotechnology may improve the laser beam targets in order to come closer to fusion ignition conditions at a lower input energy than is provided by the direct methods applied in big facilities. In this way, we hope to assist with the effects from plasmonics, from the collective motion of electrons illuminated by intense and ultrashort, well-contrasted laser pulses. After the first few years of this project, carried out with modest means, we report our initial experiences of the enhancement of laser energy absorption due to nanoplasmonic effects, stemming from doping metal (Au) nanoantennas in a transparent polymer target [2–7].

We prepare 20–160 micrometer-thick targets made of UDMA-TEGDMA copolymer, a material used in dentistry as tooth filler [2]. Our choice was motivated by the fact that while nanometals are easy to mix under a fluid, for laser shootings and transportation, a solid carrier is advantageous. The UDMA-TEGDMA can be solidified using UV light treatment after mixing. Au nanorods have been implemented into the polymer with carefully chosen size resonant to the laser wavelength of 795 nm. Furthermore, after finding the first signals for the presence of deuterium in the target remainder in molecular vibration spectra after shooting at it, for comparison and background measurements, we also use deuterized targets. This is achieved by admixing another, shorter molecule, the MMA, where all eight H atoms can be replaced by D atoms. In this way, up to 31 per cent D/H ratio can be achieved in the total copolymer molecule of the target. The width of the target is larger (160 μm), permitting us to study the crater sizes remaining after single shots, while it shall be smaller when planning multilayer targets in successive experiments.

The theory behind hoping for an enhancement of energy density due to plasmonics requires extended computer simulations on the motion of electrons on the Au nanoantennas of resonant size and various shapes. Although in the experiment, we are using cylindrical shapes, other shapes and metals sometimes promise a higher value of near field enhancement [8]. Theoretically, near-field enhancement (NFE) factors in the order of 100 can be reached, meaning an energy density enhancement of 10^4 in near atomic layers, up to cca. 10–30 nm-s.

The classical approach to studying plasmonic effects on nanoparticles involves the use of the dielectric function of the free electron gas, which often neglects important phenomena such as electron–electron interactions and spill-out effects; these are typically included by fit parameters. However, alongside such classical methods, we also use kinetic models utilizing the particle-in-cell (PIC) method. In PIC simulations, marker particles are randomly distributed on the metal surface based on the electron number density, and these particles move in a continuous phase space, while densities and currents are computed in stationary mesh cells.

The PIC method has been shown to be an efficient tool for analyzing the electron dynamics and for modeling spill-out effects in plasmon simulations [9,10].

A kinetic model simulation [11–14] reveals collectively moving electrons and protons up to momenta in the 100 MeV range—such projectiles in principle may initiate a few nuclear reactions in the surrounding polymer atomic layers (cf. Figure 1) [15,16].

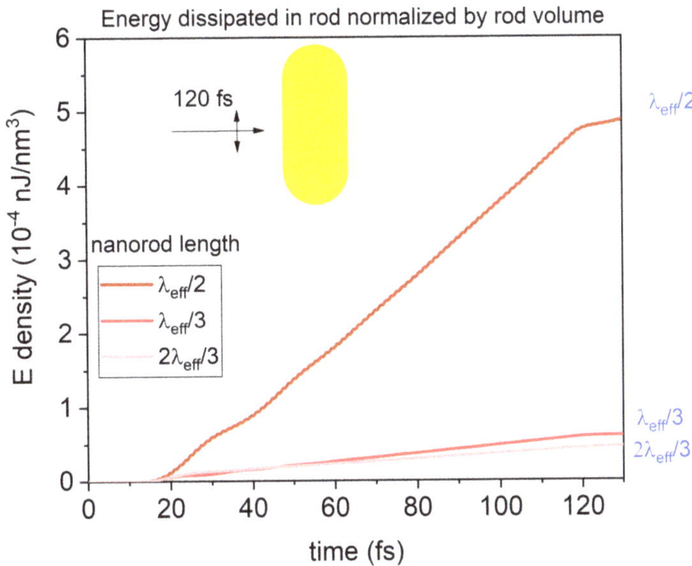

Figure 1. Result of a single nanoantenna simulation using a hydrodynamic (HDM) model implemented into FEM numerical codes. The importance of the in–medium resonant length is shown by the energy of accelerated electrons from the conducting band in the metal.

Embedding nanoantennas into the fusion remedies another obstacle inherent in present nuclear fusion techniques. Both MCF and ICF methods are fighting the Rayleigh–Taylor instabilities when the target is compressed to achieve high nuclear reaction rates. This leads to more and to less compressed domains, and consequently, only the highly compressed domains ignite. Due to the rapid expansion arising from the high pressure, the less compressed domains do not reach ignition at all.

In ultra-relativistic heavy ion reactions, this problem does not occur, since the hadronization process takes place on a timelike oriented hypersurface that is simultaneously in the whole spatial volume. It is possible to achieve a similar situation in ICF fusion with *short* laser ignition pulses of picosecond or femtosecond lengths [17]. For this purpose, one has to regulate the energy deposition in the fusion target. The implanted nanoantennas, with an adequately designed density distribution, also help to reach this goal [8,18].

In the following section, we concentrate on the results of spectral investigations and their correlation within each other.

2. Results: Sers, Crater Sizes, Libs Spectra

The study of the effect of 40 fs laser pulses ranging in energy from 1 mJ to 30 mJ on the above-described target reveals a monotonic, almost linear dependence of crater diameters on the increasing pulse energy [19,20]. All measurements are compared at the best focusing, i.e., at maximal laser field intensity. For the optimal focusing case, down to a light beam hit diameter of about 20 μm, high enough intensities can be reached (up to 10^{17} W/cm^2). At the best focus position plus–minus 1 mm, the reflection drops dramatically from 70% down to around 10%. In this situation, the effects on the target are not disturbed by much light reflection on plasma mirrors. The craters are investigated sometimes days after the shootings, since the target material conserves them in a way similar to that of the old-fashioned gel-film detectors in nuclear experiments. Hundreds of craters have been studied by our research team using white light interferometry. As an example cf. Figure 2.

Additionally, soft spectral measurements on the treated targets were also conducted after the shootings. Most prominently, SERS (surface enhanced Raman scattering) performed by illumination with infrared lasers with low-energy (ns long) pulses was used to study molecular vibrations, which are typical in organic polymers. In particular, CH_2 and NH bonds were sought, with the aim of replacing them with CD_2 and ND groups. Certainly, a standard computer model also had to be used in the background to produce and calculate the wavelength of the corresponding lines in the SERS spectrum [21–23]. Here, we have found a wide enhancement of the intensity in the ranges characteristic for CD_2 and ND, in an integrated yield way beyond that which could have been counted for the natural deuterium ratio of $D/H \approx 1/6000$ cf. Ref. [19].

Laser-Induced Breakdown Spectroscopy (LIBS) is being employed more and more in the nuclear/fusion research field of late [24–29], due to the fact that it can sensitively detect not only heavy but also the lightest elements. It is microdestructive, can be applied remotely, and in a high-vacuum environment, it is able to provide isotopic resolution for all three hydrogen isotopes. All these features make this method appealing for laser-ignited fusion research conducted in a vacuum chamber, where the microplasma generated on the surface of the target can act as an emission source for LIBS-monitoring measurements. We are studying the emission intensities of the Balmer alpha line of protium (H) at 656.240 nm and deuterium (D) at 656.123 nm in the plasma plume. The wavelengths of these lines differ with 0.117 nanometer, but in a high-vacuum environment, the line widths are smaller than this; thus, they can be separated.

We are using LIBS to detect a possible excess D being formed from H via nuclear processes. Comparative monitoring experiments are being carried out on UDMA/TEGDMA co-polymer targets with and without gold nanorods added. For optimization and calibration purposes, we also prepare and use partially deuterized targets made from a mixture of UDMA and a fully deuterized methyl-methacrylate (MMA) monomer. By changing the mixing ratio of the monomers, a series of co-polymer targets with varying D content is fabricated.

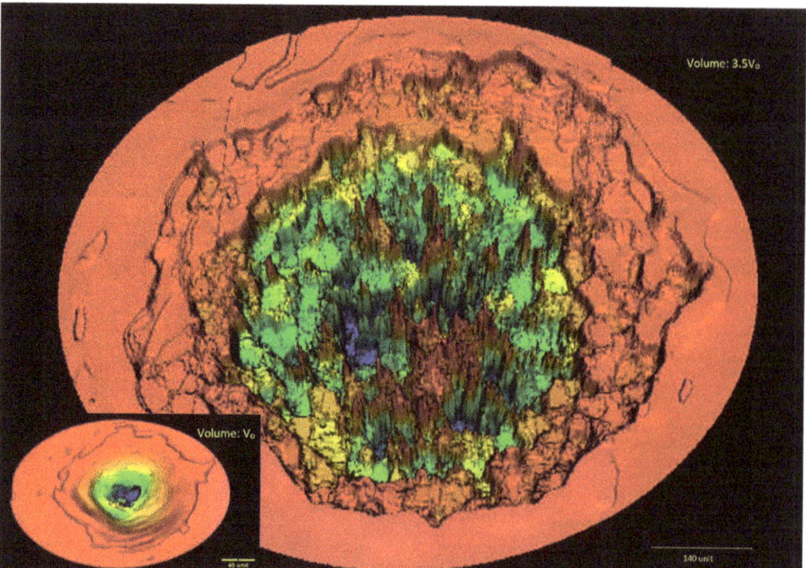

Figure 2. Comparison of crater sizes between undoped and Au-doped targets at the same laser pulse of 27 mJ energy. The pictures are made using white light interferometry and brought to their respective sizes to agree based on the real micrometer dimension of the horizontal line notations.

Complementary in situ high-resolution mass spectrometry measurements, aiming at the detection of low mass/charge ratio charged and chargeless products by controlling the ionization source, are also currently being performed during the laser irradiation of the targets. Since these measurements are still ongoing, we cannot report the final results here.

Preliminary experimental data show that the particular experimental conditions and requirements make the measurements difficult. A large number of repeated measurements are needed to obtain reliable spectra. Given the association of the laser intensity not only with the possible ignition of a few fusion reactions, but also with the LIBS emission signal generation, further increasing of the laser intensity (including better focusing of the beam and an improved contrast of the pulse with a preceding pedestal) would make this analysis even more sensitive.

3. Conclusions

In conclusion, we have presented the NAPLIFE collaboration and project with the goal of studying ultrashort laser pulse energy utilization based on their plasmonic effects in nanotechnologically manipulated polymer targets towards nuclear fusion ignition energies. The present preliminary experiments did show a drastic change in the energy absorption due to gold nanorod embedding into UDMA/TEGDMA copolymer, even at random orientation and low density. The crater sizes observed after laser shots by microscope techniques reveal a factor of 3–4-fold enhancement, while theoretical simulations of the near field enhancement (NFE) predict up to a factor of 100 in the field strength enhancement [8,20,30]. The first results from the kinetic modeling of electron and proton motion on and near to the nanoantennas are also presented [11]. Here, we find proton momenta up to the 100 MeV range; to overcome the Coulomb barrier in vacuum (without nanoparticle effects on the screening of it) one needs a factor of roughly 10 more.

Further studies are planned by varying the nanoparticle density, form and material, as well as trying different coatings, possibly delivering nuclear reactants with near-threshold resonances, such as boron and beryllium. Beyond our local activity, we plan to join experiments at facilities providing higher laser pulse energy and better contrast, e.g., at ELI-ALPS in Szeged, Hungary.

Author Contributions: Conceptualization, T.S.B., N.K. and L.P.C.; Data curation, N.K., M.A., M.Á.K., J.K., Á.N.S., R.H., I.R., A.B. (Alexandra Borók), S.Z., R.K., N.T., A.S., D.V., E.T. and M.S.; Formal analysis, T.S.B. and M.C.; Funding acquisition, T.S.B. and N.K.; Investigation, N.T. and M.S.; Methodology, T.S.B., N.K., L.P.C., M.V., R.H., A.B. (Attila Bonyár), M.C. and G.G.; Project administration, T.S.B.; Resources, T.S.B.; Software, I.P., R.H., M.C., A.S., D.V. and E.T.; Validation, T.S.B., N.K., M.A., I.R. and G.G.; Visualization, I.P.; Writing—original draft, T.S.B.; Writing—review and editing, N.K., L.P.C., M.V., Á.N.S., G.G. and M.C. All authors have read and agreed to the published version of the manuscript. Authorship has been limited to those who have contributed substantially to the work reported.

Funding: This research is funded by NKFIH (National Bureau for Research, Development and Innovation) grant Nr. 2022-NL-2.1.1.-2022-00002.

Institutional Review Board Statement: Not applicable.

Informed Consent Statement: Not applicable.

Data Availability Statement: No data can be shared beyond the already cited publications to this date.

Acknowledgments: T.S.B. and N.K. acknowledge enlightening discussions with Johann Rafelski. The administration of the project is acknowledged to Anett Szeledi. L.P.C. acknowledges support from Wigner RCP, Budapest.

Conflicts of Interest: The authors declare no conflict of interest.

References

1. Fusion Power Is Coming Back into Fashion (This Time It Might Even Work). *The Economist (Science & Technology)*, 22 March 2023.
2. Barszczewska-Rybarek, I.M. A Guide through the Dental Dimethacrylate Polymer Network Structural Characterization and Interpretation of Physico-Mechanical Properties. *Materials* **2019**, *12*, 4057. [CrossRef]
3. Ding, T.; Mertens, J.; Lombardi, A.; Scherman, O.A.; Baumberg, J.J. Light-Directed Tuning of Plasmon Resonances via Plasmon-Induced Polymerization Using Hot Electrons. *ACS Photonics* **2017**, *4*, 1453–1458. [CrossRef]
4. Li, J.; Cushing, S.K.; Meng, F.; Senty, T.R.; Bristow, A.D.; Wu, N. Plasmon-Induced Resonance Energy Transfer for Solar Energy Conversion. *Nat. Photonics* **2015**, *9*, 601–607. [CrossRef]
5. Minamimoto, H.; Toda, T.; Futashima, R.; Li, X.; Suzuki, K.; Yasuda, S.; Murakoshi, K. Visualization of Active Sites for Plasmon-Induced Electron Transfer Reactions Using Photoelectrochemical Polymerization of Pyrrole. *J. Phys. Chem. C* **2016**, *120*, 16051–16058. [CrossRef]
6. Wu, K.; Chen, J.; McBride, J.R.; Lian, T. Efficient Hot-Electron Transfer by a Plasmon-Induced Interfacial Charge-Transfer Transition. *Science* **2015**, *349*, 632–635. [CrossRef] [PubMed]
7. Wang, Z.; Kan, Z.; Shen, M. Study the plasmonic property of gold nanorods highly above damage threshold via single-pulse spectral hole-burning experiments. *Nat. Sci. Rep.* **2021**, *11*, 22232. [CrossRef] [PubMed]
8. Csete, M.; Szenes, A.; Tóth, E.; Vass, D.; Fekete, O.; Bánhelyi, B.; Papp, I.; Biró, T.; Csernai, L.P.; Kroó, N.; et al. Comparative study on the uniform energy deposition achievable via optimized plasmonic nanoresonator distributions. *Plasmonics* **2022**, *17*, 775–787. [CrossRef]
9. Maier, S.A. *Plasmonics: Fundamentals and Applications*; Springer Science and Business Media: New York, NY, USA, 2007.
10. Ding, W.J.; Lim, J.Z.J.; Do, H.T.B.; Xiong, X.; Mahfoud, Z.; Png, C.E.; Bosman, M.; Ang, L.K.; Wu, L. Particle simulation of plasmons. *Nanophotonics* **2020**, *9*, 3383. [CrossRef]
11. Papp, I.; Bravina, L.; Csete, M.; Kumari, A.; Mishustin, I.I.; Molnar, D.; Motornenko, A.; Racz, P.; Satarov, L.M.; Stocker, H.; et al. Kinetic Model Evaluation of the Resilience of Plasmonic Nanoantennas for Laser-Induced Fusion. *Phys. Rev. X Energy* **2022**, *1*, 023001. [CrossRef]
12. Nanbu, K.; Yonemura, G. Weighted particles in Coulomb collision simulations based on the theory of a cumulative scattering amplitude. *J. Comput. Phys.* **1998**, *145*, 639–654. [CrossRef]
13. Pérez, F.; Gremillet, L.; Decoster, A.; Drouin, M.; Ledfebvre, E. Improved modeling of relativistic collisions and collisional ionization in particle-in-cell codes. *Phys. Plasmas* **2012**, *19*, 083104. [CrossRef]
14. Arber, T.; Bennett, K.; Brady, C.; Lawrence-Douglas, A.; Ramsay, M.; Sircombe, N.; Gillies, P.; Evans, R.; Schmitz, H.; Bell, A.; et al. Contemporary particle-in-cell approach to laser-plasma modeling. *Plasma Phys. Control. Fusion* **2015**, *57*, 113001. [CrossRef]
15. Papp, I.; Bravina, L.; Csete, M.; Mishustin, I.N.; Molnár, D.; Motornenko, A.; Satarov, L.M.; Stöcker, H.; Strottman, D.D.; Szenes, A.; et al. Laser Wake Field Collider. *Phys. Lett. A* **2021**, *396*, 12724. [CrossRef]
16. István, P.; Larissa, B.; Mária, C.; Archana, K.; Anton, M.I.N.; Péter, R.; Horst, S.L.M.S.; András, S.D.S.; Dávid, V.; Nagyné, S.; et al. Kinetic model of resonant nanoantennas in polymer for laser induced fusion. *Front. Phys.* **2023**, *11*, 1116023.
17. Csernai, L.P.; Strottman, D.D. Volume ignition via time-like detonation in the pellet fusion. *Laser Part. Beams* **2015**, *33*, 279–282. [CrossRef]
18. Csernai, L.P.; Kroo, N.; Papp, I. Radiation dominated implosion with nano-plasmonics. *Laser Part. Beams* **2018**, *36*, 171–178. [CrossRef]
19. Bonyár, A.; Szalóki, M.; Borók, A.; Rigó, I.; Kámán, J.; Zangana, S.; Veres, M.; Rácz, P.; Aladi, M.; Kedves, M.; et al. The Effect of Femtosecond Laser Irradiation and Plasmon Field on the Degree of Conversion of a UDMA-TEGDMA Copolymer Nanocomposite Doped with Gold Nanorods. *Int. J. Mol. Sci.* **2022**, *23*, 13575. [CrossRef] [PubMed]
20. Rigó, I.; Kámán, J.; Szokol, Ń.; Bonyár, A.; Szalóki, M.; Borók, A.; Zangana, S.; Rácz, P.; Aladi, M.; Kedves, M.; et al. Raman spectroscopic characterization of crater walls formed upon single-shot high energy femtosecond laser irradiation of dimethacrylate polymer doped with plasmonic gold nanorods. *arXiv* **2022**, arXiv:2210.00619. [CrossRef]
21. Becke, A.D. Density-functional exchange-energy approximation with correct asymptotic behavior. *Phys. Rev. A* **1998**, *38*, 3098–3100. [CrossRef]
22. Lee, C.; Yang, W.; Parr, R.G. Development of the Colle-Salvetti correlation-energy formula into a functional of the electron density. *Phys. Rev. B* **1998**, *37*, 785–789. [CrossRef]
23. Rassolov, V.A.; Pople, J.A.; Ratner, M.A.; Windus, T.L. 6-31G* basis set for atoms K through Zn. *J. Chem. Phys.* **1998**, *109*, 1223–1229. [CrossRef]
24. Csernai, L.P.; Mishustin, I.N.; Satarov, L.M.; Stoecker, H.; Bravina, L.; Csete, M.; Kámán, J.; Kumari, A.; Motornenko, A.; Papp, I.; et al. Crater Formation and Deuterium Production in Laser Irradiation of Polymers with Implanted Nano-antennas. *arXiv* **2022**, arXiv:2211.14031. [CrossRef]
25. Galbács, G.; Kovács-Széles, É. Nuclear Applications of Laser-Induced Breakdown Spectroscopy. In *Laser Induced Breakdown Spectroscopy (LIBS)*; Singh, V.K., Tripathi, D.K., Deguchi, Y., Wang, Z., Eds.; Wiley: Hoboken, NJ, USA, 2023. [CrossRef]
26. Li, C.; Feng, C.L.; Oderji, H.Y.; Luo, G.N.; Ding, H.B. Review of LIBS application in nuclear fusion technology. *Front. Phys.* **2016**, *11*, 114–214. [CrossRef]
27. Wu, J.; Qiu, Y.; Li, X.; Yu, H.; Zhang, Z.; Qiu, A. Progress of laser-induced breakdown spectroscopy in nuclear industry applications. *J. Phys. Appl. Phys.* **2020**, *53*, 023001. [CrossRef]

8.	Kurniawan, K.H.; Kagawa, K. Hydrogen and deuterium analysis using laser-induced plasma spectroscopy. *Appl. Spectrosc. Rev.* **2006**, *41*, 99–130. [CrossRef]
9.	Craners, D.A.; Chinni, R.C. Lares-Induced Breakdown Spectroscopy—Capabilities and Limitations. *Appl. Spectrosc. Rev.* **2009**, *44*, 457–506. [CrossRef]
0.	Vass, D.; Szenes, A.; Tóth, E.; Bánhelyi, B.; Papp, I.; Bíró, T.; Csernai, L.P.; Kroó, N.; Csete, M. Plasmonic nanoresonator distributions for uniform energy deposition in active targets. *Opt. Mater. Express* **2023**, *13*, 9–27. [CrossRef]

MDPI

St. Alban-Anlage 66

4052 Basel

Switzerland

www.mdpi.com

Universe Editorial Office

E-mail: universe@mdpi.com

www.mdpi.com/journal/universe